Innovative Robot Designs and Approaches

Innovative Robot Designs and Approaches

Editors

Giuseppe Carbone
Med Amine Laribi

MDPI • Basel • Beijing • Wuhan • Barcelona • Belgrade • Manchester • Tokyo • Cluj • Tianjin

Editors
Giuseppe Carbone
Department of Mechanical, Energy and Management Engineering, Università della Calabria, Rende, Italy

Med Amine Laribi
Department GMSC, Institut PPRIME, CNRS, Université de Poitiers, ENSMA, Poitiers, France

Editorial Office
MDPI
St. Alban-Anlage 66
4052 Basel, Switzerland

This is a reprint of articles from the Special Issue published online in the open access journal *Applied Sciences* (ISSN 2076-3417) (available at: https://www.mdpi.com/journal/applsci/special_issues/Innovative_Robot).

For citation purposes, cite each article independently as indicated on the article page online and as indicated below:

LastName, A.A.; LastName, B.B.; LastName, C.C. Article Title. *Journal Name* **Year**, *Volume Number*, Page Range.

ISBN 978-3-0365-6714-3 (Hbk)
ISBN 978-3-0365-6715-0 (PDF)

Contents

About the Editors

Giuseppe Carbone

Giuseppe Carbone graduated cum laude with a Master's degree at the University of Cassino (Italy), where he also completed his PhD studies. He served as a Key Member of LARM (Laboratory of Robotics and Mechatronics) for around 20 years. From 2015 to 2017, he was a Senior Lecturer at Sheffield Hallam University (UK) and a member of the executive board of Sheffield Robotics. Since Dec. 2018, he has worked as an Associate Professor at DIMEG, University of Calabria, Italy. Prof. Carbone has received several Awards, including three IFToMM "Young Delegate" Awards and two JSPS Awards in Japan. His research interests cover aspects of mechanics of manipulation and grasp, mechanics of robots and mechanics of machinery, and he has published more than 300 papers and over 10 patents relating to these topics. He has edited/co-edited four books with Springer International Publisher and one book with Elsevier. He has participated in or coordinated more than 20 research projects at national and international levels, including the 7th European Framework and Horizon 2020. Currently, he is the Chair of the IFToMM Technical Committee on Robotics and Mechatronics, a Member of the Executive Board of Directors of the International Society of Bionic Engineering and a Treasurer of the IFToMM Italy Society.

Med Amine Laribi

Med Amine Laribi is an Associate Professor in the Fundamental and Applied Sciences Faculty of the University of Poitiers (UP), where he teaches robotics and mechanics. He obtained his Mechanical Engineer degree from École Nationale d'Ingénieurs de Monsatir (E.N.I.M.) in 2001 and M.S. in Mechanical Design in 2002. He received his Ph.D. in Mechanics from the University of Poitiers in 2005 and his Ph.D. in National Habilitation in Mechanics from the University of Poitiers in 2018. He serves as an Associate Editor in several internationals journals, including the *ASME Journal of Medical Devices*, *Robotic Intelligence and Automation* (Emerald Publishing), *Mechanical Sciences* (Copernicus Publications), *MDPI Robotics* and *MDPI Machines*. He is a member of the international scientific committee of MESROB, RAAD, ISRM, METRAPP and ROMANSY. His research interests include aspects of robot design, mechanism synthesis, cable-driven robots, parallel robots, haptic interfaces and collaborative robots. He has published 57 peer reviewed journal papers, 8 edited books, 10 edited journal Special Issues and assisted six PhD students and two postdoctoral students. He is currentlyleading several national and international research projects in the fields of medical robotics and biomimetics.

applied sciences

MDPI

Editorial

Recent Trends on Innovative Robot Designs and Approaches

Giuseppe Carbone [1,*] and Med Amine Laribi [2]

[1] Department of Mechanical Engineering, Energy Engineering and Management, University of Calabria, 87036 Rende, Italy
[2] Department GMSC, Institut PPRIME, Université de Poitiers, CNRS, ENSMA, UPR 3346, 86962 Poitiers, France
* Correspondence: giuseppe.carbone@unical.it

1. Introduction

The use and function of robots are evolving at a fast pace, sparking interest in creative solutions within a quickly expanding potential market in cutting-edge industries with applications including service robotics, surgical and rehabilitative robotics, and assistive robotics. In this context, fresh ideas, approaches, and applications still require considerable attention. For example, assistive robotics, surgical and rehabilitative robots, service robotics and other cutting-edge application domains are becoming increasingly important, not only from a technological and financial standpoint but also in terms of their consequences for daily life and society as reported for example in [1–12]. Even the use and function of robots on assembly lines and in other conventional frameworks are being extensively altered in favor of innovative flexible and agile manufacturing methods. Novel designs are also being extensively researched, including cable-driven parallel robots (CDPRs), as their conceptual design can provide a key performance in terms of large workspace, reconfigurability, large payload capacity, and dynamics [13–16].

This Special Issue aims at attracting cutting-edge research and review articles on any innovative robot design or modelling/control approach. The published papers in this Special Issue cover a wide range of topics, including robot manipulation, variable stiffness actuation, mobile system, social robotics, task optimization, robot compliance, biomedical devices, collaborative robotics, trajectory planning, and wearable robotics. The first published paper presents the concept of a robotic system for the aliquoting of biomaterials, consisting of a serial manipulator in combination with a parallel Delta-like robot. This is particularly valuable to avoid the risks of contaminations as reported with a design solution and simulation models in [17]. The second paper addresses the concept of "Industry 4.0" as based on the utilization of collaborative robots [18]. In particular, authors present a gestural framework for controlling a collaborative robotic manipulator using pointing gestures. A unique robotic collaborative workspace called the Complex Collaborative HRI Workplace (COCOHRIP) was designed around the gestural framework to evaluate the method and provide a basis for the future development of HRI applications [18]. The third paper focuses on robots for rehabilitation tasks by presenting the development of an internal torque monitoring system for ASPIRE, a parallel robot designed for shoulder rehabilitation. A complete analysis regarding the components of the robotic system is carried out with the purpose of determining the dynamic behavior of the system, as reported in [19]. The fourth paper addresses the structural-parametric synthesis and kinematic analysis of the RoboMech class of parallel mechanisms (PM) having two sliders. The proposed methods allow the synthesis of a PM with its structure and geometric parameters of the links to obtain the given laws of motions of the input and output links (sliders). The paper outlines a possible application of the proposed approach to design a PM for a cold-stamping technological line, as reported in [20].

The fifth paper presents novel models of reconfigurable parallel mechanisms (RPMs) with a single active degree-of-freedom (1-DOF). The mechanisms contain three to six

Citation: Carbone, G.; Laribi, M.A. Recent Trends on Innovative Robot Designs and Approaches. *Appl. Sci.* **2023**, *13*, 1388. https://doi.org/10.3390/app13031388

Received: 17 January 2023
Accepted: 18 January 2023
Published: 20 January 2023

identical kinematic chains, which provide three (for the tripod) to zero (for the hexapod) uncontrollable DOFs. Each kinematic chain in the synthesized mechanism consists of planar and spatial parts. Such a design provides them with reconfiguration capabilities even when the driving link is fixed. This allows the reproduction of diverse output trajectories without using additional actuators, as reported in [21]. The sixth paper presents a mechanical design of a four degrees of freedom (DOF) wheelchair-mounted upper limb exoskeleton. The design takes advantage of a non-back-drivable mechanism that can hold the output position without energy consumption and assist the completely paralyzed users. Preliminary results are provided to show the effectiveness and reliability of using the proposed design for physically disabled people, as reported in [22]. The seventh paper deals with the optimal design of a planar cable-driven parallel robot (CDPR), with three degrees of freedom, intended for assisting the patient's affected upper limb along a prescribed movement. A prototype of the optimal design of the CDPR was developed and validated experimentally, as reported in [23]. The eighth paper addresses the development of ResQbot 2.0—a mobile rescue robot designed for performing casualty extraction. The proposed design and development of the mechanical system as well as the method for safely loading a full-body casualty onto the robot's 'stretcher bed', are described in detail, as reported in [24]. The nineth paper deals with the Instant center that is an important kinematic characteristic which can be used for velocity and singularity analysis, configuration synthesis and dynamic modeling of multi-degree of freedom (multi-DOF) planar linkages. The paper proposes two criteria to convert single-loop multi-DOF planar linkages into a two-loop virtual linkage by adding virtual links. The proposed method can be applied to a wide range of single-loop multi-DOF N-bar ($N \geq 5$) planar linkages, as reported in [25]. The tenth paper deals with the synthesis of the kinematic structure of a robotic manipulator to determine the optimal manipulator for a given task by proposing four different algorithms using the standard Denavit–Hartenberg convention and Bézier splines approximation and vector algebra. The results are demonstrated with three chosen example poses and are evaluated by measuring the manipulability and total link length of the final kinematic structures, as reported in [26].

The eleventh paper refers to robotic deburring by proposing a mechanism that can automatically reduce cutting forces in the event that the burr is too high, and is able to return to the baseline configuration when the burr thickness is acceptable again. The effectiveness of the proposed mechanism is verified by means of dynamic simulations using selected test cases. A reduction of 60% of the cutting forces is obtained, considering a steel burr 6 mm in height, as reported in [27]. The twelfth paper introduces a novel kinematic model for a tendon-driven compliant torso mechanism for humanoid robots, which describes the complex behavior of a system integrating rigid bodies with flexible actuation tendons. Inspired by the human spine, the proposed mechanism is based on a flexible backbone whose shape is controlled by two pairs of antagonistic tendons. Preliminary tests are reported to show the accuracy and efficiency of the proposed torso mechanism, as reported in [28]. The last paper introduces an innovative robotic foot design inspired by the functionality and anatomy of the human foot for a humanoid design. The proposed foot mechanism consists of three main bodies, to represent the heel, plant and toes, connected by compliant joints for improved balancing and impact absorption. The proposed design was validated with a numerical simulation, as reported in [29].

The Guest Editors of this Special Issue would like to express their gratitude to the authors and reviewers for their efforts and time spent in the valuable scientific contributions and useful feedback that have confirmed the high-scientific-quality of the recent trends in innovative robot design and approaches.

Conflicts of Interest: The authors declare no conflict of interest.

References

1. Tai, K.; El-Sayed, A.-R.; Shahriari, M.; Biglarbegian, M.; Mahmud, S. State of the Art Robotic Grippers and Applications. *Robotics* **2016**, *5*, 11. [CrossRef]
2. Muñoz, J.; López, B.; Quevedo, F.; Barber, R.; Garrido, S.; Moreno, L. Geometrically constrained path planning for robotic grasping with Differential Evolution and Fast Marching Square. *Robotica* **2023**, *41*, 414–432. [CrossRef]
3. Gull, M.A.; Bai, S.; Bak, T. A Review on Design of Upper Limb Exoskeletons. *Robotics* **2020**, *9*, 16. [CrossRef]
4. Zhang, P.; Zhang, J. Motion generation for walking exoskeleton robot using multiple dynamic movement primitives sequences combined with reinforcement learning. *Robotica* **2022**, *40*, 2732–2747. [CrossRef]
5. Gherman, B.; Birlescu, I.; Plitea, N.; Carbone, G.; Tarnita, D.; Pisla, D. On the singularity-free workspace of a parallel robot for lower-limb rehabilitation. *Proc. Rom. Acad. Ser. A Math. Phys. Tech. Sci. Inf. Sci.* **2019**, *20*, 383–391.
6. Carbone, G.; Gherman, B.; Ulinici, I.; Vaida, C.; Pisla, D. Design Issues for an Inherently Safe Robotic Rehabilitation Device. In *Advances in Service and Industrial Robotics, Proceedings of the 26th International Conference on Robotics in Alpe-Adria-Danube Region, RAAD 2017, Torino, Italy, 21–23 June 2017*; Springer: Cham, Switzerland, 2018; pp. 1025–1032. [CrossRef]
7. Vaida, C.; Birlescu, I.; Pisla, A.; Ulinici, I.-M.; Tarnita, D.; Carbone, G.; Pisla, D. Systematic Design of a Parallel Robotic System for Lower Limb Rehabilitation. *IEEE Access* **2020**, *8*, 34522–34537. [CrossRef]
8. BEN Hamida, I.; Laribi, M.A.; Mlika, A.; Romdhane, L.; Zeghloul, S. Dimensional Synthesis and Performance Evaluation of Four Translational Parallel Manipulators. *Robotica* **2021**, *39*, 233–249. [CrossRef]
9. Mishra, S.K.; Kumar, C.S. Compliance modeling of a full 6-DOF series–parallel flexure-based Stewart platform-like micromanipulator. *Robotica* **2022**, *40*, 3435–3462. [CrossRef]
10. Han, Q.; Ji, A.; Jiang, N.; Hu, J.; Gorb, S.N. A climbing robot with paired claws inspired by gecko locomotion. *Robotica* **2022**, *40*, 3686–3698. [CrossRef]
11. Nekoo, S.R.; Acosta, J.; Ollero, A. Quaternion-based state-dependent differential Riccati equation for quadrotor drones: Regulation control problem in aerobatic flight. *Robotica* **2022**, *40*, 3120–3135. [CrossRef]
12. Hoffmann, T.; Prause, G. On the Regulatory Framework for Last-Mile Delivery Robots. *Machines* **2018**, *6*, 33. [CrossRef]
13. Ennaiem, F.; Chaker, A.; Arévalo, J.; Laribi, M.; Bennour, S.; Mlika, A.; Romdhane, L.; Zeghloul, S. Sensitivity Based Selection of an Optimal Cable-Driven Parallel Robot Design for Rehabilitation Purposes. *Robotics* **2021**, *10*, 7. [CrossRef]
14. Ennaiem, F.; Chaker, A.; Sandoval, J.; Mlika, A.; Romdhane, L.; Bennour, S.; Zeghloul, S.; Laribi, M.A. A hybrid cable-driven parallel robot as a solution to the limited rotational workspace issue. *Robotica* **2023**, *41*, 1–19. [CrossRef]
15. Laribi, M.A.; Carbone, G.; Zeghloul, S. On the Optimal Design of Cable Driven Parallel Robot with a Prescribed Workspace for Upper Limb Rehabilitation Tasks. *J. Bionic Eng.* **2019**, *16*, 503–513. [CrossRef]
16. Qian, S.; Zi, B.; Shang, W.-W.; Xu, Q. A Review on Cable-driven Parallel Robots. *Chin. J. Mech. Eng.* **2018**, *31*, 66. [CrossRef]
17. Malyshev, D.; Rybak, L.; Carbone, G.; Semenenko, T.; Nozdracheva, A. Optimal Design of a Parallel Manipulator for Aliquoting of Biomaterials Considering Workspace and Singularity Zones. *Appl. Sci.* **2022**, *12*, 2070. [CrossRef]
18. Čorňák, M.; Tölgyessy, M.; Hubinský, P. Innovative Collaborative Method for Interaction between a Human Operator and Robotic Manipulator Using Pointing Gestures. *Appl. Sci.* **2022**, *12*, 258. [CrossRef]
19. Pisla, D.; Tarnita, D.; Tucan, P.; Tohanean, N.; Vaida, C.; Geonea, I.D.; Bogdan, G.; Abrudan, C.; Carbone, G.; Plitea, N. A Parallel Robot with Torque Monitoring for Brachial Monoparesis Rehabilitation Tasks. *Appl. Sci.* **2021**, *11*, 9932. [CrossRef]
20. Baigunchekov, Z.; Laribi, M.A.; Carbone, G.; Mustafa, A.; Amanov, B.; Zholdassov, Y. Structural-Parametric Synthesis of the RoboMech Class Parallel Mechanism with Two Sliders. *Appl. Sci.* **2021**, *11*, 9831. [CrossRef]
21. Fomin, A.; Petelin, D.; Antonov, A.; Glazunov, V.; Ceccarelli, M. Virtual and Physical Prototyping of Reconfigurable Parallel Mechanisms with Single Actuation. *Appl. Sci.* **2021**, *11*, 7158. [CrossRef]
22. Gull, M.; Thoegersen, M.; Bengtson, S.; Mohammadi, M.; Struijk, L.A.; Moeslund, T.; Bak, T.; Bai, S. A 4-DOF Upper Limb Exoskeleton for Physical Assistance: Design, Modeling, Control and Performance Evaluation. *Appl. Sci.* **2021**, *11*, 5865. [CrossRef]
23. Ennaiem, F.; Chaker, A.; Laribi, M.; Sandoval, J.; Bennour, S.; Mlika, A.; Romdhane, L.; Zeghloul, S. Task-Based Design Approach: Development of a Planar Cable-Driven Parallel Robot for Upper Limb Rehabilitation. *Appl. Sci.* **2021**, *11*, 5635. [CrossRef]
24. Saputra, R.P.; Rakicevic, N.; Kuder, I.; Bilsdorfer, J.; Gough, A.; Dakin, A.; de Cocker, E.; Rock, S.; Harpin, R.; Kormushev, P. ResQbot 2.0: An Improved Design of a Mobile Rescue Robot with an Inflatable Neck Securing Device for Safe Casualty Extraction. *Appl. Sci.* **2021**, *11*, 5414. [CrossRef]
25. Nie, L.; Ding, H.; Ting, K.-L.; Kecskeméthy, A. Instant Center Identification of Single-Loop Multi-DOF Planar Linkage Using Virtual Link. *Appl. Sci.* **2021**, *11*, 4463. [CrossRef]
26. Huczala, D.; Kot, T.; Pfurner, M.; Heczko, D.; Oščádal, P.; Mostýn, V. Initial Estimation of Kinematic Structure of a Robotic Manipulator as an Input for Its Synthesis. *Appl. Sci.* **2021**, *11*, 3548. [CrossRef]
27. Bottin, M.; Cocuzza, S.; Massaro, M. Variable Stiffness Mechanism for the Reduction of Cutting Forces in Robotic Deburring. *Appl. Sci.* **2021**, *11*, 2883. [CrossRef]

28. Russo, M.; Ceccarelli, M.; Cafolla, D. Kinematic Modelling and Motion Analysis of a Humanoid Torso Mechanism. *Appl. Sci.* **2021**, *11*, 2607. [CrossRef]
29. Russo, M.; Chaparro-Rico, B.D.M.; Pavone, L.; Pasqua, G.; Cafolla, D. A Bioinspired Humanoid Foot Mechanism. *Appl. Sci.* **2021**, *11*, 1686. [CrossRef]

applied sciences

Article

A Bioinspired Humanoid Foot Mechanism

Matteo Russo [1], Betsy D. M. Chaparro-Rico [2], Luigi Pavone [2], Gabriele Pasqua [3] and Daniele Cafolla [2,*]

[1] Faculty of Engineering, University of Nottingham, Nottingham NG8 1BB, UK; matteo.russo@nottingham.ac.uk
[2] IRCCS Neuromed, Via dell'Elettronica, 86077 Pozzilli (IS), Italy; betsychaparro@hotmail.com (B.D.M.C.-R.); bioingegneria@neuromed.it (L.P.)
[3] Department of Medicine and Health Science, University of Molise, 86100 Campobasso (CB), Italy; ing.gabrielepasqua@gmail.com
* Correspondence: contact@danielecafolla.eu

Abstract: This paper introduces an innovative robotic foot design inspired by the functionality and the anatomy of the human foot. Most humanoid robots are characterized by flat, rigid feet with limited mobility, which cannot emulate the physical behavior of the foot–ground interaction. The proposed foot mechanism consists of three main bodies, to represent the heel, plant, and toes, connected by compliant joints for improved balancing and impact absorption. The functional requirements were extracted from medical literature, and were acquired through a motion capture system, and the proposed design was validated with a numerical simulation.

Keywords: robotics; humanoids; foot mechanism; prosthetics; neurorehabilitation

Citation: Russo, M.; Chaparro-Rico, B.D.M.; Pavone, L.; Pasqua, G.; Cafolla, D. A Bioinspired Humanoid Foot Mechanism. *Appl. Sci.* **2021**, *11*, 1686. https://doi.org/10.3390/app11041686

Academic Editors: Manuel Armada and Giuseppe Carbone

Received: 21 January 2021
Accepted: 9 February 2021
Published: 13 February 2021

1. Introduction

Humanoid robotics has fascinated and challenged scientists for decades [1]. The design of humanoid robots has evolved from the serial design of the first humanoids, such as WABOT (WAseda roBOT) [2], to more refined architectures [3–12]. However, despite decades of research, the mobility of most of these robots is still limited to the legs, arms, hands, and head only, with very few examples including torso [10] and foot [13–23] mobility, despite the feet and torso playing a key role in human motion [24]. The reduced mobility of humanoid robots in those regions hinders their capability to achieve balance and to perform complex dynamic tasks. Human foot architecture is characterized by three main segments—namely the heel, mid-foot, and toes—that fulfill distinct roles during bipedal locomotion [25].

However, several humanoid robots use flat, one-segment feet [2–12]. Some famous examples include Honda's ASIMO (Advanced Step in Innovative Mobility) [3], the iCub design and its variants from the Italian Institute of Technology [7], and Softbank Robotics NAO [9]. Several research groups have tried to improve the foot performance by compensating a limited mechanical design with an enhanced sensing/control capability, which can be obtained by adding force/pressure sensors to the plant of the foot [26–30], with mixed results. Others kept a one-segment design, but moved from a flat rigid sole to a compliant one [13–15], obtaining significant results and increased performance (e.g., Boston Dynamics' Atlas [14]). Soft options have been also explored through designs with a highly elastic foot sole directly connected to the ankle joint [16].

Some research groups investigated the possibility of using two-segment feet, with a heel segment and a toe segment [16–20]. Foot designs with three segments have been proposed to better mimic human-like motion [21,22], but are usually characterized by active mechanisms that require a complex control and motion coordination with the rest of the body [21], or by planar rigid-body mechanisms [22] that achieve balance through spring-loaded joints only. While successful, the lack of compliance in these three-segment designs limits their behavior when obstacles introduce 3D torques and loads.

In this paper, an innovative passive mechanism for robotic feet is presented, which is able to adapt to the environment without the need of external actuation thanks to joint elasticity. First, the mobility requirements to achieve human-like motion are extracted from medical literature and experimental gait analysis using a motion capture system. Then, the proposed design is introduced, along with its main characteristics and technical details. The foot mechanism is validated through numerical simulation with both Finite Element Analysis (FEA) and Multibody Dynamics, and a prototype is manufactured with 3D printing in order to prove the feasibility of the proposed mechanism on a range of terrains and obstacles.

2. Materials and Methods

2.1. Requirements for Humanoid Foot Mechanisms

In order to obtain the mobility requirements for humanoid foot mechanisms, the human gait was analyzed in terms of absolute and angular motion [31–35] looking retrospectively at clinical outcomes from patients who outcome a healthy gait cycle. In accordance with the medical standards, a variation of the Davis protocol [31]—with a marker on the second metatarsus instead of fifth—was used to acquire the motion of 11 subjects (details in Table 1) over 10 distinct gait sessions of four walking steps each. The motion was acquired through a VICON Motion Capture system, with the following retroreflective marker set configuration and alignment (Figures 1 and 2):

- RA, right ankle marker: Placed at the level of the lateral malleolus;
- RT, right toe marker: Placed on the lateral aspect of the foot at the second metatarsal head;
- RQ, right heel marker: Positioned so that the heel–toe marker vector is parallel to (but offset from) the sole of the foot, and is aligned with the foot progression line (i.e., the line from the ankle center to the space between the second and third metatarsal heads).
- LA, left ankle marker: Placed at the level of the lateral malleolus;
- LT, left toe marker: Placed on the lateral aspect of the foot at the second metatarsal head;
- LQ, left heel marker: Positioned so that the heel–toe marker vector is parallel to (but offset from) the sole of the foot, and is aligned with the foot progression line.

Table 1. Data of the subjects for the gait acquisition.

Subject	Age	Gender	Weight (kg)	Height (cm)	Pelvis Width (cm)	Pelvis Height (cm)	Right Leg Length (cm)	Left Leg Length (cm)	Right Knee Diameter (cm)	Left Knee Diameter (cm)	Right Ankle Diameter (cm)	Left Ankle Diameter (cm)
1	27	F	67	172	26.5	10	34	34	11	11	6	6
2	37	M	75	175	30	8	34	34	10	10	6	6
3	33	M	80	173	24.5	8	35	35	10	10	6.5	6.5
4	31	F	63	161	27.5	8.5	32	32	10.5	10.5	5	5
5	28	F	56	169	25	8	35	34	8	8.5	4.5	5
6	29	M	65	178	20	7	35	35	10	10	6.5	6.5
7	22	F	56	157	24.5	6	30	30	8.5	8.5	5	5
8	25	F	55	168	24	7.5	31	31	9	8.5	5	5
9	27	M	72	172	21.5	8	35	35	8	8	6	6
10	23	M	75	180	23	8.5	33	34	8.5	8.5	7	6.5
11	21	M	70	179	24	11	35	35	8.5	9	7	7

The BTW Gaitlab software (BTS Bioengineering Corp., Quincy, MA, USA) [32] was used for the data processing, according to the following steps:

- For each four-step acquisition, each step of the gait was isolated.
- The time of each single step was normalized in a 0 to 100 scale, in which 0 represents the beginning of the step cycle (stance with dual foot support) and 100 represents the end of that step cycle, and corresponds to the beginning of the following step.
- The normalized step data were filtered and compared for all of the acquisitions in order to remove out-of-phase outliers, whereas the outliers on the ankle angle acquired value were left in the graph in Figure 3 for completeness, but were removed from further calculations.

- The remaining dataset was used to obtain a normalized average gait in terms of ankle angle, which is illustrated by the black line in Figure 3.

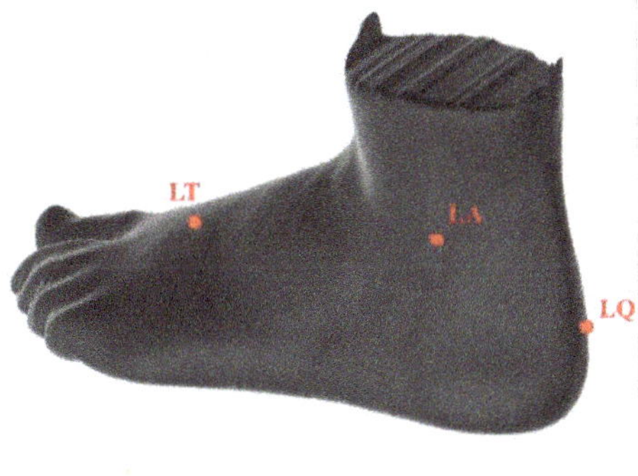

Figure 1. Example of the marker placements on the left foot (CAD model), with the left ankle (LA), left toe (LT) and left heel (LQ) markers.

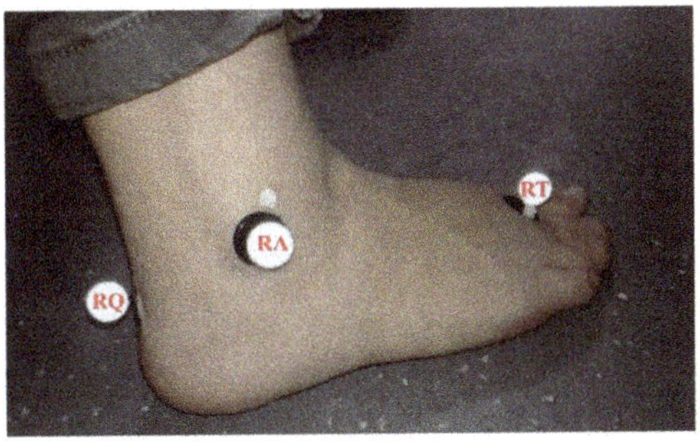

Figure 2. Example of the marker placements on the right foot of a patient, with the right ankle (RA), right toe (RT) and right heel (RQ) markers.

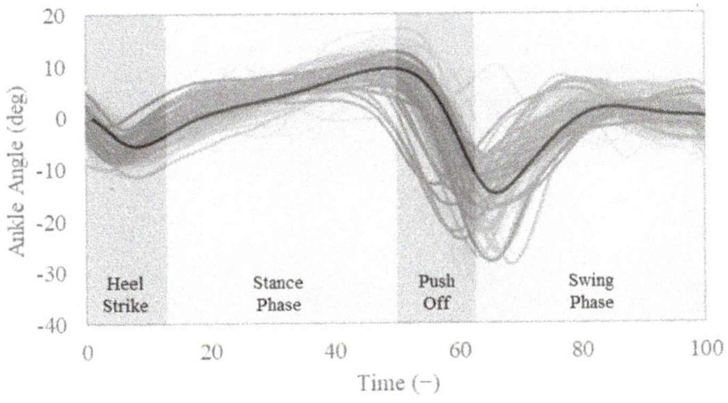

Figure 3. Normalized average gait as the ankle angle, in normalized time.

The resulting average angular displacement for the whole motion is approximately 25 degrees, compared to a peak value of 45 degrees of the ankle angular displacement for some acquisitions. By expressing the angular motion range of the heel part of the foot, the values in Figure 3 identify all of the possible foot configurations in which the foot must be in stable contact with the ground in order to support the body, and will be used both for dimensioning and simulating the proposed humanoid foot mechanism design.

2.2. An Underactuated Humanoid Foot Mechanism

The proposed foot mechanism aims to enhance robot mobility and efficiency by enabling body support in all of the stances acquired in Section 2. In order to do so, a three-segment design was considered, based on the diagram in Figure 4. During the gait, the foot motion is ruled by four main centers of rotation, which correspond to the ankle joint, the heel bone (or calcaneus), the navicular tuberosity, and the metatarsophalangeal (MTP) joint, as shown in Figure 4. The first segment of the foot can be associated to the calcaneus-talus complex, which includes the heel and ankle, and can be approximated to a rigid body. The motion of the central segment of the foot is enabled by the active stiffening behavior of the intrinsic foot muscles in the plantar arches, which connect the calcaneus to the MTP joint. The plantar fascia and several ligaments support the plantar arches, which store mechanical energy during weight bearing by deforming. This behavior can be reproduced with the elasticity of a compliant body. The motion of the last segment of the foot can be approximated to a flexion/extension movement of the toes, which rotate around the MTP joint and the interphalangeal joints. The MTP joint is usually described as a condyloid joint, while the interphalangeal joints resemble revolute joints.

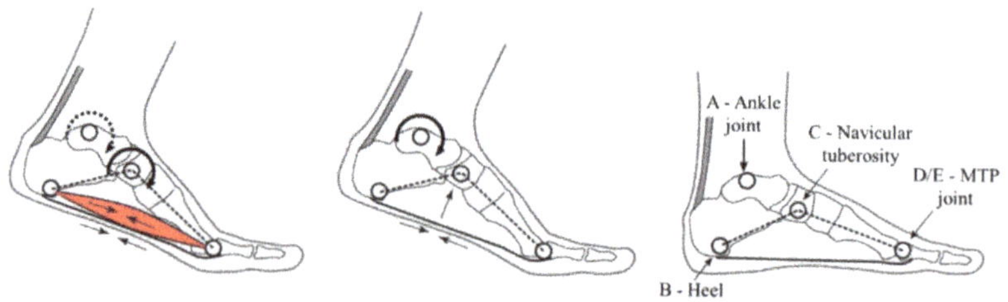

Figure 4. Human foot mobility [23], with a focus on the main joints of the foot: the ankle joint (**A**), the heel bone (or calcaneus (**B**)), the navicular tuberosity (**C**), and the metatarsophalangeal (MTP) joint (**D**/**E**).

In order to replicate the behavior of the human foot, a new compliant foot mechanism based on the kinematic scheme in Figure 5 is here presented. The proposed mechanism is based on a compliant mechanism in the plant, which can be approximated with an RRPR (revolute-revolute-prismatic-revolute) architecture, with revolute joints in points B, C, and D, and a link of varying length between B and D. Furthermore, a second compliant mechanism is added in the MTP joint in point E, with a rigid revolute joint controlled by a torsional spring.

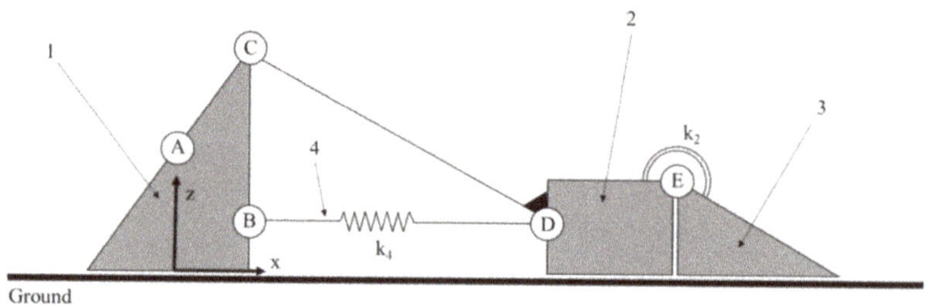

Figure 5. Kinematic scheme of the proposed foot mechanism based on the corresponding points (A to E) in the foot mobility diagram in Figure 4: (1) hindfoot; (2) midfoot; (3) forefoot; (4) compliant body.

The mechanism works mainly on the sagittal plane, even if the compliant part can be designed to balance 3D loads as well. When analyzing the planar behavior of the proposed foot design, its degrees of freedom can be evaluated from its rigid mechanism equivalent with the Chebychev-Grübler-Kutzbach criterion as two, controlled respectively by the plantar compliant element and by the torsional spring of the MTP joint. An exemplar design of the proposed foot mechanism is shown in Figure 6.

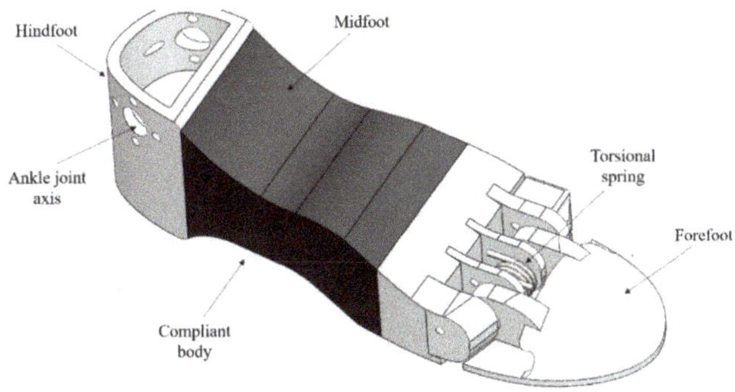

Figure 6. CAD model of the proposed foot mechanism, based on the equivalent kinematic scheme of Figure 5.

3. Results

3.1. Simulation and Results

The behavior of the humanoid foot was checked through a nonlinear, dynamic study using FEA. Iterative schemes to solve Equation (1) through the Newton–Raphson method (NR), already integrated in the software, were used for each node (i).

$$[M]^{t+\Delta t}\{U''\}^{(i)} + [C]^{t+\Delta t}\{U'\}^{(i)} + {}^{t+\Delta t}[K]^{(i)}\,{}^{t+\Delta t}[\Delta U]^{(i)} = {}^{t+\Delta t}\{R\} - {}^{t+\Delta t}\{F\}^{(i-1)} \quad (1)$$

where:

M = Mass matrix of the system;
C = Damping matrix of the system;

- ${}^{t+\Delta t}[K]^{(i)}$ = Stiffness matrix of the system;
- ${}^{t+\Delta t}[R]$ = Vector of externally-applied nodal loads;
- ${}^{t+\Delta t}[F]^{(i-1)}$ = Vector of internally-generated nodal forces at iteration (i−1);
- ${}^{t+\Delta t}[\Delta U]^{(i)}$ = Vector of incremental nodal displacements at iteration (i);
- ${}^{t+\Delta t}\{U\}^{(i)}$ = Vector of total displacements at iteration (i);
- ${}^{t+\Delta t}\{U'\}^{(i)}$ = Vector of total velocities at iteration (i);
- $[M]^{t+\Delta t}\{U''\}^{(i)}$ = Vector of total accelerations at iteration (i).

In nonlinear static analysis, equations have to be solved at any time step $t+\Delta t$. Because the internal nodal forces ${}^{t+\Delta t}\{F\}^{(i-1)}$ depend on the nodal displacements at time $t+\Delta t$, ${}^{t+\Delta t}[\Delta U]^{(i)}$, an iterative method must be used to find a converging solution. The above-mentioned iterations have several methodologies in place. The NR scheme was used for this simulation. Within this method, the tangential stiffness matrix is formed and decomposed in a certain step in every iteration. The NR method using Newmark integration is used because it has a high rate of convergence, and is quadratic. However, each iteration generates and decomposes the tangential rigidity, which is prohibitively costly for larger models. Thus, a different iterative approach may be advantageous.

The humanoid foot was placed above a solid body, simulating the ground. The ground was fixed, giving it 0 degrees of freedom. A gravity of 9.807 m/s^2 was set, acting on the entire assembly. The temperature loads were included, setting a constant temperature of 298 Kelvin. The mean angular displacement of the ankle acquired from the subject was used as the input motion for the nonlinear dynamic simulation. The definition of the time curve for the angle displacement application was set to simulate the mean step duration for 0.63 s, using a force control technique. The input given to the simulation was extracted by the gait analysis data, and it corresponds to the average gait of Figure 3. The input motion was applied to the ankle joint axis, which is highlighted in blue in Figure 7.

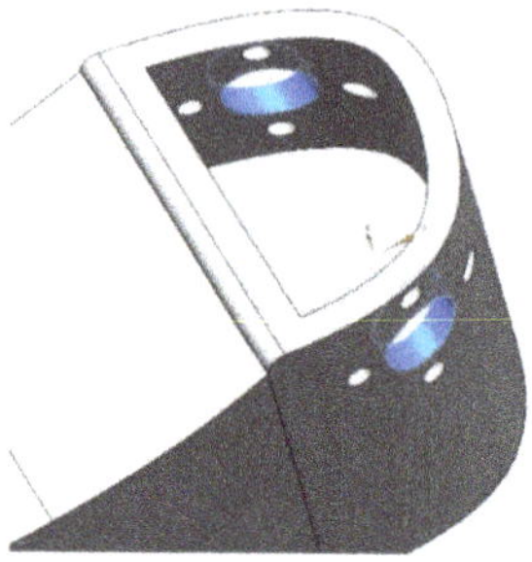

Figure 7. Ankle joint axis of the CAD model, on which the input motion is applied.

The humanoid foot is composed of four bodies, three of which are in ABS rigid plastic (Table 2), and the middle of which is set to thermoplastic polyurethane (TPU) for his flexible mechanical properties, simulating the muscle part. Another key component is the spring—shown in Figure 8—in the forefoot, which passively improves the ability of the foot to adapt to the ground; the spring was modelled with a 2 mm diameter, five coils, and an 8 mm length, and alloy steel was chosen as material in order to simulate a commercial part. After setting all of the required parameters, a mesh was generated with the parameters in Table 3 to perform the FEA.

Table 2. Mechanical properties of the materials in the FEA.

Property	ABS	TPU	Alloy Steel
Yield strength	4.50×10^7 N/m^2	5.41×10^7 N/m^2	2.76×10^7 N/m^2
Tensile strength	7.30×10^7 N/m^2	7.79×10^7 N/m^2	6.89×10^7 N/m^2
Elastic modulus	3.00×10^9 N/m^2	1.48×10^7 N/m^2	6.90×10^{10} N/m^2
Poisson's ratio	0.35	0.55	0.33
Mass density	1200.00 kg/m^3	1217.00 kg/m^3	2700.00 kg/m^3

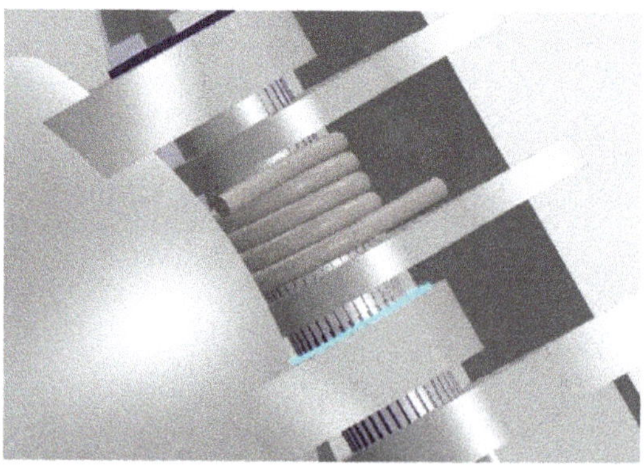

Figure 8. Detailed view of the forefoot torsional spring.

Table 3. Mesh parameters.

Mesh information	Values
Jacobian points	4 Points
Maximum element size	2.00 mm
Tolerance	0.10 mm
Total Nodes	175,296
Total Elements	100,015

Figure 9 shows the FEM nonlinear dynamic analysis stress results computed by the use of the Von Mises formulation. A maximum value of 2.57×10^6 N/m^2 is observed in the upper part of the rigid midfoot, next to the connection to the compliant body. The maximum value is significantly smaller than the yield limit of the material, thus validating the proposed humanoid foot for an average walking gait. The behavior of the impact of the foot on the ground is obtained by applying the mean angle displacement taken from the analyzed subjects (see Section 2). The results can be observed in the step-by-step effect of the load on the different bodies of the humanoid foot, as is shown in the linear displacement distribution reported in Figure 10:

- Starting from the rigid body of the hindfoot, the angular displacement is applied to the ankle joint. The hindfoot is connected to the flexible body of the midfoot, which starts to deform.
- When the angular displacement causes the foot to hit the ground, the midfoot is characterized by a stress concentration in the upper part, as per Figure 9.
- The arc bottom part of the foot is loaded with a maximum tension of 1.57×10^6 N/m^2, behaving as the human foot muscle when stepping.
- The spring connecting the two bodies of the forefoot is stretched. This helps the foot to adapt to the ground, and the potential energy of the spring restores the neutral 'flat' position when the contact is released.

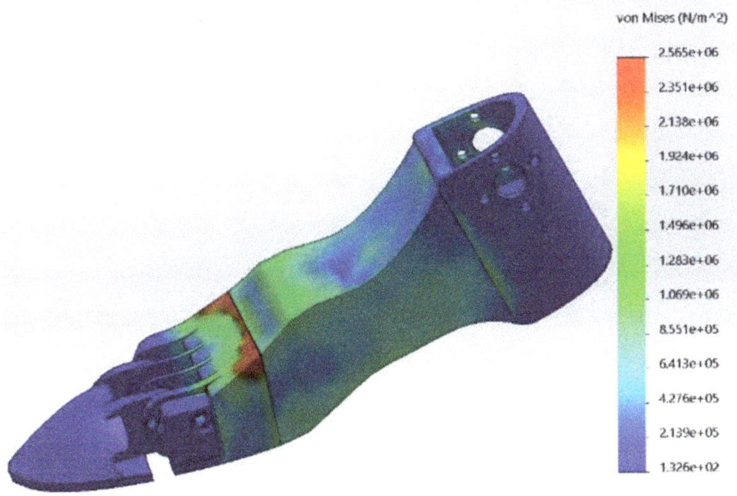

Figure 9. Von Mises stress distribution.

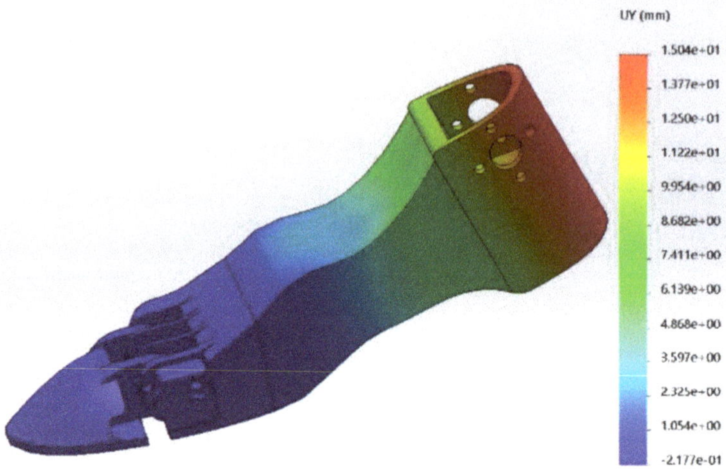

Figure 10. Displacement values in the humanoid foot.

These results validate the behavior of the proposed humanoid foot, showing that it acts similarly to the human foot in terms of stress while walking. Furthermore, the results prove that the proposed design can safely withstand the loads associated with an average human gait, including the impact of the foot on the ground.

3.2. Experimental Tests

Due to the promising simulation results, a preliminary prototype was built at the Biomechatronics Lab of IRCSS Neuromed in Pozzilli, Italy. The prototype is composed of six parts: five were manufactured by 3D printing, using the rapid prototyping techniques explained by Cafolla et al. [36], whereas the last one is a commercial mechanical component. The printing materials were chosen from among a wide variety of commercial resins and plastics, in order to ensure the desired foot behaviour: the four rigid bodies were printed in PLA filament; the flexible component, representing the midfoot of the mechanism, was manufactured using a TPU filament with properties matching the ones of the FEA. The properties of both materials are reported in Table 4, for reference. Different infill percentages and patterns were tested for the midfoot, and the final infill was selected for its optimal TPU stiffness. Both the value of the infill and the 3D support pattern with which the infill was printed were optimized according to the surface curve of the compliant component through an FEA with constrained motion (as per the acquired gait).

The correct relative positioning of the components is defined by male and female rectangular pins. Then, the components are locked together with interlocking shapes. Additionally, epoxy glue was added to the coupled surfaces in order to improve the strength of the connection, as it is the most loaded component of the system, as shown in Figure 9. The assembly of the entire foot is shown in Figure 11. The prototype of the humanoid foot is composed of a rigid hind foot, a flexible mid foot, and a rigid forefoot, which is made of three 3D printed rigid parts, namely the connection pin, the sole and the toe, and a commercial part, the spring. The prototype was scaled down from human size in order to fit a service humanoid robot design [10–12], and thus the whole foot can fit into a box of ($146 \times 52 \times 41$) mm; it weighs 57.80 g.

Table 4. Mechanical properties of the materials of the 3D printed prototype.

Property	ABS	ABS Test Method	TPU	TPU Test Method
Specific gravity	1.24 g/cc	ASTM D1505	1.14 g/cc	ISO 1183
Flow rate	6.0 g/10 min	-	39 cm^3/10 min	ISO 1133
Tensile strength	110 MPa (MD)	ASTM D882	-	-
	145 MPa (TD)	ASTM D882	-	-
Strain at break	160% (MD)	ASTM D882	530%	ISO 527
	100% (TD)	ASTM D882	-	-
Tensile modulus	3310 MPa (MD)	ASTM D882	95 MPa	ISO 527
	3860 MPa (TD)	ASTM D882	-	-
Impact Strength	7.5 KJ/m^2	-	Notched	ISO 179
			No break	Charpy 23C
Hardness	-	-	45 Shore D	ISO 868

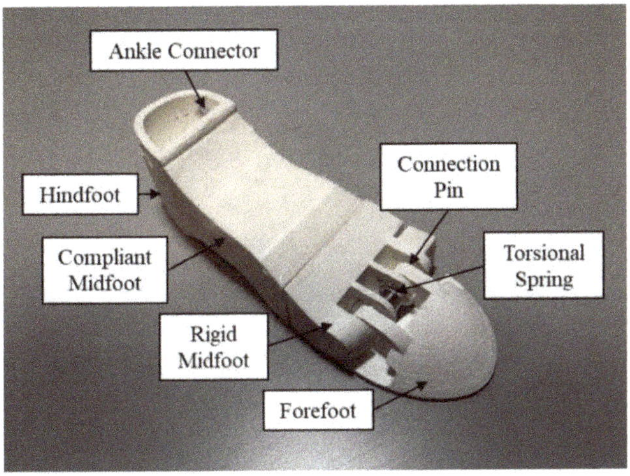

Figure 11. Humanoid foot prototype assembly.

As a preliminary test, the adaptability of the foot to different step heights was tested. Figures 12 and 13 show the foot adapting to a 22 mm high step and a 54 mm high step, respectively. The combination of the forefoot spring and the compliant midfoot actions allow the proposed foot design to balance on step heights up to approximately 35% of the length of the foot.

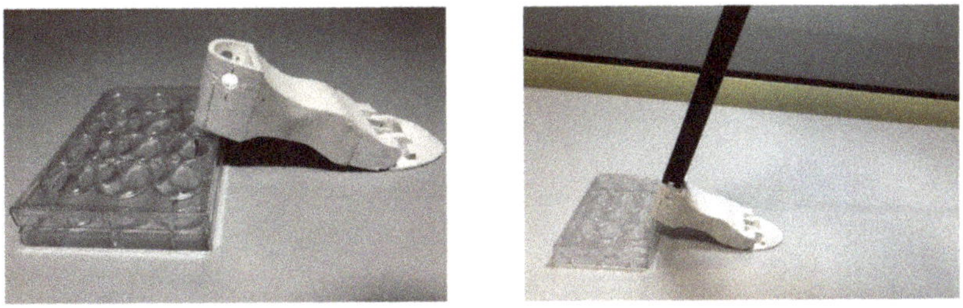

(a) (b)

Figure 12. Humanoid foot prototype assembly on a 22 mm step: (**a**) unloaded test; (**b**) loaded test.

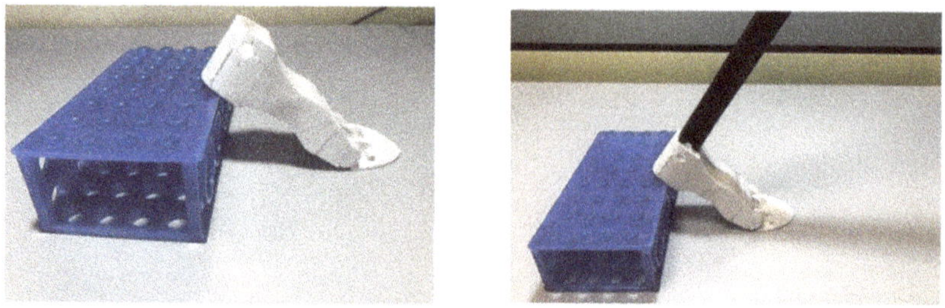

(a) (b)

Figure 13. Humanoid foot prototype assembly on a 54 mm step: (**a**) unloaded test; (**b**) loaded test.

Further experimental tests were conducted in order to measure the reaction force between the foot and the ground, and to validate the FEA results of Figures 9 and 10. By using the experimental setup shown in Figure 14, a load cell was calibrated with a set of calibration weights (accuracy 1%) before the experimental test, and was then used to measure the force, according to the scheme in Figure 15. A camera was used to measure the foot motion by tracking the marker in Figure 14 through the Kinovea software. By using the real dimension of the marker, i.e., 10 × 10 mm, the motion in pixels is converted to real-world one with a semi-automated tracking process, which has an estimated accuracy of 5% (measured through a checkerboard calibration).

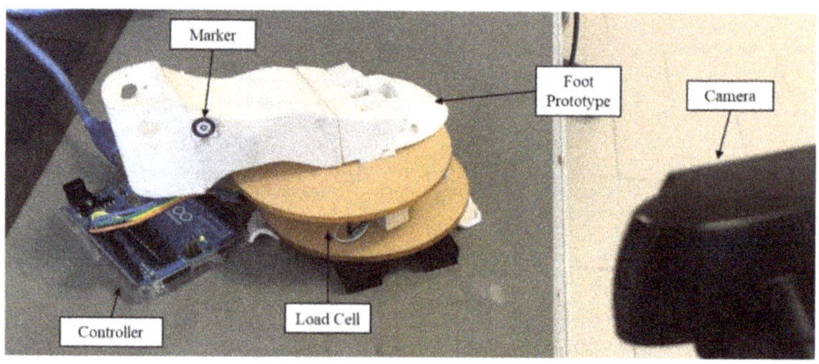

Figure 14. Experimental setup for weight and displacement acquisition.

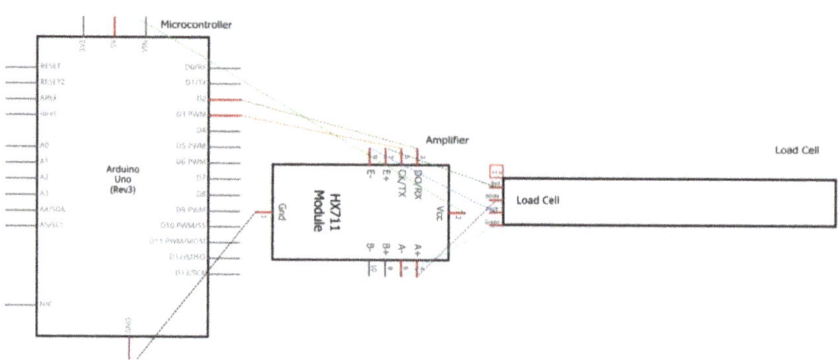

Figure 15. Experimental setup for weight and displacement acquisition.

A 1-second stance cycle was manually reproduced, moving according to the motion in Figure 16, which was measured with the camera during the experiment. During the load cycle, the marker moved 9.5 mm, with a comparable motion to the corresponding point in the analysis in Figure 10. The action of the foot on the ground, measured with the load cell, was reported in Figure 17.

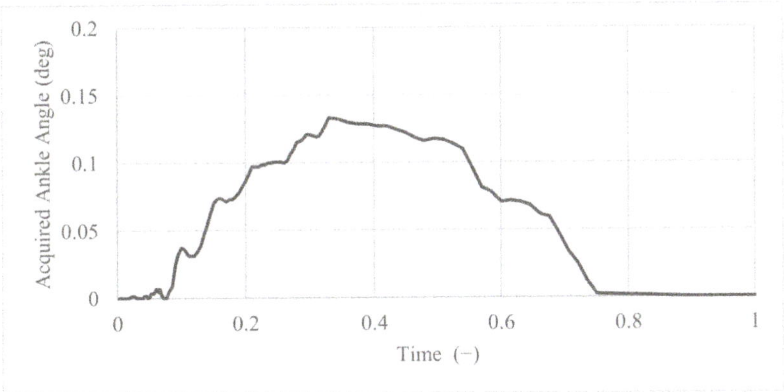

Figure 16. Ankle angle, as acquired with motion capture by the camera during a gait cycle.

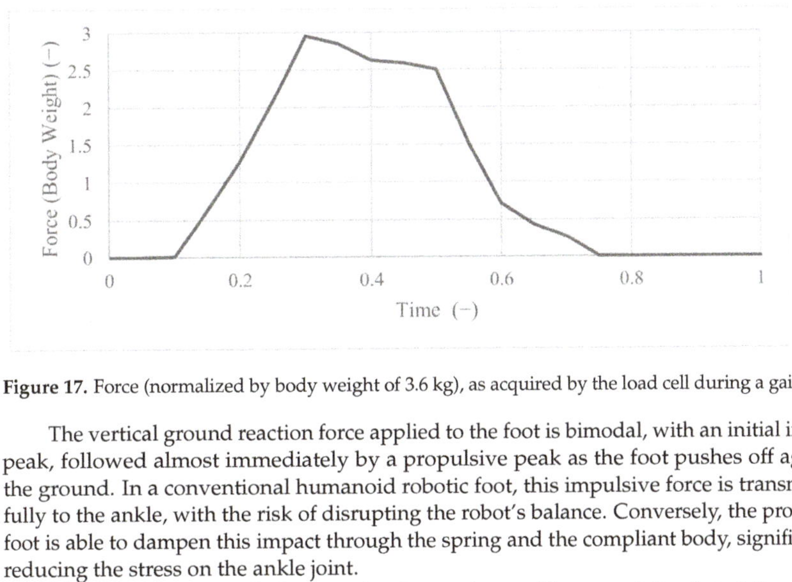

Figure 17. Force (normalized by body weight of 3.6 kg), as acquired by the load cell during a gait cycle.

The vertical ground reaction force applied to the foot is bimodal, with an initial impact peak, followed almost immediately by a propulsive peak as the foot pushes off against the ground. In a conventional humanoid robotic foot, this impulsive force is transmitted fully to the ankle, with the risk of disrupting the robot's balance. Conversely, the proposed foot is able to dampen this impact through the spring and the compliant body, significantly reducing the stress on the ankle joint.

Hall [37] and Munro [38] report that for a gait speed between 3.0 m/s to 5.0 m/s, the impact forces range from 1.6 to 2.3 times the body weight, and the propulsive forces range from 2.5 to 2.8 times the body weight. In order to compare their results to the experimental ones, an overall body weight of 3.6 kg was estimated from humanoid robots of similar sizes and proportions [10]. The acquired results, reported in Figure 16, are in a comparable range (6% error on displacement) of the FEA simulation, and show a comparable behaviour to that of the human locomotion system, with a propulsive peak of approximately 3.0 times body weight.

4. Discussion and Conclusions

In this paper, an innovative robotic foot design was introduced, based on the mobility of the human foot in walking gaits. Whereas most other humanoid robot feet use a single-body or two-segment design, the proposed system is characterized by three underactuated segments that replicate the behavior of the human foot. The motion of the human foot was studied with gait analysis, which was performed on a sample of 110 human gaits in order to acquire the motion requirements for the proposed design. A passive foot mechanism with compliant elements was then introduced in order to improve humanoid robots' impact absorption and stability during walking gaits. The proposed design was validated through a simulation with FEA and Multibody Dynamics, which demonstrated its functionality. Experiments with a 3D printed prototype of the proposed foot mechanism were also reported to prove the feasibility of this study.

Overall, this work introduced a novel design that increases foot mobility thanks to a three-segment foot mechanism, without introducing additional degrees of freedom and control complexity into the system. In future works, the foot will be assembled on a humanoid robot for full validation. In addition, the proposed humanoid foot is feasible for future developments and studies in the field of prosthetics and neurorehabilitation.

Author Contributions: Conceptualization, M.R. and D.C.; methodology, B.D.M.C.-R. and D.C.; validation, B.D.M.C.-R. and D.C.; formal analysis, B.D.M.C.-R. and D.C.; investigation, M.R., B.D.M.C.-R. and D.C.; resources, L.P. and D.C.; data curation, M.R., B.D.M.C.-R., L.P., G.P., D.C.; writing—original draft preparation, M.R. and D.C.; writing—review and editing, B.D.M.C.-R. and D.C.; visualization, M.R. and D.C.; supervision, D.C. All authors have read and agreed to the published version of the manuscript.

Funding: This work was funded by a grant from Ministero della Salute (Ricerca Corrente 2021).

Institutional Review Board Statement: This is a retrospective observational study where data were used from clinical outcomes of 11 patients who outcome a healthy gait cycle. The patients were diagnosed and treated at IRCCS NEUROMED—the Mediterranean Neurological Institute (Italy), according to the national guidelines and agreements that govern its hospital center. All data were collected as part of routine diagnosis and treatment. This study does not report on the use of experimental or new protocols. Although the rehabilitation program carried out by the 11 selected patients is described in this study, the rehabilitation program was neither designed nor modified for the purposes of this study.

Informed Consent Statement: Patients entering IRCCS NEUROMED gave a generic consent to use their data for future scientific research purposes according to GDPR (General Data Protection Regulation) regulation.

Data Availability Statement: The data presented in this study are available on request at the discretion of the corresponding author. The data are not publicly available due to privacy reasons.

Conflicts of Interest: The authors declare no conflict of interest.

References

1. Siciliano, B.; Khatib, O. *Springer Handbook of Robotics*; Springer: Berlin/Heidelberg, Germany, 2016.
2. Lim, H.-O.; Takanishi, A. Biped walking robots created at waseda university: Wl and wabian family. *Philos. Trans. R. Soc. A Math. Phys. Eng. Sci.* **2007**, *365*, 49–64. [CrossRef] [PubMed]
3. Hirose, M.; Ogawa, K. Honda humanoid robots development. *Philos. Trans. R. Soc. A Math. Phys. Eng. Sci.* **2007**, *365*, 11–19. [CrossRef] [PubMed]
4. Kajita, S.; Nagasaki, T.; Kaneko, K.; Yokoi, K.; Tanie, K. A hop towards running humanoid biped. In Proceedings of the IEEE International Conference on Robotics and Automation, ICRA'04, New Orleans, LA, USA, 26 April–1 May 2004; Volume 1, pp. 629–635.
5. Park, I.-W.; Kim, J.-Y.; Lee, J.; Oh, J.-H. Mechanical design of humanoid robot platform khr-3 (kaist humanoid robot 3: Hubo). In Proceedings of the 5th IEEE-RAS International Conference on Humanoid Robots, Tsukuba, Japan, 5–7 December 2005; pp. 321–326.
6. Tajima, R.; Honda, D.; Suga, K. Fast running experiments involving a humanoid robot. In Proceedings of the 2009 IEEE International Conference on Robotics and Automation, Kobe, Japan, 12–17 May 2009; pp. 1571–1576.
7. Tsagarakis, N.G.; Li, Z.; Saglia, J.; Caldwell, D.G. The design of the lower body of the compliant humanoid robot iCub. In Proceedings of the 2011 IEEE International Conference on Robotics and Automation, Shanghai, China, 9–13 May 2011; pp. 2035–2040.

8. Zyu, H.Q.; Ma, G.; Chen, X.; Zhang, W.; Li, J.; Gao, J. Design and development of the humanoid robot bhr-5. *Adv. Mech. Eng.* **2014**, *6*, 852937.
9. Gouaillier, D.; Hugel, V.; Blazevic, P.; Kilner, C.; Monceaux, J.; Lafourcade, P.; Marnier, B.; Serre, J.; Maisonnier, B. Mechatronic design of NAO humanoid. In Proceedings of the 2009 IEEE International Conference on Robotics and Automation, Kobe, Japan, 12–17 May 2009; pp. 769–774.
10. Russo, M.; Cafolla, D.; Ceccarelli, M. Design and experiments of a novel humanoid robot with parallel architectures. *Robotics* **2018**, *7*, 79. [CrossRef]
11. Cafolla, D.; Wang, M.; Carbone, G.; Ceccarelli, M. Larmbot: A new humanoid robot with parallel mechanisms. In Proceedings of the Symposium on Robot Design, Dynamics and Control, Udine, Italy, 20–23 June 2016; Springer: Berlin/Heidelberg, Germany, 2016; pp. 275–283.
12. Russo, M.; Ceccarelli, M. Dynamics of a humanoid robot with parallel architectures. In Proceedings of the IFToMM World Congress on Mechanism and Machine Science, Krakow, Poland, 15–18 July 2019; Springer: Berlin/Heidelberg, Germany, 2019; pp. 1799–1808.
13. Li, J.; Huang, Q.; Zhang, W.; Yu, Z.; Li, K. Flexible foot design for a humanoid robot. In Proceedings of the 2008 IEEE International Conference on Automation and Logistics, Qingdao, China, 1–3 September 2008; pp. 1414–1419.
14. Boston Dynamics. ATLAS. Available online: https://www.bostondynamics.com/atlas (accessed on 15 June 2020).
15. Lapeyre, M.; Rouanet, P.; Oudeyer, P.-Y. The poppy humanoid robot: Leg design for biped locomotion. In Proceedings of the 2013 IEEE/RSJ International Conference on Intelligent Robots and Systems, Tokyo, Japan, 3–7 November 2013; pp. 349–356.
16. Piazza, C.; della Santina, C.; Gasparri, G.M.; Catalano, M.G.; Grioli, G.; Garabini, M.; Bicchi, A. Toward an adaptive foot for natural walking. In Proceedings of the 2016 IEEE-RAS 16th International Conference on Humanoid Robots (Humanoids), Cancun, Mexico, 15–17 November 2016; pp. 1204–1210.
17. Nishiwaki, K.; Kagami, S.; Kuniyoshi, Y.; Inaba, M.; Inoue, H. Toe joints that enhance bipedal and full body motion of humanoid robots. In Proceedings of the Proceedings 2002 IEEE International Conference on Robotics and Automation (Cat. No. 02CH37292), Washington, DC, USA, 11–15 May 2002; Volume 3, pp. 3105–3110.
18. Buschmann, T.; Lohmeier, S.; Ulbrich, H. Humanoid robot lola: Design and walking control. *J. Physiol.* **2009**, *103*, 141–148. [CrossRef] [PubMed]
19. Borovac, B.; Slavnic, S. Design of multi-segment humanoid robot foot. In Proceedings of the International Conference on Research and Education in Robotics, Heidelberg, Germany, 22–24 May2008; Springer: Berlin/Heidelberg, Germany, 2008; pp. 12–18.
20. Agrawal, A.; Banala, S.K.; Agrawal, S.K.; Binder-Macleod, S.A. Design of a twodegree-of-freedom ankle-foot orthosis for robotic rehabilitation. In Proceedings of the 9th International Conference on Rehabilitation Robotics, ICORR, Chicago, IL, USA, 28 June–1 July 2005; pp. 41–44.
21. Narioka, K.; Homma, T.; Hosoda, K. Humanlike ankle-foot complex for a bipedrobot. In Proceedings of the 2012 12th IEEE-RAS International Conference on Humanoid Robots (Humanoids 2012), Osaka, Japan, 29 November–1 December 2012; pp. 15–20.
22. Hashimoto, K.; Motohashi, H.; Takashima, T.; Lim, H.; Takanishi, A. Shoes-wearable foot mechanism mimicking characteristics of human's foot arch and skin. In Proceedings of the 2013 IEEE International Conference on Robotics and Automation, Karlsruhe, Germany, 6–10 May 2013; pp. 686–691.
23. Torricelli, D.; Gonzalez, J.; Weckx, M.; Jimenez-Fabian, R.; Vanderborght, B.; Sartori, M.; Dosen, S.; Farina, D.; Lefeber, D.; Pons, J.L. Human-like compliant locomotion: State of the art of robotic implementations. *Bioinspir. Biomim.* **2016**, *11*, 051002. [CrossRef] [PubMed]
24. Knudson, D. *Fundamentals of Biomechanics*; Springer Science & Business Media: Berlin/Heidelberg, Germany, 2007.
25. Holowka, N.B.; Lieberman, D.E. Rethinking the evolution of the human foot: Insights from experimental research. *J. Exp. Biol.* **2018**, *221*, jeb174425. [CrossRef] [PubMed]
26. Takahashi, Y.; Nishiwaki, K.; Kagami, S.; Mizoguchi, H.; Inoue, H. High-speed pressure sensor grid for humanoid robot foot. In Proceedings of the 2005 IEEE/RSJ International Conference on Intelligent Robots and Systems, Edmonton, AB, Canada, 2–6 August 2005; pp. 3909–3914.
27. Kim, G.-S.; Shin, H.-J.; Yoon, J. Development of 6-axis force/moment sensor for a humanoid robot's intelligent foot. *Sens. Actuators A Phys.* **2008**, *141*, 276–281. [CrossRef]
28. Wu, B.; Luo, J.; Shen, F.; Ren, Y.; Wu, Z. Optimum design method of multi-axis force sensor integrated in humanoid robot foot system. *Measurement* **2011**, *44*, 1651–1660. [CrossRef]
29. Yuan, C.; Luo, L.-P.; Yuan, Q.; Wu, J.; Yan, R.-J.; Kim, H.; Shin, K.-S.; Han, C.-S. Development and evaluation of a compact 6-axis force/moment sensor with a serial structure for the humanoid robot foot. *Measurement* **2015**, *70*, 110–122. [CrossRef]
30. Nishiwaki, K.; Murakami, Y.; Kagami, S.; Kuniyoshi, Y.; Inaba, M.; Inoue, H. Asix-axis force sensor with parallel support mechanism to measure the ground reaction force of humanoid robot. In Proceedings of the 2002 IEEE International Conference on Robotics and Automation (Cat. No. 02CH37292), Washington, DC, USA, 11–15 May 2002; Volume 3, pp. 2277–2282.
31. Davis, R.B.; Ounpuu, S.; Tyburski, D.; Gage, J.R. A Gait Analysis Data Collection and Reduction Technique. *Hum. Mov. Sci.* **1991**, *10*, 575–587. [CrossRef]
32. Gaitlab, B. Multifactorial Gait Analysis with Davis Protocol. Available online: https://www.btsbioengineering.com/products/bts-gaitlab-gait-analysis (accessed on 14 June 2020).
33. Baker, R. Gait analysis methods in rehabilitation. *J. Neuroeng. Rehabil.* **2006**, *3*, 4. [CrossRef] [PubMed]

34. Kressig, R.W.; Beauchet, O.; European GAITRite Network Group. Guidelines for clinical applications of spatiotemporal gait analysis in older adults. *Aging Clin. Exp. Res.* **2006**, *18*, 174–176. [CrossRef] [PubMed]
35. Perry, J.; Burnfield, J.M. Gait analysis: Normal and pathological function. *J. Pediatr. Orthop.* **1992**, *12*, 815. [CrossRef]
36. Cafolla, D.; Ceccarelli, M.; Wang, M.; Carbone, G. 3d printing for feasibility check of mechanism design. *Int. J. Mech. Control* **2016**, *17*, 3–12.
37. Hall, S. *Basic Biomechanics*; McGraw-Hill Higher Education: New York, NY, USA, 2014.
38. Munro, C.F.; Miller, D.I.; Fuglevand, A.J. Ground reaction forces in running: A reexamination. *J. Biomech.* **1987**, *20*, 147–155. [CrossRef]

applied sciences

MDPI

Article

Kinematic Modelling and Motion Analysis of a Humanoid Torso Mechanism

Matteo Russo [1], Marco Ceccarelli [2] and Daniele Cafolla [3,*]

[1] Faculty of Engineering, University of Nottingham, Nottingham NG81BB, UK; matteo.russo@nottingham.ac.uk
[2] LARM2: Laboratory of Robot Mechatronics, University of Rome "Tor Vergata", 00133 Rome, Italy; marco.ceccarelli@uniroma2.it
[3] Biomechatronics Lab, IRCCS Neuromed, 86077 Pozzilli, Italy
* Correspondence: contact@danielecafolla.eu

Featured Application: Model-based motion planning of cable-driven compliant mechanisms with a flexible backbone.

Abstract: This paper introduces a novel kinematic model for a tendon-driven compliant torso mechanism for humanoid robots, which describes the complex behaviour of a system characterised by the interaction of a complex compliant element with rigid bodies and actuation tendons. Inspired by a human spine, the proposed mechanism is based on a flexible backbone whose shape is controlled by two pairs of antagonistic tendons. First, the structure is analysed to identify the main modes of motion. Then, a constant curvature kinematic model is extended to describe the behaviour of the torso mechanism under examination, which includes axial elongation/compression and torsion in addition to the main bending motion. A linearised stiffness model is also formulated to estimate the static response of the backbone. The novel model is used to evaluate the workspace of an example mechanical design, and then it is mapped onto a controller to validate the results with an experimental test on a prototype. By replacing a previous approximated model calibrated on experimental data, this kinematic model improves the accuracy and efficiency of the torso mechanism and enables the performance evaluation of the robot over the reachable workspace, to ensure that the tendon-driven architecture operates within its wrench-closure workspace.

Keywords: humanoid robotics; assistive robotics; service robotics; mechanism design; kinematics; cable-driven robots; compliant mechanisms; underactuated mechanisms; motion analysis; workspace

Citation: Russo, M.; Ceccarelli, M.; Cafolla, D. Kinematic Modelling and Motion Analysis of a Humanoid Torso Mechanism. *Appl. Sci.* **2021**, *11*, 2607. https://doi.org/10.3390/app11062607

Academic Editor: Manuel Armada

Received: 1 March 2021
Accepted: 12 March 2021
Published: 15 March 2021

1. Introduction

Mobile robots are often limited in their tasks by an environment designed to be inhabited by humans. A wheeled robot, for example, cannot climb stairs, while an animal-like robot, such as hexapods or quadrupeds, is often stopped by a closed door. Conversely, humanoid robots can interact with human-sized items and navigating in human-sized environments. Furthermore, people tend to react better to humanoid robots than to other architectures [1].

Most humanoid robots are developed to optimise their limb mobility with a "black box" approach, in which the locomotion system is developed independently from the manipulation system and then integrated only through control software. Thus, complex leg and arm designs can be found in the literature [2], whereas the robot torso usually consists of a single body with no mobility. This kind of architecture can be observed in the most successful humanoid robots, including Honda's ASIMO [3], University of Waseda's WABIAN family [4], and Softbank's NAO [5] and Pepper [6].

However, the human torso plays a key role in both locomotion and manipulation tasks, by supporting dynamic balance and increasing reach and dexterity [7–9]. Thus, recent humanoid robot designs have started introducing torso motion control [10] or one or more degrees of mobility in the torso to enhance performance. These designs range from simple serial mechanisms [11–13] to parallel architectures [14–16], which have demonstrated how much torso mobility can improve robot performance.

Among these robots, the LARMbot humanoid [17–20] is equipped with a cable-driven compliant torso mechanism with a complex behaviour inspired by the human spine: a central compliant element represents the backbone; actuated cables act as muscle tendons to define the shape of the central element, and rigid disk-like vertebrae route the cables in parallel to the backbone [15].

In previous studies, the behaviour of this mechanism has been numerically determined by linearly interpolating motion calibration data experimentally [16]. However, while a satisfying performance can be obtained with this procedure, the result is limited to a specific design and geometry and high precision can be ensured only in a limited motion range. Conversely, a model-based control would ensure correct behaviour of the mechanism over its whole range of motion and for any given geometry.

While an analytical kinematic model would significantly improve torso performance, modelling the complex cable-driven compliant mechanism poses several challenges. While hybrid mechanisms with a rigid joint in parallel to two or more actuation cables have been kinematically modelled in the past [21,22], previous results can only be partially applied to the proposed mechanism due to the increased mobility and compliance of the compliant backbone, which is capable of three main motion modes: bending, compression, and axial torsion.

A solution to this can be found in continuum and snake-like robot literature. These systems are characterised by a compliant backbone with bending capabilities [23,24] that are described through constant [25] or variable [26,27] curvature kinematics. These models use two variables to define the bending mode, namely bending angle and direction of the bending plane, and they are also capable of describing backbone elongation [28].

In this paper, an extension of the constant curvature kinematic model which includes torsion and elongation is proposed as a novel solution to define a kinematic model that accurately describes the complex behaviour of the LARMbot's torso mechanism. First, the mechanical design of the torso is introduced with its main constructive details. Then, the behaviour of the different components of the system—tendons, compliant backbone, and rigid disks—is kinematically modelled to obtain a closed-form solution for an efficient control system. Finally, the model is validated with numerical results.

2. Materials and Methods

In this section, the architecture of the LARMbot's torso mechanism is briefly introduced with its main constructive elements and its functioning. Then, the mechanism's behaviour is modelled with a constant curvature model to describe its bending compliance, expanded to include both backbone torsion and compression due to cable tension. The proposed model achieves a closed-form analytical formulation for the robot kinematics, for accurately and efficiently controlling the system as well as enabling a complete characterization of its behaviour.

2.1. Mechanical Design of LARMbot's Torso Mechanism

The LARMbot humanoid robot was developed between 2012 and 2018 [15–20] at the LARM laboratory of the University of Cassino. It was designed as a low-cost user-friendly humanoid robot for service tasks, and it features parallel mechanisms both in its legs [29] and torso [30] to achieve a high kinematic and dynamic performance.

As shown in Figure 1, this humanoid robot is characterised by 22 active degrees of freedom (DoFs), a height of 850 mm, and an overall mass of 3.600 kg only, thanks to its lightweight 3D-printed frame [17]. Its torso mechanism, CAUTO (CAssino hUmanoid TOrso), is bioinspired, reproducing Functional Spine Units (FSU). The torso is at the core of the LARMbot performance, as it contributes to both walking balance, general navigation, and manipulation tasks thanks to the additional three active DoFs that it provides to the system. The main geometrical parameters of this mechanism are reported in Table 1, while the main components are illustrated on a 3D-printed prototype in Figure 2.

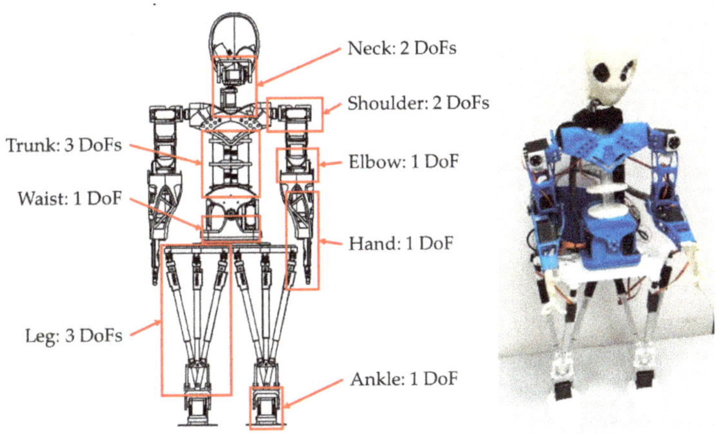

Figure 1. The LARMbot humanoid robot [17,18]: (**a**) Architecture and mobility; (**b**) A prototype.

Table 1. Technical specifications of the LARMbot's torso mechanism (CAUTO) [15,16].

Width [mm]	**Depth [mm]**	**Height [mm]**
200	150	300
Mass [kg]	**Backbone Mobility**	**Actuators**
1.200	4 Degrees of Freedom [1]	4 Servomotors [2]

[1] Bending angle, direction of bending, axial torsion, axial elongation. [2] The tendon-driven architecture means that four motors can actively control three degrees of freedom.

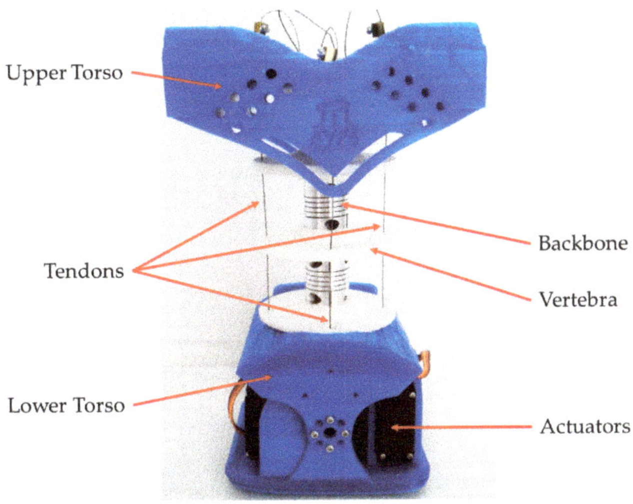

Figure 2. A prototype of the LARMbot's torso mechanism with its main components [15,16].

The main structural element of the torso is the compliant backbone, made of commercial flexible shaft couplers that are characterised by a compliant response to bending and a stiffer response to axial torsion, compression, and elongation. As these components connect the rigid vertebrae of the spine, they determine the mobility of the system. The intrinsic stiffness of the backbone also dampens the effect of unexpected wrenches on the system and restores a straight torso position in case of structural failure of the tendons. The configuration of the couplers is actuated by four tendons that determine the pose of the upper torso with respect to the lower torso.

2.2. Kinematic Modelling of a Compliant Tendon-Driven Torso Mechanism

When compared to other similar tendon-driven mechanisms with four cables driving a joint, such as the rehabilitation device in [22], the LARMbot's torso mechanism poses additional challenges due to the compliance of the backbone. First, the backbone is not an idle joint, and, because of its stiffness, it exerts a wrench on the upper platform in any non-straight configuration. This wrench results in a higher actuation force required to move the torso mechanism but improves the dynamic behaviour of the mechanism by restoring a stable position when force closure is not achieved [22]. Furthermore, the torso mechanism, which is illustrated in the kinematic scheme in Figure 3, is also characterised by a degree of underactuation.

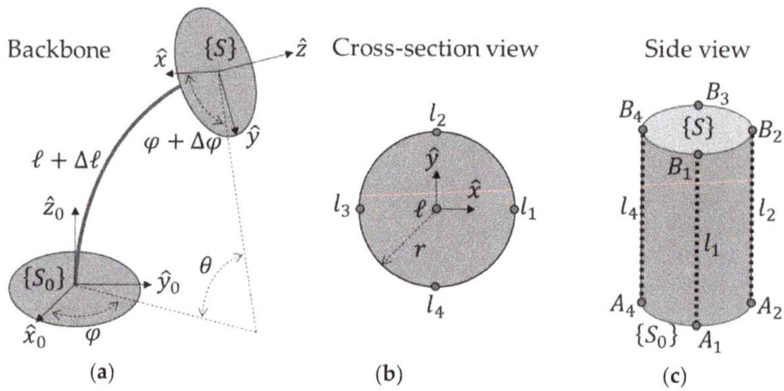

Figure 3. A kinematic scheme of the LARMbot's torso mechanism: (**a**) Main geometrical and motion parameters of the compliant backbone; (**b**) Cross-section view of a vertebra with actuation tendon routing geometry; (**c**) Side representation of the entire system in a straight configuration with tendons routing points on the lower and upper torso.

The backbone has four DoFs, which can be characterised by the following parameters with reference to the kinematic schemes in Figure 3:

- Backbone bending angle θ, which is defined as the angle between the x_0y_0 plane of the lower torso frame $\{S_0\}$ and the xy plane of the upper torso frame $\{S\}$.
- Direction of bending φ, which is defined as the angle between the plane of bending and the x_0y_0 plane of the lower torso frame $\{S_0\}$.
- Axial torsion $\Delta\varphi$, which is defined as the angle between the x_0z_0 plane of the lower torso frame $\{S_0\}$ and the xz plane of the upper torso frame $\{S\}$.
- Axial elongation/compression $\Delta\ell$, which is defined as the variation of the neutral backbone length ℓ caused by the wrench acting on the component.

However, only three of these DoFs can be actively controlled by the four actuators, as the tendons (routed as per Figure 3b,c) can only pull, but not push [22]. Thus, the configuration of the mechanism cannot be determined with motor position only, as its torsional mobility is underactuated and depends on the passive spring-like elements in the system, that is, the torso's backbone. Thus, kinematics must be integrated with a stiffness model of the backbone for complete characterization.

As previously mentioned, a piecewise constant curvature kinematic (PCCK) model is used to describe the main bending of the backbone. PCCK models have been conceived as a convenient tool to analyse and control continuum robots, as explained in [25], even though they are recently being replaced by more complex models with variable curvature [26,27] to accurately describe the hyper-redundant architecture of those systems, with complex motion coupling between independently bending sections. However, as the proposed torso mechanism can be represented by a single bending section, a constant curvature model can appropriately model the compliant backbone with limited approximation errors. To describe the pose of the lower and upper torso, the reference frames $\{S_0\}$ and $\{S\}$ can be defined at the base and at the end of the backbone, respectively. By assuming constant curvature along the backbone, the translation along its central curve $\boldsymbol{t} \in \mathbb{R}^3$ can be expressed by an arc of a circle as:

$$\boldsymbol{t}(\theta, \varphi, \Delta\ell) = \frac{\ell + \Delta\ell}{\theta} \begin{bmatrix} \cos\varphi(1-\cos\theta) \\ \sin\varphi(1-\cos\theta) \\ \sin\theta \end{bmatrix}, \tag{1}$$

while the rotation from $\{S_0\}$ to $\{S\}$ can be written as a rotation matrix $\boldsymbol{R} \in SO(3)$ that is:

$$\boldsymbol{R}(\theta, \varphi, \Delta\varphi) = \boldsymbol{R}_z(\varphi) \cdot \boldsymbol{R}_y(\theta) \Delta \boldsymbol{R}_z(\Delta\varphi - \varphi), \tag{2}$$

where $\boldsymbol{R}_z(\varphi)$ represents a rotation of φ around the z-axis, $\boldsymbol{R}_y(\theta)$ represents a rotation of θ around the y-axis, and $\boldsymbol{R}_z(\Delta\varphi - \varphi)$ represents a rotation of $\Delta\varphi - \varphi$ around the z-axis. These equations expand the conventional PCCK that is described in [25] to include both the elongation/compression and the axial torsion that are characteristic of the mechanism under examination. The forward kinematics of the torso mechanism can be thus defined with the homogeneous transformation $\boldsymbol{T} \in SE(3)$ from $\{S_0\}$ to $\{S\}$ as:

$$\boldsymbol{T}(\theta, \varphi, \Delta\ell, \Delta\varphi) = \begin{bmatrix} \boldsymbol{R}(\theta, \varphi, \Delta\varphi) & \boldsymbol{t}(\theta, \varphi, \Delta\ell) \\ \boldsymbol{0} & \boldsymbol{1} \end{bmatrix}. \tag{3}$$

Equation (3) represents only the first part of the kinematics of the structure, as it relates the pose of the upper torso (with respect to the lower torso) only to its configuration, not to the length of the actuating tendons. To compute tendon length $\boldsymbol{l} = (l_1;\ l_2;\ l_3;\ l_4)^T$, the four tendons can be modelled as circle arcs that bend in parallel to the backbone, with the same bending angle and direction. Thus, their lengths can be evaluated as:

$$l_i = \ell + \Delta\ell + \theta r \cos\left(\varphi + \Delta\varphi + \frac{(i-1)\pi}{2}\right); i = \{1,\ 2,\ 3,\ 4\}. \tag{4}$$

By using Equations (3) and (4), the relationship between actuation vector $\boldsymbol{l}$ and upper platform pose $\boldsymbol{T}$ can be obtained, defining the kinematic input-output function of the robotic system. However, as previously mentioned, the stiffness of the backbone must be considered for full characterization. In order to do so, a linear elastic behaviour of the backbone is assumed, with a torsional stiffness k_t, a compression module of k_c, and a bending stiffness k_b that can be evaluated from the material properties and are related to motion parameters as:

$$M_z = k_t \Delta\varphi;\ F_z = k_c \Delta\ell;\ M_\theta = k_b \theta, \tag{5}$$

where M_z is the moment around the z-axis of the backbone, representing axial torsion; F_z is the force along the z-axis of the backbone, representing compression; and $M_\theta = M_y \sin\varphi - M_x \cos\varphi$ is the bending moment acting on the backbone, which can be written as a combination of the moments around the x- and y-axes of the backbone, M_x and M_y. Given the tensions in the tendons F_1, F_2, F_3, and F_4, these values can be computed as:

$$\begin{aligned} M_x &= (F_2 - F_4)r; \\ M_y &= (F_3 - F_1)r; \\ M_\theta &= M_y \sin\varphi - M_x \cos\varphi; \\ F_z &= \sum_{i=1}^{4} F_i. \end{aligned} \tag{6}$$

Thus, as expressed in Equation (6) and with reference to Figure 4, the four actuation cables can actively control the two DoFs related to bending, whereas the compression/elongation and torsion of the backbone are coupled and cannot be actuated independently.

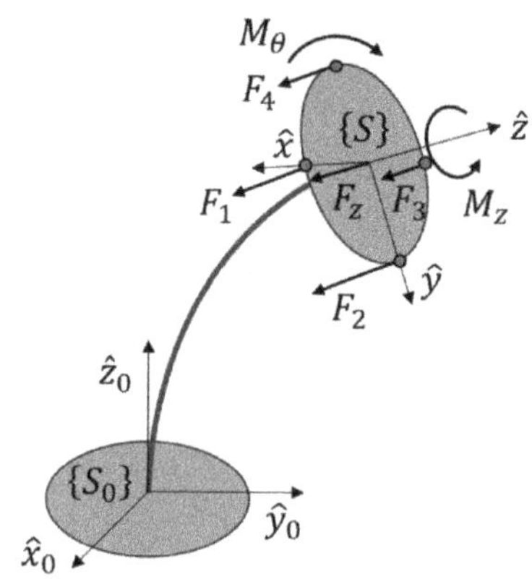

Figure 4. A free-body diagram of the upper torso without any external wrench. The reactions of the backbone are represented by M_z, F_z, and M_θ, whereas F_1, F_2, F_3, and F_4 are the tensions in the actuation tendons.

If a general external wrench $(P_x;\ P_y;\ P_z;\ W_x;\ W_y;\ W_z)^T$ is applied to the upper torso, the elongation and torsion of the backbone can be estimated from:

$$\begin{aligned} M_x &= (F_2 - F_4)r + \tfrac{\ell+\Delta\ell}{\theta}(P_z \sin\varphi(1-\cos\theta) - P_y \sin\theta) + W_x; \\ M_y &= (F_3 - F_1)r + \tfrac{\ell+\Delta\ell}{\theta}(P_x \sin\theta - P_z \cos\varphi(1-\cos\theta)) + W_y; \\ F_z &= P_z + \sum_{i=1}^{4} F_i; \end{aligned} \tag{7}$$

and decoupled through a pose measurement with an onboard motion sensor (for example, with the Inertial Measurement Unit (IMU) installed on the LARMbot's upper torso).

In summary, the model in this section can be used to accurately move the CAUTO torso by integrating the feedback from the motors' sensors and the IMUs on the upper torso and the head of the humanoid robot to enable closed-loop control.

3. Results

The kinematic model defined in the previous section is here used to evaluate the workspace of the torso mechanism and its characteristics. Furthermore, a simple mapping of the kinematic variables to an input joystick device is discussed for smooth motion planning with experimental validation of a prototype moving in the torso's workspace that is evaluated with the proposed model.

3.1. Workspace of the Proposed Torso Mechanism

Whereas the original motion planning was based on a regression between calibration points, the new analytical kinematic model enables a full characterization of the reachable workspace of the robot by defining all the reachable points of the mechanism. By using the forward kinematics in Equation (3), the reachable workspace of the mechanism was evaluated, given the size and motion parameters of the torso prototype in Figure 2, which are reported in Table 2. Once defined the motion limits on angle and direction of bending, as per Table 2, and by assuming a negligible compression of the backbone, the minimum and maximum tendon lengths were computed through Equation (4). The reachable workspace was defined as the geometrical locus of all the points that can be reached by the upper endpoint of the backbone's centreline, and it was computed in MATLAB R2021a as plotted in Figure 5. As expected, for a negligible compression, the operational workspace shape resembles a convex surface.

Table 2. Size and motion parameters of the torso mechanism prototype in Figure 2.

Backbone Length ℓ	Bending Angle θ	Bending Direction φ
129.0 mm	[0; 15] deg	[0; 360] deg
Backbone compression $\Delta\ell$	**Tendon length l_i**	**Tendon radius r**
0.0 mm	[120.0; 138.0] mm	34.0 mm

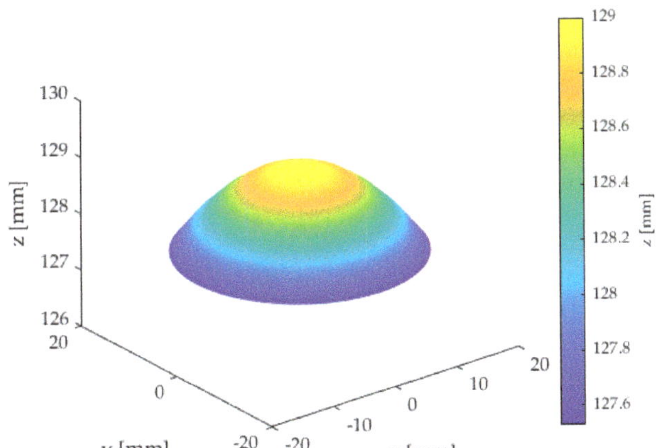

Figure 5. Workspace of the LARMbot's torso mechanism evaluated through the proposed PCCK model with elongation and torsion.

3.2. Experimental Validation

To validate the proposed kinematic model, the workspace result was compared to the reachable workspace of the prototype as per Figure 6, which was evaluated in the configuration space of the robot by measuring the orientation of the upper torso. The experimental setup is illustrated in Figure 7 with its main components, which include the torso prototype with an embedded IMU on the upper torso on the last Functional Spinal Unit (FSU) to extract the orientation data, an Arduino microcontroller with the kinematic model, a power supply, and a joystick (spring-loaded to the centre) to teleoperate the system. All the motors are embedded in the waist of the robot not to hinder spine motion. While a wired control was used in this experimental validation, wireless communication can enable autonomous navigation with the humanoid robot. The motion of the torso was mapped to the joystick motion according to the proposed kinematic model, with the direction of the joystick controlling the direction of bending φ and the position of the joystick linearly mapped to the bending angle θ. The extreme positions of this mapping are illustrated in Figure 6 with the corresponding upper torso pose.

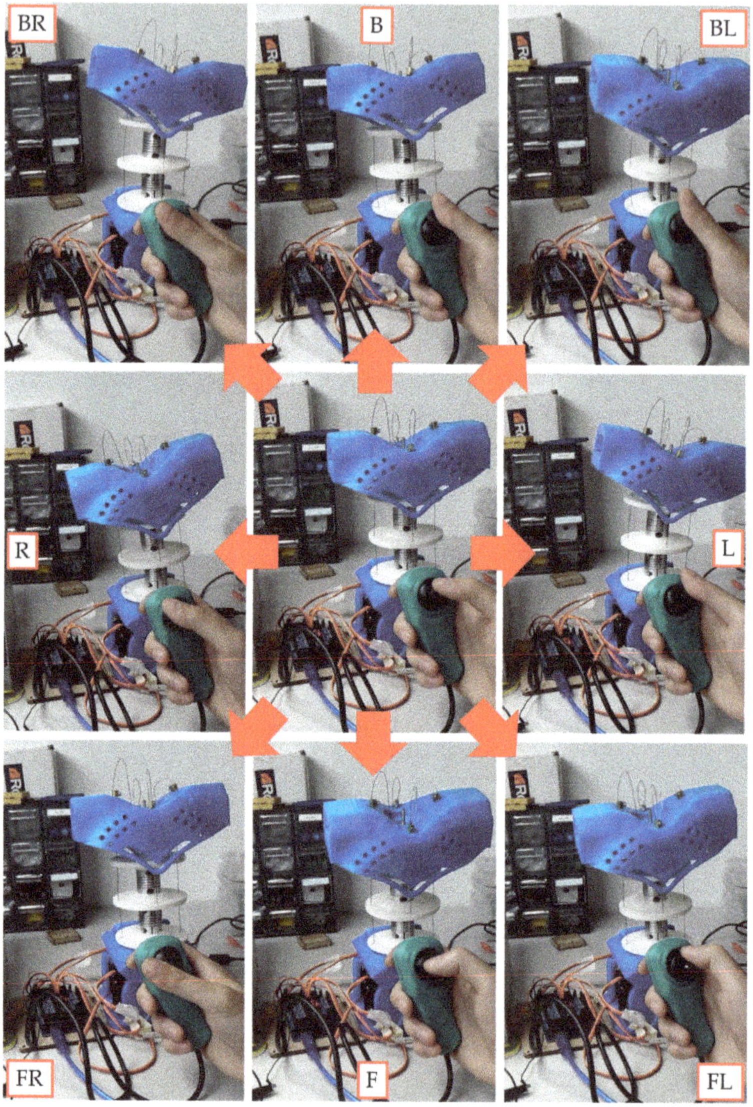

Figure 6. Motion control of the LARMbot's torso mechanism through a joystick, with motion mapping to angle of direction and bending angle. F: forward; FL: forward-left; L: left; BL: backward-left; B: backward; BR: backward-right; R: right; FR: forward-right.

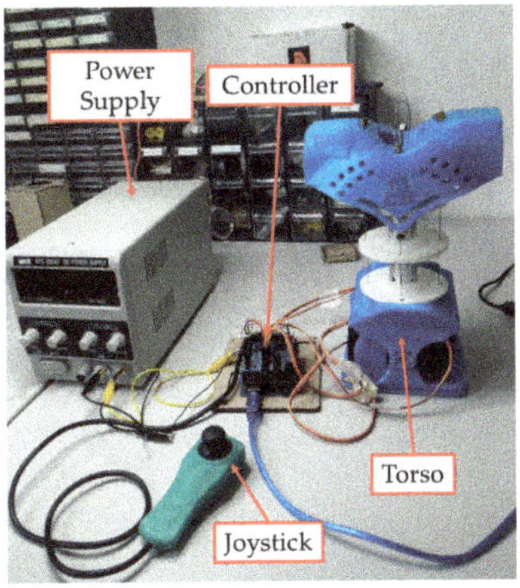

Figure 7. Experimental setup for the validation of the proposed kinematic model, including the torso mechanism, an Arduino microcontroller, a joystick for teleoperation, and a power supply.

In the reported experiment, the robot was driven from a central position (straight backbone) to a point on the outer border of its workspace and then was moved around the entire border once before going back to the central position (see Figure 6). This motion was repeated four times, and the motion data of the upper torso were reported in Figures 8 and 9 as acquired by the IMU. Figure 8 illustrates the acquired motion as bending around the x- and y-axes of the torso mechanism, whereas Figure 9 reports the overall bending angle and axial torsion of the torso. The proposed torso mechanism can reach and move around the limits on its workspace, as illustrated in Figure 10, which shows the acquired IMU data points during a manually operated motion around the reachable workspace that was computed with the new analytical model. The experiments also show a negligible backbone elongation and torsion in absence of any external wrench, as motion is here obtained by tendon action only. The proposed model achieves a smooth motion, with acceleration values always below 36 rad/s^2, as reported in Figure 11. The power consumption is also fairly low, with a maximum value below 15 W, as reported in Figure 12. The experimental results are summarised in Table 3.

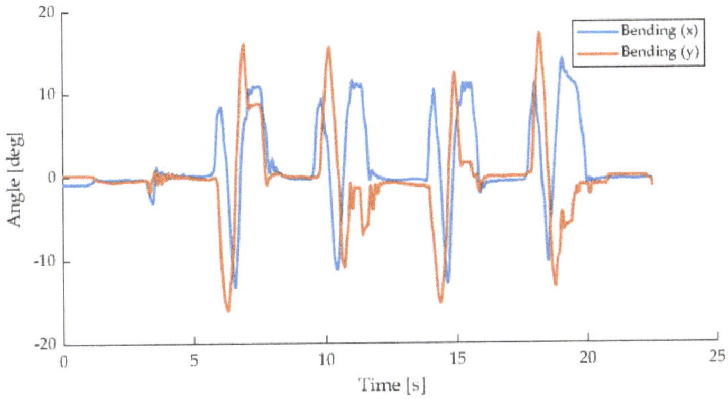

Figure 8. Raw orientation data acquired by the IMU on the upper torso expressed as bending around the x- and y-axes of the upper torso of the robot in Figure 6.

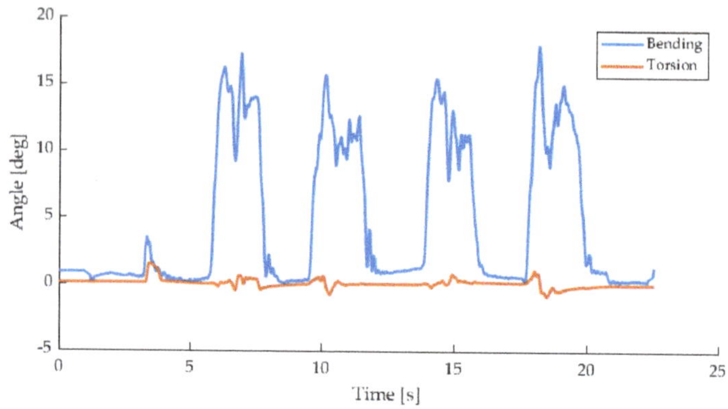

Figure 9. Orientation data acquired by the IMU on the upper torso and mapped as torso bending angle θ and axial torsion $\Delta\varphi$.

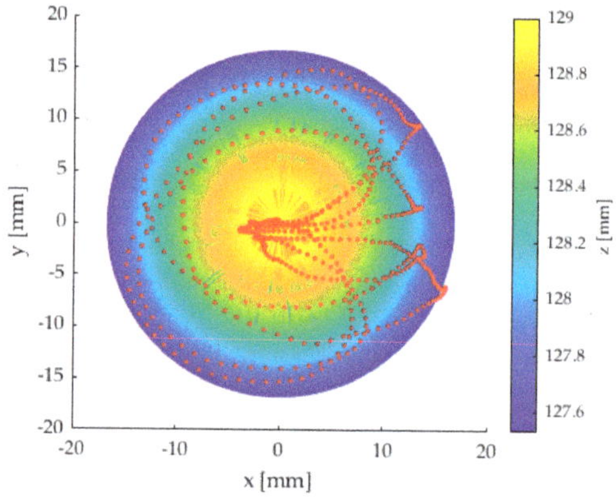

Figure 10. Top view of the acquired teleoperated torso motion and comparison with the computed workspace in Figure 5.

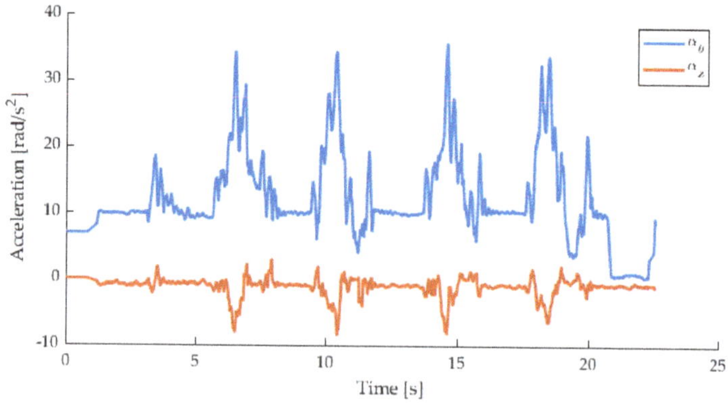

Figure 11. Acceleration data acquired by the IMU on the upper torso mapped as angular acceleration around the instantaneous bending axis α_θ and around the backbone axis α_z.

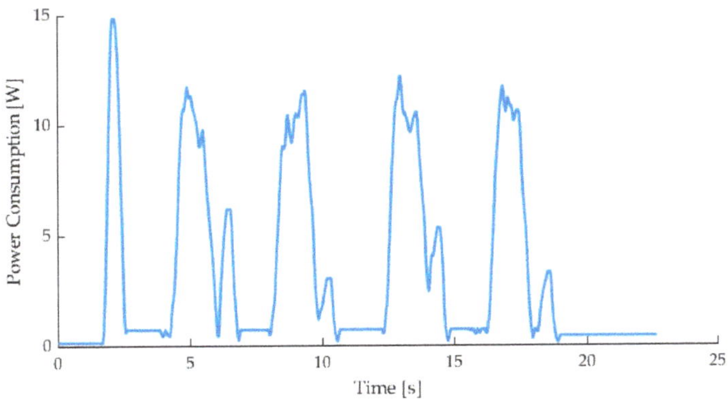

Figure 12. Power consumption acquired with a current sensor during the torso motion reported in Figures 8 and 9.

Table 3. Experimental results.

	Maximum Angular Value	Maximum Acceleration
Backbone bending	$\theta = 17.95$ deg	$\alpha_\theta = 35.75\ \mathrm{rad/s^2}$
Backbone torsion	$\Delta\varphi = 1.54$ deg	$\alpha_z = 8.32\ \mathrm{rad/s^2}$

4. Discussion

In this paper, a kinematic model was proposed for a four-DoF tendon-driven compliant torso mechanism for humanoid robots. The proposed model is presented with a closed-form analytical formulation and it is used to compute the workspace of the mechanism under examination. The results are validated on a prototype, which is operated with a joystick on which the proposed model is mapped. The main findings of this paper can be summarised as:

- *Mobility analysis:* The mobility of a four-DoF tendon-driven compliant torso mechanism for humanoid robots is analysed to identify its main modes of motion. The degrees of mobility of the robot are further classified as active degrees of freedom, which can be controlled by the action of one or more actuators, and passive degrees of freedom, which depend on the intrinsic stiffness of the system only.
- *Kinematic modelling:* The proposed constant curvature kinematic model, which is usually used for continuum robots, is used to describe the main bending mode of motion $(\theta,\ \varphi)$ of the spine of the torso. The conventional constant curvature kinematic model is characterised for the analysed system, and an expanded model is here proposed to include backbone elongation $\Delta\ell$ and torsion $\Delta\varphi$.
- *Stiffness modelling:* A linearised stiffness model is introduced to create an efficient framework with lumped parameters to relate the deformation of the backbone to the wrench acting on the upper torso of the mechanism. This model also outlines how the axial torsion of the backbone cannot be controlled by the actuation tendons, as it can only be caused by external wrenches.
- *Workspace analysis:* The behaviour of the torso mechanism is characterised by evaluating its motion limit as a reachable workspace, bending and direction angles, and tendon displacement.
- *Joystick mapping:* A joystick mapping of the main motion parameters of the proposed kinematic model is proposed, with direction and angle of bending linearly mapped to joystick orientation and magnitude respectively.

- *Experimental validation:* The workspace computed with the proposed model is confirmed by experiments with the prototype, whose motion is acquired by an onboard inertial measurement unit.

The main advantages of the proposed model, which replaces a previous approximated model calibrated on experimental data, can be identified as higher accuracy and efficiency and a quick response thanks to the closed-form analytical formulation. Furthermore, this model can be also used to evaluate the operating performance of the robot over the reachable workspace and to ensure that the tendon-driven architecture operates within its wrench-closure workspace.

Future works will focus on refining the current model by introducing a dynamic and stiffness model of the system, which will be validated with the addition of load cells on the tendons; by defining the performance of the torso through numerical indices that can be used to optimise motion planning; by analysing the compliant behaviour of the backbone with Finite Element analysis simulations in order to evaluate the error introduced by the lumped parameter model that is proposed in this manuscript; and finally by integrating this model with the kinematics of the whole humanoid, in order to implement a dynamic control of the system.

Author Contributions: Conceptualization, M.R., M.C., and D.C.; methodology, M.R.; software, M.R. and D.C.; validation, D.C.; formal analysis, M.R. and M.C.; investigation, M.R.; resources, M.C. and D.C.; data curation, M.R.; writing—original draft preparation, M.R.; writing—review and editing, M.R., M.C., and D.C.; visualization, M.R. and D.C.; supervision, M.C. and D.C.; project administration, M.C. and D.C.; funding acquisition, D.C. All authors have read and agreed to the published version of the manuscript.

Funding: This work was funded by a grant from Ministero della Salute (Ricerca Corrente 2021).

Institutional Review Board Statement: Not applicable.

Informed Consent Statement: Not applicable.

Data Availability Statement: Not applicable.

Conflicts of Interest: The authors declare no conflict of interest.

References

1. Fitzpatrick, P.; Harada, K.; Kemp, C.C.; Matsumoto, Y.; Yokoi, K.; Yoshida, E. Humanoids. In *Springer Handbook of Robotics*; Springer: Cham, Switzerland, 2016; pp. 1789–1818.
2. Gulletta, G.; Erlhagen, W.; Bicho, E. Human-like arm motion generation: A Review. *Robotics* **2020**, *9*, 102. [CrossRef]
3. Sakagami, Y.; Watanabe, R.; Aoyama, C.; Matsunaga, S.; Higaki, N.; Fujimura, K. The intelligent ASIMO: System overview and integration. In Proceedings of the IEEE/RSJ International Conference on Intelligent Robots and Systems, Lausanne, Switzerland, 30 September–4 October 2002; IEEE: Piscataway, NJ, USA, 2002; Volume 3, pp. 2478–2483.
4. Ogura, Y.; Aikawa, H.; Shimomura, K.; Kondo, H.; Morishima, A.; Lim, H.O.; Takanishi, A. Development of a new humanoid robot WABIAN-2. In Proceedings of the 2006 IEEE International Conference on Robotics and Automation, 2006. ICRA 2006, Orlando, FL, USA, 15–19 May 2006; IEEE: Piscataway, NJ, USA, 2006; pp. 76–81.
5. Gouaillier, D.; Hugel, V.; Blazevic, P.; Kilner, C.; Monceaux, J.; Lafourcade, P.; Maisonnier, B. Mechatronic design of NAO humanoid. In Proceedings of the 2009 IEEE International Conference on Robotics and Automation, Kobe, Japan, 12–17 May 2009; IEEE: Piscataway, NJ, USA, 2009; pp. 769–774.
6. Pandey, A.K.; Gelin, R. A mass-produced sociable humanoid robot: Pepper: The first machine of its kind. *IEEE Robot. Autom. Mag.* **2018**, *25*, 40–48. [CrossRef]
7. Simonidis, C.; Stelzner, G.N.; Seemann, W. A kinematic study of human torso motion. In Proceedings of the International Design Engineering Technical Conferences and Computers and Information in Engineering Conference, Las Vegas, NV, USA, 4–7 September 2007; Volume 48094, pp. 773–780.
8. Kljuno, E.; Williams, R.L. Humanoid walking robot: Modeling, inverse dynamics, and gain scheduling control. *J. Robot.* **2010**, *2010*, 278597. [CrossRef]
9. Knudson, D. *Fundamentals of Biomechanics*; Springer Science & Business Media: Berlin, Germany, 2007.
10. Li, Q.; Yu, Z.; Chen, X.; Meng, F.; Meng, L.; Huang, Q. Dynamic Torso Posture Compliance Control for Standing Balance of Position-Controlled Humanoid Robots. In Proceedings of the 2020 5th International Conference on Advanced Robotics and Mechatronics (ICARM), Shenzhen, China, 18–21 December 2020; IEEE: Piscataway, NJ, USA, 2020; pp. 529–534.

11. Kaneko, K.; Kaminaga, H.; Sakaguchi, T.; Kajita, S.; Morisawa, M.; Kumagai, I.; Kanehiro, F. Humanoid robot HRP-5P: An electrically actuated humanoid robot with high-power and wide-range joints. *IEEE Robot. Autom. Lett.* **2019**, *4*, 1431–1438. [CrossRef]
12. Yao, P.; Li, T.; Luo, M.; Zhang, Q.; Tan, Z. Mechanism design of a humanoid robotic torso based on bionic optimization. *Int. J. Hum. Robot.* **2017**, *14*, 1750010. [CrossRef]
13. Cao, B.; Sun, K.; Jin, M.; Huang, C.; Zhang, Y.; Liu, H. Design and development of a two-DOF torso for humanoid robot. In Proceedings of the 2016 IEEE International Conference on Advanced Intelligent Mechatronics (AIM), Banff, AB, Canada, 12–15 July 2016; IEEE: Piscataway, NJ, USA, 2016; pp. 46–51.
14. Fiorio, L.; Scalzo, A.; Natale, L.; Metta, G.; Parmiggiani, A. A parallel kinematic mechanism for the torso of a humanoid robot: Design, construction and validation. In Proceedings of the 2017 IEEE/RSJ International Conference on Intelligent Robots and Systems (IROS), Vancouver, BC, Canada, 24–28 September 2017; IEEE: Piscataway, NJ, USA, 2017; pp. 681–688.
15. Cafolla, D.; Ceccarelli, M. Design and simulation of a cable-driven vertebra-based humanoid torso. *Int. J. Hum. Robot.* **2016**, *13*, 1650015. [CrossRef]
16. Cafolla, D.; Ceccarelli, M. An experimental validation of a novel humanoid torso. *Robot. Auton. Syst.* **2017**, *91*, 299–313. [CrossRef]
17. Russo, M.; Cafolla, D.; Ceccarelli, M. Design and experiments of a novel humanoid robot with parallel architectures. *Robotics* **2018**, *7*, 79. [CrossRef]
18. Ceccarelli, M.; Cafolla, D.; Russo, M.; Carbone, G. LARMBot Humanoid Design Towards a Prototype. *Moj Int. Jnl Appl. Bionics Biomech.* **2017**, *1*, 00008.
19. Cafolla, D.; Marco, C. Design and simulation of humanoid Spine. *Mech. Mach. Sci.* **2015**, *24*, 585–593.
20. Cafolla, D.; Ceccarelli, M. Design and FEM analysis of a novel humanoid torso. *Mech. Mach. Sci.* **2015**, *25*, 477–488.
21. Li, C.; Gu, X.; Ren, H. A cable-driven flexible robotic grasper with lego-like modular and reconfigurable joints. *IEEE/ASME Trans. Mechatron.* **2017**, *22*, 2757–2767. [CrossRef]
22. Russo, M.; Ceccarelli, M. Analysis of a Wearable Robotic System for Ankle Rehabilitation. *Machines* **2020**, *8*, 48. [CrossRef]
23. Amanov, E.; Nguyen, T.D.; Burgner-Kahrs, J. Tendon-driven continuum robots with extensible sections—A model-based evaluation of path-following motions. *Int. J. Robot. Res.* **2019**. [CrossRef]
24. Russo, M.; Raimondi, L.; Dong, X.; Axinte, D.; Kell, J. Task-oriented optimal dimensional synthesis of robotic manipulators with limited mobility. *Robot. Comput. Integr. Manuf.* **2021**, *69*, 102096. [CrossRef]
25. Webster, R.J., III; Jones, B.A. Design and kinematic modeling of constant curvature continuum robots: A review. *Int. J. Robot. Res.* **2010**, *29*, 1661–1683. [CrossRef]
26. Gonthina, P.S.; Kapadia, A.D.; Godage, I.S.; Walker, I.D. Modeling variable curvature parallel continuum robots using euler curves. In Proceedings of the 2019 International Conference on Robotics and Automation (ICRA), Montreal, QC, Canada, 20–24 May 2019; IEEE: Piscataway, NJ, USA, 2019; pp. 1679–1685.
27. Mahl, T.; Hildebrandt, A.; Sawodny, O. A variable curvature continuum kinematics for kinematic control of the bionic handling assistant. *IEEE Trans. Robot.* **2014**, *30*, 935–949. [CrossRef]
28. Garriga-Casanovas, A.; Rodriguez y Baena, F. Kinematics of continuum robots with constant curvature bending and extension capabilities. *J. Mech. Robot.* **2019**, *11*, 011010. [CrossRef]
29. Russo, M.; Ceccarelli, M.; Takeda, Y. Force transmission and constraint analysis of a 3-SPR parallel manipulator. *Proc. Inst. Mech. Eng. Part C J. Mech. Eng. Sci.* **2018**, *232*, 4399–4409. [CrossRef]
30. Cafolla, D.; Ceccarelli, M. Characteristics and Performance of CAUTO (CAssino hUmanoid TOrso) Prototype. *Inventions* **2017**, *2*, 17. [CrossRef]

applied sciences

MDPI

Article

Variable Stiffness Mechanism for the Reduction of Cutting Forces in Robotic Deburring

Matteo Bottin, Silvio Cocuzza * and Matteo Massaro

Department of Industrial Engineering, University of Padova, Via Venezia 1, 35131 Padova, Italy; matteo.bottin@unipd.it (M.B.); matteo.massaro@unipd.it (M.M.)
* Correspondence: silvio.cocuzza@unipd.it; Tel.: +39-049-8276793

Featured Application: The proposed variable stiffness mechanism will be used in robotic deburring applications, in particular in applications with high and irregular burr profiles, ensuring a precision cut without damaging the cutting tool or the robot.

Abstract: One of the main issues related to robotic deburring is that the tool can get damaged or stopped when the burr thickness exceeds a certain threshold. The aim of this work is to devise a mechanism that can reduce cutting forces automatically, in the event that the burr is too high, and is able to return to the baseline configuration when the burr thickness is acceptable again. On the one hand, in normal cutting conditions, the mechanism should have high stiffness to ensure high cutting precision. On the other hand, when the burr is too high the mechanism should exploit its compliance to reduce the cutting forces and, as a consequence, a second cutting cycle will be necessary to completely remove the burr. After the conceptual design of the mechanism and the specification of the desired stiffness curve, the main design parameters of the system are derived thanks to an optimization method. The effectiveness of the proposed mechanism is verified by means of dynamic simulations using selected test cases. A reduction up to 60% of the cutting forces is obtained, considering a steel burr up to 6 mm high.

Keywords: deburring; robot; cutting forces; compliant mechanism

Citation: Bottin, M.; Cocuzza, S.; Massaro, M. Variable Stiffness Mechanism for the Reduction of Cutting Forces in Robotic Deburring. *Appl. Sci.* **2021**, *11*, 2883. https://doi.org/10.3390/app11062883

Academic Editor: Giuseppe Carbone

Received: 18 February 2021
Accepted: 18 March 2021
Published: 23 March 2021

1. Introduction

The deburring process is mainly performed by using manual labor or computerized numerical control (CNC) machines, but in the research field also robotic solutions have been considered [1]. Workers can easily adapt to the unpredictability of the burr, but the dangerous working conditions and the difficulty of the task suggest using dedicated machines. To do so, special CNC machines have been developed, even if a comparable robotic solution costs less than 1/3 than the cost of a CNC machine [2]; moreover, when compared to a CNC machine, an industrial robot can provide a wider workspace, is usually more flexible (since it has a higher number of degrees of freedom), and can provide a higher dexterity, adapting its movement to complex geometries. The drawback of robotic solutions relies on the fact that robots are generally less stiff and accurate than CNC machines [3]: this affects the finishing of the workpiece, can cause chatter [4], and increases the programming and setup time.

The literature related to robotic deburring is limited [1], when compared to the literature related to robotic machining. This can be related to the high variability of the burr profile, which may behave differently depending on the workpiece material [5,6] and may influence burr removal time [7]. In other fields, the effect of burr properties and grinding tool on the final result has been studied. In particular, it has been of great interest in the biomedical field, where the bone grinding is crucial for surgical operations [8]. In fact, the shape of the tool itself provides different deburring results both in terms of temperature and tool wear [9], which may be fundamental for industrial applications, in which the burr to be removed is made of a material sensible to temperature (e.g., plastics or aluminum) or in intensive and repetitive tasks (e.g., robotic deburring). In depth analysis of the grinding process has been developed for yttrium aluminum garnet (YAG) single crystals, in which it is shown that strain rate plays an important role in the final deformation [10]. This knowledge can be integrated in robotic deburring for specific applications.

In industrial robotics, most of the research effort has focused on obtaining a precise control of the robot motion: Tao et al. [11] developed a sliding mode control method based on radial basis function (RBF) neural network, which is able to learn uncertain control actions. Quian et al. [12] proposed a sensorless force controller that uses the information retrieved only from collaborative robot internal motor sensors to estimate the burr dimensions. Other methods used to handle the high variability of the burr profile include the use of sensors [13,14], but these approaches are not suitable in many deburring applications since the working environment is usually dirty and dusty. If the ambient is clean, however, sensors can be used effectively [15]. El Naser et al. [16] and Gou et al. [17] have developed a robotic arm designed purposely for deburring. This approach is similar to the development of a special CNC machine. On the other hand, Lai et al. [18] have presented a novel design that includes an additional manipulator to be linked to an existing industrial robot. In this way, the kinematic chain becomes similar to the one of a parallel robot, thus increasing stiffness. These last two approaches, however, requires a complex external equipment, which may be expensive and time-consuming for industrial applications.

To increase or decrease the stiffness of the mechanical system during specific tasks, variable stiffness actuators (VSA) and mechanisms have been developed. They can be useful in many fields of application, such as in machining operations, since the stiffness can be changed to improve the surface finishing or reduce vibrations. During the years different solutions have been proposed. Vuong et al. [19] proposed a mechanism that can be tuned to change its stiffness in a wide range, from zero to (theoretically) infinite. It uses multiple linear guides in combination with a moving pivot and linear springs to achieve this functionality. Another example of VSA can be found in [20], in which the authors focused on the compactness of the system and obtained a stiffness inversely proportional to the gear displacement. A variable stiffness robotic arm has also been designed [21]: in this case a 7-degrees-of-freedom manipulator takes advantage of VSA to perform different tasks in the SHERPA mission. Moreover, Petit et al. [22] developed an ad-hoc controller to take advantage of VSA capabilities.

Deburring is a redundant task: the workpiece can be machined on any point of the grinding surface of the grinding wheel. Therefore, it is possible to achieve functional redundancy [23], and the robot can perform the same task using different configurations. For example, if the stiffness and inertial properties of the robot are known or experimentally identified [24,25] it is possible to choose a proper configuration so that the kinematic chain of the robot is stiff enough to perform the grinding operation. However, this field of research has only been partially explored. To overcome the functional redundancy, Nemec et al. [26] proposed to create robot recipes by means of programming by demonstration: by analyzing the movements of an expert human demonstrator, the robot can perform the same task (in this case polishing) exploiting redundancy. Moreover, as can be seen in [27], the stiffness map of a robot can be used to improve the accuracy of the assigned task. In particular, in this work, the authors introduced a compensation strategy to improve a roll hemming process. A stiffness model of a parallel haptic mechanism suitable for real-time computation has been proposed in [28].

When traditional industrial robots are concerned, it is usually difficult to implement force feedback controllers for deburring control purposes. Moreover, industrial applications require high flexibility and reduced implementation costs.

In sum, the literature review suggests that novel deburring techniques are needed to improve flexibility and precision and reduce costs. Indeed, CNC machines provide the best performance, but are generally more expensive and less flexible when compared to robotic solutions. One of the main issues related to robotic deburring is that the tool can get damaged or stopped when the burr thickness exceeds a certain threshold. A simple system suitable for industrial application is thus required to automatically compensate cutting force peaks. A passive mechanical system is a possible solution: the robot programming remains unaffected, and, if unexpected deburring forces appear, the mechanical system will automatically adapt to reduce the forces applied to the robot.

To achieve this objective, in this paper a simple mechanism is presented that can be used to reduce the peaks of deburring forces that could harm the robot structure or deburring tool. The mechanism will reduce cutting forces automatically if the burr is too high thanks to its compliance, while it will return to the baseline configuration when the burr thickness is acceptable again. The proposed system has also the advantage not to need a high setup time, when compared to a force feedback controller implemented in an industrial robot.

The work is organized as follows: in Section 2 the formulation of deburring forces is covered; in Section 3 the proposed mechanism and its mathematical model are presented; in Section 4 the design of the mechanism is addressed; Section 5 presents the dynamic analysis and simulation of the system in selected test cases; finally, Section 6 concludes the paper.

2. Deburring Forces

Limited information is available about the estimation of deburring forces. However, a simple model has been developed starting from the grinding process [6,29,30]. In the deburring configuration considered (Figure 1), the material is removed by the cylindrical surface of the deburring wheel, with the workpiece moving from the right to the left (along the y axis) and the deburring wheel rotating counterclockwise.

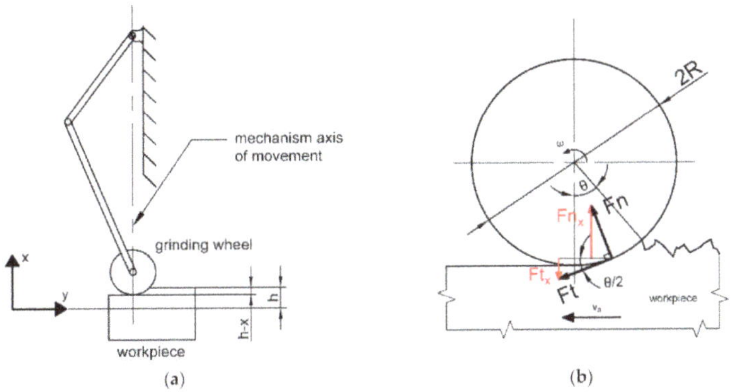

Figure 1. Scheme of the workpiece position in relation to the mechanism axis of movement (**a**); deburring forces (**b**).

The material removing rate (MRR) of a single chunk of material can be calculated from the burr dimensions and workpiece movement speed:

$$MRR = b_w \cdot h \cdot v_a \quad \left[\frac{m^3}{s}\right] \tag{1}$$

where b_w and h are the width and height of the material chunk, and v_a is the linear workpiece movement speed. If the material and its specific removal energy u are known and the rotational speed ω of the grinding wheel is defined, it is possible to calculate the tangential cutting force needed to remove the slice of material as follows:

$$F_t = \frac{b_w \cdot h \cdot v_a \cdot u}{\omega \cdot R} \quad [N] \tag{2}$$

where R is the radius of the grinding wheel.

In practice, (2) depends on the position of the mechanism since the burr height that the grinding wheel removes varies with its x position. As a result, F_t can be expressed as:

$$F_t = \frac{b_w \cdot v_a \cdot u}{\omega} \cdot \frac{(h-x)}{R} \tag{3}$$

It is worthy to point out that the other main force component involved in the cutting process is the force F_n normal to the burr surface. The ratio ($\mu = F_t/F_n$) between F_t and F_n depends on the material, but its value is usually between 0.2 (ceramics and high strength steel) and 0.8 (mild steel) [31]

If the material is removed by a small portion of the grinding wheel surface (Figure 1):

$$h - x = R - Rcos(\theta) \quad \rightarrow \quad \theta = \arccos\left(1 - \frac{h-x}{R}\right) \tag{4}$$

Since F_t and F_n are applied evenly on the surface of the material being removed, it is possible to consider them applied in the middle of the surface, so they are inclined of $\theta/2$ with respect to the y and x axis, respectively. The resulting force along the mechanism x axis is:

$$\begin{gathered} F_{tx} = F_t \sin\left(\frac{\theta}{2}\right), \quad F_{nx} = F_n \cos\left(\frac{\theta}{2}\right) = \frac{F_t}{\mu}\cos\left(\frac{\theta}{2}\right) \\ F_{tot} = F_{nx} - F_{tx} = \frac{b_w \cdot v_a \cdot u}{\omega} \cdot \frac{(h-x)}{R}\left(\frac{1}{\mu}\cos\left(\frac{\theta}{2}\right) - \sin\left(\frac{\theta}{2}\right)\right) \end{gathered} \tag{5}$$

It is worth to notice that in practice $R \gg h$, so usually $\theta \rightarrow 0$. Equation (5) is still valid for a generic deburring process, but in this case it is possible to simplify it to calculate the actual normal component of the deburring force applied to the mechanism F_b:

$$F_b \approx \frac{b_w \cdot v_a \cdot u}{\mu\, \omega} \cdot \frac{(h-x)}{R} \tag{6}$$

3. Proposed Mechanism

It is assumed that the robot arm is holding the workpiece, while the grinding wheel is attached to the ground through the proposed mechanism (Figure 2). The robot end-point trajectory assures the desired depth of cut and feed rate of the piece with respect to the grinding wheel. To develop a mechanism that is as simple as possible, the main compliance direction, assumed normal to the burr (i.e., parallel to the direction of the normal component of the deburring force (6)), should be fixed. The cutting forces that arise during the deburring can be projected along the tangent and normal directions to the burr profile [32], as discussed in Section 2, and the tangential component of the deburring force is balanced by the manipulator.

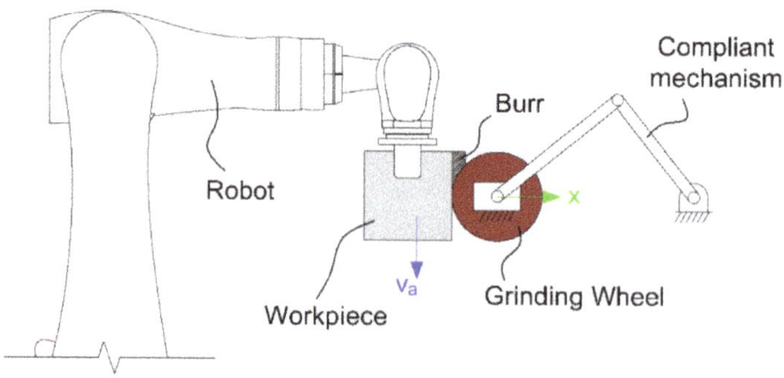

Figure 2. System configuration: the robot holds the workpiece whilst the compliant mechanism holds the grinding wheel.

The main objective is to obtain a mechanism that does not change configuration in normal conditions (assuring a precision cut thanks to its high stiffness), while it changes configuration and reduces its stiffness (and thus the forces applied to the grinding wheel) when the burr exceeds a certain threshold value. In this case, a second cutting cycle is necessary to completely remove the burr.

The proposed mechanism is a slider-crank, where the crank is attached to the ground and the grinding wheel is fixed to the slider. To introduce some forces that allow the system to return to the baseline configuration, some springs are connected to different parts of the mechanism. To obtain a suitable mechanism behavior, it is chosen to place the mechanism close to a singularity, namely close to the top-dead-center (TDC) point. As a result, when the mechanism is close to the TDC, the slider-crank mechanism can react with a maximum force value (see Section 3.1), below which the slider (which supports the grinding wheel) does not move. Of course, if the mechanism is exactly at the TDC, the force value becomes the one giving the structural break of the structure. If this threshold is exceeded, the grinding wheel moves away from the workpiece, thus removing less material, and consequently reducing the cutting force. The slider-crank is ideal for the task at hand since it is easy to build and is suitable for working conditions near the TDC. To avoid to reach the singular configuration, a mechanical stop is applied to the slider (described in more detail in Section 5).

3.1. First Temptative Mechanism

The possibility to add three translational springs to the classic slider-crank mechanism is explored (Figure 3): one for the y-translation of the crank-end (k_1); one for the x-translation of the crank-end (k_2); one for the x-translation of the slider (k_3). In this first stage the springs are connected to sliders, so that their elongations remain parallel to the main axes x and y. The free lengths of the springs are such that they do not generate any forces when the mechanism is in the TDC (angle of the crank $\theta = 0$, and angle of the rocker arm link $\alpha = 0$), i.e., the free length of k_1 is zero, the free length of k_2 is l_1, while the free length of k_3 depends on the anchor point.

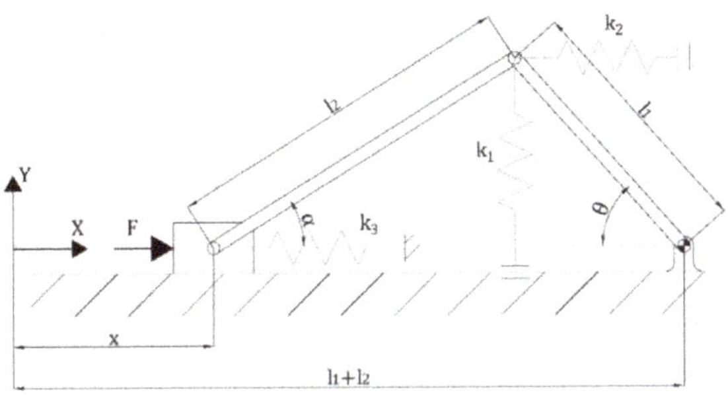

Figure 3. Slider-crank mechanism with three translational springs, two of which are connected to sliders so that their movement is purely along one axis. l_1 is the crank length, l_2 is the rocker arm link length.

When neglecting friction forces, the principle of virtual work gives:

$$\delta L = 0 \rightarrow \sum_i \delta L_i = \sum_i F_i \cdot \delta x_i = 0 \tag{7}$$

where F_i is the i-th force applied to the mechanism, δx_i is the associated virtual displacement, and δL_i the related virtual work. From the mechanism configuration of Figure 3 and (7) it is found that:

$$F \cdot \delta x - k_3 x \cdot \delta x - k_1 s_1 \cdot \delta s_1 - k_2 s_2 \cdot \delta s_2 = 0 \tag{8}$$

where:

$$
\begin{array}{ll}
s_1 = l_1 \sin\theta \rightarrow & \delta s_1 = l_1 \cos\theta \cdot \delta\theta \\
s_2 = l_1(1-\cos\theta) \rightarrow & \delta s_2 = l_1 \sin\theta \cdot \delta\theta
\end{array}
\tag{9}
$$

Substituting s_1, s_2, and their differentials in (8) results in:

$$
(F - k_3 x)\delta x - \left(k_1 l_1^2 \sin\theta\cos\theta + k_2 l_1^2 (1-\cos\theta)\sin\theta\right)\delta\theta = 0 \tag{10}
$$

It is possible to calculate the magnitude of the force F that is balanced by the springs in each configuration of the mechanism starting from the speed ratios of the slider-crank mechanism:

$$
\tau_{\theta x} = \frac{\dot{\theta}}{\dot{x}} = \frac{\cos(\alpha)}{l_1 \sin(\alpha+\theta)} = \frac{\delta\theta}{\delta x}\,, \quad \tau_{\alpha x} = \frac{\dot{\alpha}}{\dot{x}} = \frac{\cos(\theta)}{l_1 \sin(\alpha+\theta)} = \frac{\delta\alpha}{\delta x} \tag{11}
$$

and using them in (10), which results in:

$$
F = k_3 x + \frac{k_1 l_1 \sin\theta\cos\theta\cos\alpha + k_2 l_1 (1-\cos\theta)\sin\theta\cos\alpha}{\sin(\alpha+\theta)} \tag{12}
$$

This equation is only a function of the variable x, since α and θ can be written as:

$$
\begin{array}{l}
\theta = \arccos\left(\frac{l_1^2 + (l_1+l_2-x)^2 - l_2^2}{2 l_1 (l_1+l_2-x)}\right) \\
\alpha = \arcsin\left(\frac{l_1}{l_2}\sin\theta\right)
\end{array}
\tag{13}
$$

Starting from (12) it is possible to obtain different behaviors of the mechanism by changing the stiffnesses of springs and link lengths. As an example, in Figure 4 two configurations are shown: to the left only the springs k_1 and k_2 are used, while to the right only k_1 and k_3 are used. In the first case the mechanism can withstand higher forces with the increase of x, whereas in the second case the opposite happens.

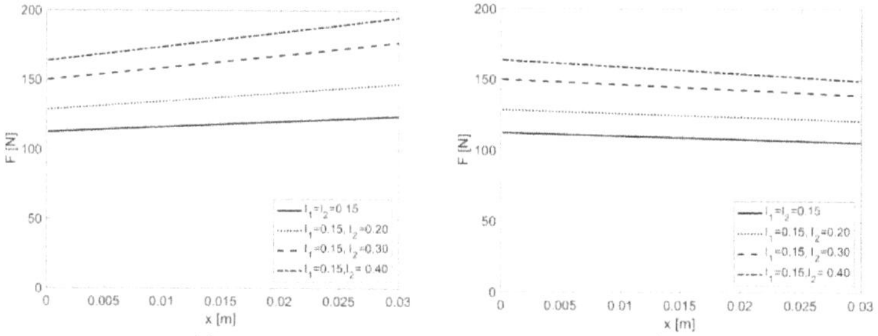

Figure 4. Variation of the force necessary to maintain the mechanism in static equilibrium as a function of x. (**a**) Only k_1 and k_2 are used (k_1 =1500 N/m, k_2 = 3000 N/m). (**b**) Only k_1 and k_3 are used (k_1 = 1500 N/m, k_3 = 150 N/m).

It is important to notice that the force F results in a finite value when the mechanism is close to the TDC configuration, and this force is a function of k_1 only, i.e., (12) when $x \rightarrow 0$ and using (13) becomes:

$$
\lim_{x\rightarrow 0} F = k_1 \frac{l_1 l_2}{l_1 + l_2} \tag{14}
$$

In sum, both progressive and regressive forces as a function of the slider displacement can be obtained with the proposed mechanism and layout, together with a finite value force for configurations close to the TDC. These last two are precisely the ingredients necessary to build a system capable of reducing the cutting forces when the deburring height exceeds a given threshold. Indeed, during deburring the grinding wheel has to remove a certain amount of material, while preserving the structural integrity of the robot. The mechanism is able to withstand a certain deburring force (finite value around the TDC). When this threshold is exceeded, the mechanism becomes compliant (with regressive curve), thus reducing the cutting forces.

3.2. Final Mechanism

The mechanism proposed in Section 3.1, although theoretically suitable for the target application, is difficult to build. In particular, it would be quite impractical to include sliders at the at the anchor points with the frame, in order to keep the springs aligned with the reference axes. Therefore, a new mechanism is proposed with the springs fixed at some distance from the crank fixed hinge (Figure 5). Even with this modification, the basic principles described in Section 3.1 hold valid.

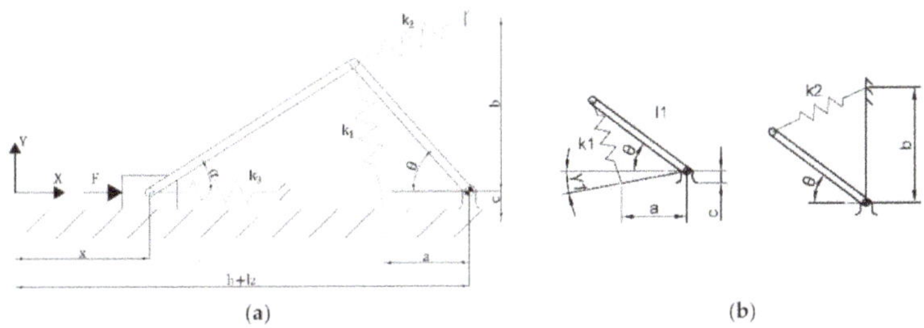

Figure 5. Slider-crank mechanism with translational springs connected to fixed ends (**a**). Two particulars of the mechanism are shown in the right figure (**b**).

Additionally, in this modified design, the springs are unloaded when the mechanism is close to the TDC ($x \rightarrow 0$). The principle of virtual work is to be applied in the same way as in (7), but the virtual displacements of k_1 and k_2 (s_1 and s_2, respectively) have to be modified according to the new mechanism. Starting from the dimensions of Figure 5, the spring displacements are:

$$
\begin{aligned}
s_1 &= \sqrt{l_1^2 + d^2 - 2l_1 d\cos(\gamma_1 + \theta)} - \sqrt{(a - l_1)^2 + c^2} \\
s_2 &= \sqrt{l_1^2 + b^2} - \sqrt{l_1^2 + b^2 - 2l_1 b\sin(\theta)}
\end{aligned} \tag{15}
$$

where d is the distance between the anchor point of k_1 and the crank hinge, and γ_1 is the angle of the vector crank hinge-anchor point with respect to the x-axis:

$$
d = \sqrt{a^2 + c^2}, \quad \gamma_1 = \operatorname{atan}\left(\frac{c}{a}\right) \tag{16}
$$

The differentials of s_1 and s_2 can be calculated as:

$$
\begin{aligned}
\delta s_1 &= \frac{\partial s_1}{\partial \theta} \cdot \delta\theta = \frac{l_1 d\sin(\gamma_1+\theta)}{\sqrt{l_1^2+d^2-2l_1 d\cos(\gamma_1+\theta)}} \cdot \delta\theta \\
\delta s_2 &= \frac{\partial s_2}{\partial \theta} \cdot \delta\theta = \frac{l_1 b\cos(\theta)}{\sqrt{l_1^2+b^2-2l_1 b\sin(\theta)}} \cdot \delta\theta
\end{aligned} \tag{17}
$$

As a result, the force F at the slider becomes:

$$
F = k_3 x + \left(k_1 s_1 \frac{l_1 d\sin(\gamma_1 + \theta)}{\sqrt{l_1^2 + d^2 - 2l_1 d\cos(\gamma_1 + \theta)}} + k_2 s_2 \frac{l_1 b\cos(\theta)}{\sqrt{l_1^2 + b^2 - 2l_1 b\sin(\theta)}} \right) \tau_{\theta x} \tag{18}
$$

The expression (18) for $x \to 0$ simplifies to:

$$\lim_{x \to 0} F = \frac{k_1 l_1 l_2 c^2 (b^2 + l_1^2) + k_2 l_1 l_2 b^2 \left((a - l_1)^2 + c^2 \right)}{(l_1 + l_2) \left((a - l_1)^2 + c^2 \right) (b^2 + l_1^2)} \tag{19}$$

Differently from (14), which depends on k_1 only, (19) depends both on k_1 and k_2. Similarly to the mechanism of Figure 3, the new mechanism behaviour depends on its main dimensions. Nevertheless, in this case also the dimensions a, b, and c have a great impact on F. As an example, two different mechanisms are shown in Figure 6, whose main parameters are listed in Table 1. Similarly to the mechanism presented in Section 3.1., both progressive and regressive forces as a function of the x displacement can be obtained, with a finite value of the force around the TDC. The regressive curve is preferred for the application at hand since it is related to lower cutting forces.

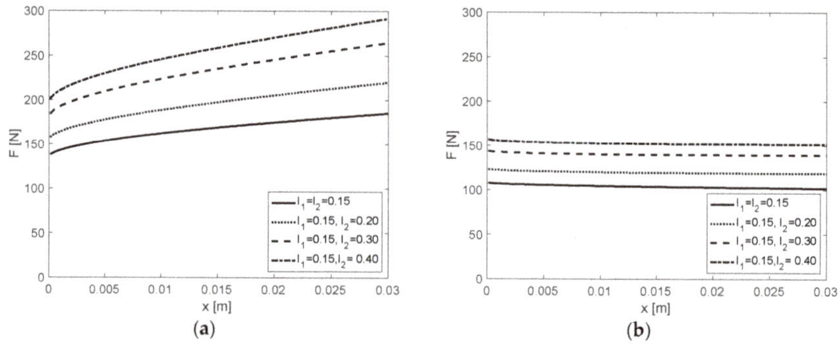

Figure 6. Variation of the force necessary to maintain the mechanism in static equilibrium as a function of x. ((**a**) Mechanism #1, (**b**) Mechanism #2).

Table 1. Main mechanism parameters.

Parameter	Mechanism #1	Mechanism #2
l_1	0.15 m	
l_2	[0.15, 0.2, 0.3, 0.4] m	
a	0.05 m	0.14 m
b	0.15 m	0.03 m
c	0.05 m	
k_1	1500 N/m	
k_2	3000 N/m	0
k_3	200 N/m	0

4. Design

In normal cutting conditions, the mechanism should have high stiffness to ensure high cutting precision. On the other hand, when the burr is too high the mechanism should exploit its compliance to reduce the cutting forces and, as a consequence, a second cutting cycle is necessary to completely remove the burr.

The desired regressive behavior of the mechanism is related to its geometrical parameters and to the springs stiffness. If the link lengths (l_1 and l_2) are defined by the scale of the application (related to the workpiece dimensions and maximum burr height), there are 5 other parameters to be chosen (see Figure 5): k_1, k_2, k_3, a, and b. It is not straightforward to find a good set of values by trial-and-error, so an optimization algorithm has been used. The algorithm aims at finding the parameters whose resulting mechanism forces F fit a desired (regressive) curve.

In this work the desired force profile is set to $F = 150 - 300x\ [N]$ (blue dotted line in Figure 7) with $l_1 = 0.15$ m and $l_2 = 0.3$ m. The *fmincon* MATLAB function has been used to find the optimal parameters that minimize the error between the actual and desired force profiles. The optimization algorithm yields the values listed in Table 2. It is worth to notice from Figure 7 that the system is stable: the normal component of the deburring force (F_b) decreases with x faster than the elastic force on the slider (F):

$$\frac{\partial F_b}{\partial x} < \frac{\partial F}{\partial x} \tag{20}$$

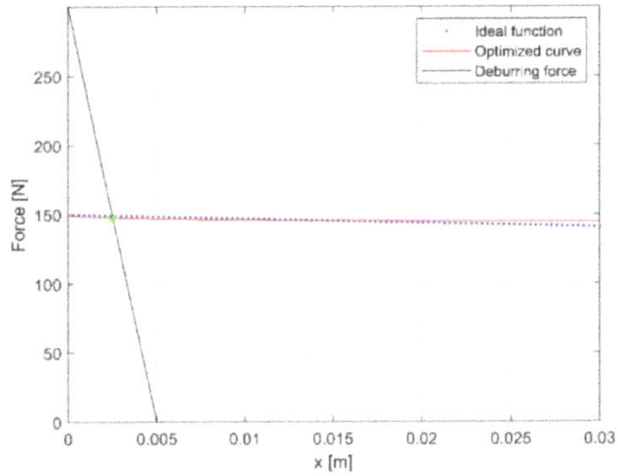

Figure 7. The optimization algorithm finds the mechanism parameters whose $F(x)$ (red line) fits at best the objective curve (blue dotted line). Black line shows how $F_b(x)$ decreases with the movement of the mechanism (with h = 5 mm).

Table 2. Parameters computed by the optimization algorithm.

k_1	k_2	k_3	a	b
1560 N/m	47 N/m	≈0 N/m	0.14 m	0.028 m

The black line of Figure 7 is the deburring force (normal component, as discussed in Section 2). This curve moves to the right for higher burr heights (with higher cutting forces for $x = 0$), and vice versa. The intersection between this line and the optimized curve (red line in figure) yields the equilibrium point, i.e., the x value at which the mechanism moves to balance the deburring force and the elastic force of the mechanism. For very low values of h the black line does not intersect the optimized curve, so the mechanism does not move. In this case, a precision cut is ensured, with cutting forces below the force threshold that activates the mechanism compliance. For the parameters of Table 2 and Figure 7 (see Table 3 in Section 5.3 for the complete list of parameters) the burr height (steel) that activates the mechanism compliance is 2.5 mm.

Table 3. Simulation parameters.

Parameter	Value
b_w	$1 * 10^{-3}$ m
v_a	$15 * 10^{-3}$ m/s
u (steel)	$30 * 10^{9}$ J/m^3
ω	100 rad/s
R	0.1 m
μ	0.75
c_{eq}	2000 Ns/m
k_1	1560 N/m
k_2	47 N/m
k_3	0 N/m
a	0.14 m
b	0.028 m
c	0.05 m
l_1	0.15 m
l_2	0.3 m
m	30 kg
m_r	1.0 kg
m_c	0.5 kg
k_{stop}	10^7 N/m
$x_{stop} = x_0$	$3 * 10^{-4}$ m
x_1	0 m
h_0	0 Ns/m
h_1	10^4 Ns/m

It is important to notice that it is possible to perform the optimization starting from a different desired force curve and, in particular, different force thresholds can be set depending on the application. This will result in a different set of springs and anchor points.

From a mechanical point of view, it is important to include a mechanical end stop for the slider to avoid possible changes in configuration of the slider-crank (i.e., to keep the rocker arm link always above the slider axis) and to avoid that the mechanism works in the TDC. Indeed, the mechanism is designed to work near the TDC, where it can generate a finite elastic force, but not exactly in the TDC, which could be dangerous for the structural integrity of the mechanism if the burr height is very large. The detailed model of the end stop will be presented in Section 5.2.

5. Dynamic Analysis

5.1. Dynamic Model

The proposed mechanism has one degree of freedom and can thus be modeled using a single equivalent mass m^* that is excited via a single equivalent force F^*. The equivalent mass and force clearly depend on the mechanism configuration (x) and on its main parameters.

The equations of motion are obtained by recalling that the power equals the derivative of the kinetic energy:

$$P = \frac{dE_k}{dt} \tag{21}$$

Since friction between the members of the mechanism and their potential energy can be neglected. The kinetic energy of the equivalent mass equals by definition the kinetic energy of all the members of the mechanism:

$$E_k = \frac{1}{2}m^*\dot{x}^2 = \frac{1}{2}\left(m\dot{x}^2 + m_r\left(\dot{x}_{Gr}^2 + \dot{y}_{Gr}^2\right) + I_r\dot{\alpha}^2 + m_c\left(\dot{x}_{Gc}^2 + \dot{y}_{Gc}^2\right) + I_c\dot{\theta}^2\right) \tag{22}$$

where $\dot{x}$ is the velocity of the equivalent mass (equal to the one of the slider); m is the mass of the slider (included the spindle and the grinding wheel); m_r, I_r, $\dot{x}_{Gr}$, $\dot{y}_{Gr}$ are the mass, barycentric moment of inertia, barycenter velocity along the x axis, and barycenter velocity along the y axis of the rocker arm link, respectively; m_c, I_c, $\dot{x}_{Gr}$, $\dot{y}_{Gr}$ are the mass, barycentric moment of inertia, barycenter velocity along the x axis, and barycenter velocity along the y axis of the crank, respectively. The coordinates of the centers of mass of the rocker arm link and of the crank are (respectively):

$$\left\{ \begin{array}{c} x_{Gr} \\ y_{Gr} \end{array} \right\} = \left\{ \begin{array}{c} x + \frac{l_2}{2}\cos(\alpha) \\ \frac{l_2}{2}\sin(\alpha) \end{array} \right\}, \left\{ \begin{array}{c} x_{Gc} \\ y_{Gc} \end{array} \right\} = \left\{ \begin{array}{c} l_1 + l_2 - \frac{l_1}{2}\cos(\theta) \\ \frac{l_1}{2}\sin(\theta) \end{array} \right\} \tag{23}$$

The corresponding velocities are:

$$\left\{ \begin{array}{c} \dot{x}_{Gr} \\ \dot{y}_{Gr} \end{array} \right\} = \left\{ \begin{array}{c} \dot{x} - \frac{l_2}{2}\sin(\alpha)\dot{\alpha} \\ \frac{l_2}{2}\cos(\alpha)\dot{\alpha} \end{array} \right\}, \left\{ \begin{array}{c} \dot{x}_{Gc} \\ \dot{y}_{Gc} \end{array} \right\} = \left\{ \begin{array}{c} \frac{l_1}{2}\sin(\theta)\dot{\theta} \\ \frac{l_1}{2}\cos(\theta)\dot{\theta} \end{array} \right\} \tag{24}$$

By introducing (24) and the speed ratios of (11), the final form of (22) is:

$$\begin{array}{c} \frac{1}{2}m^*\dot{x}^2 = \frac{1}{2}\left(m + m_r\left(1 + \frac{l_2^2}{4}\tau_{\alpha x}^2 - l_2\sin(\alpha)\tau_{\alpha x}\right) + I_r\tau_{\alpha x}^2 + m_c\left(\frac{l_1^2}{4}\tau_{\theta x}^2\right)\right. \\ \left. + I_c\tau_{\theta x}^2\right)\dot{x}^2 \end{array} \tag{25}$$

The power P of F^* can be calculated from the deburring force and the springs forces:

$$P = F^*\dot{x} = -(k_1 s_1)\dot{s}_1 - (k_2 s_2)\dot{s}_2 - (k_3 x)\dot{x} - (c_{eq}\dot{x})\dot{x} + F_b\dot{x} \tag{26}$$

in which the force of a damper to be connected in parallel to spring k_3 (with damping c_{eq}) is introduced in order to avoid undesired vibrations.

The equation of motion is found by differentiating (25) and putting it equal to (26):

$$\begin{array}{c} m^*\ddot{x} + \frac{1}{2}\left(\left(2I_r + \frac{l_2^2}{2}m_r\right)\tau_{\alpha x}\dot{\tau}_{\alpha x} + \left(2I_c + \frac{l_1^2}{2}m_c\right)\tau_{\theta x}\dot{\tau}_{\theta x}\right. \\ \left. - m_r l_2\left(\tau_{\alpha x}\,\alpha\,.\cos(\alpha) + \dot{\tau}_{\alpha x}\sin(\alpha)\right)\right)\dot{x}^2 = F^*\dot{x} \end{array} \tag{27}$$

The differential Equation (27) is implemented and solved numerically in MATLAB for different working scenarios. The simulation results and the simulation parameters used are presented in Section 5.

5.2. Mechanical Stop Model

To avoid the singular configuration at the TDC and changes in the mechanism configuration, a mechanical stop has been introduced. The mechanical stop should be made of soft material to reduce unnecessary shocks to the mechanism structure. The mechanical stop is modeled as a one degree of freedom system that is activated as soon as $x_{sup} < x_{stop}$, in which x_{sup} is the coordinate of the surface of the slider and x_{stop} is the coordinate of the mechanical stop when the slider is not in contact with it (see Figure 8).

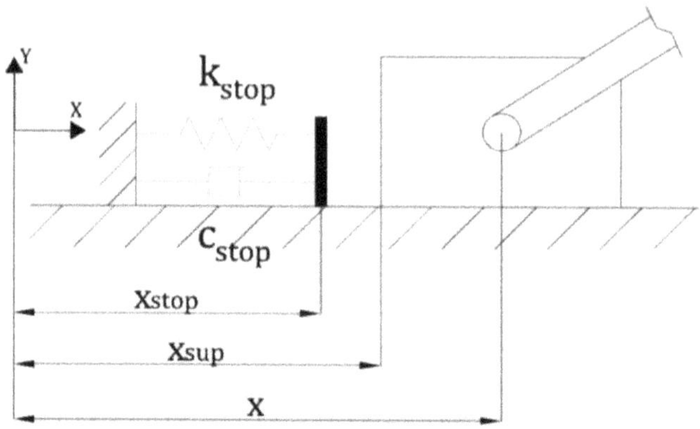

Figure 8. Model of the mechanical stop of the mechanism.

The force transferred by the mechanical stop to the mechanism is:

$$F_{stop} = k_{stop}(x_{sup} - x_0)^e + f(x_{sup}, x_0, h_0, x_1, h_1)\dot{x} \tag{28}$$

where:

- k_{stop} is the stiffness of the mechanical stop;
- x_{sup} is the coordinate of the surface of the slider that impacts the mechanical stop;
- x_0 $(=x_{stop})$ and x_1 are the coordinates of the limits of the mechanical stop;
- e is the exponent in (28), which is null in the case of the Kelvin-Voigt model;
- h_0 and h_1 are the limits of the function f.

Function f is described by a polynomial function, whose normalized shape is shown in Figure 9:

$$f(x_{sup}, x_0, h_0, x_1, h_1) = \begin{cases} h_0 & x_{sup} \geq x_0 \\ h_0 + a\Delta^2(3 - 2\Delta) & x_1 \leq x_{sup} \leq x_0 \\ h_1 & x_{sup} \leq x_1 \end{cases} \tag{29}$$

where:

$$a = h_1 - h_0 \quad , \quad \Delta = \frac{x_{sup} - x_0}{x_1 - x_0} \tag{30}$$

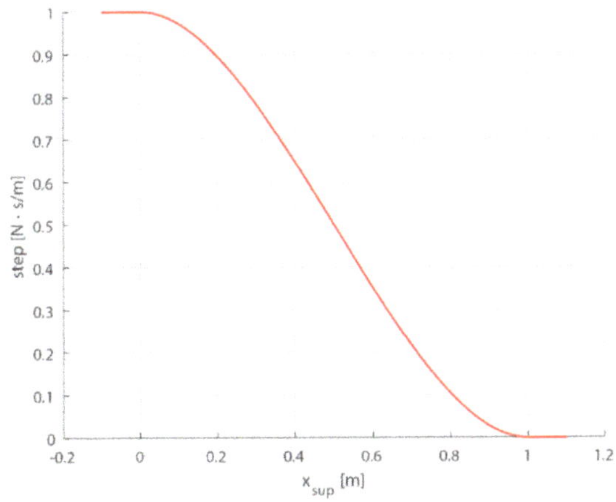

Figure 9. Normalized shape of function f (29).

5.3. Simulation Results

In this section the results of the dynamic simulations of the proposed system in different working conditions are presented. The main aim of the simulations is to verify that the mechanism works properly and, in particular, it is important to check the following conditions:

1. If the projection of the cutting force along the sliding direction is less than the elastic force threshold of the mechanism, then the grinding wheel support should not move. This happens for small burr heights. With the considered setup (specified in Table 3), the elastic force threshold of the mechanism, which is equal to the elastic reaction force for $x \to 0$, is 149.01 N). This force threshold corresponds to a burr height of 2.5 mm with the considered setup.
2. If the projection of the cutting force along the sliding direction exceeds the elastic force threshold, the grinding wheel support should move backwards according to the compliance of the mechanism and the dynamics of (27); moreover, for a constant burr height, an equilibrium condition should be achieved (after a transient) between the cutting force projection and the elastic reaction force and, therefore, a cut profile with constant height should be obtained.
3. If the burr height decreases, the compliant mechanism should move forward, and return towards the TDC. This condition is very important since it assures the stability of the system. Of course, even if the burr becomes very small, the mechanism can never reach the TDC due to the mechanical stop.

Condition 1 assures the cutting precision of the system, if the cutting force (and the burr height) is below a given threshold. Condition 2 assures that the mechanism is compliant if the cutting forces are too high, so that they are reduced to an acceptable value; nevertheless, the cutting precision is jeopardized and an additional cutting cycle is necessary to completely remove the burr. Condition 3 assures that the mechanism returns back (up to the initial configuration) if the burr height decreases.

In the simulations the parameters listed in Table 3 are used. A steel burr is considered with $\mu = 0.75$ [31]. The mechanical stop has been modeled as discussed in Section 5.2.

Different burr profiles are used as an input to simulations. The outputs of simulations are the position x (directly related to the cut profile) and velocity $\dot{x}$ of the grinding wheel support as a function of time. In order to verify the conditions 1–3, simulations are carried out with step variations of burr height. Between different steps the burr height is constant in order to verify that an equilibrium condition (cut profile with constant height) is obtained after an initial transient. Different test cases with different profiles of step burr height are analyzed as detailed below (Sections 5.3.1–5.3.5). Finally, this study will include a simulation in which the burr height is defined by a random function (Section 5.3.6).

The dynamic simulations start (for $t = 0$ s) with the grinding wheel support in contact with the mechanical stop. For this reason, the position x is 0.3 mm for $t = 0$ s in all the simulations performed, which corresponds to the position of the mechanical stop. This means that the mechanism initial configuration is very near to the TDC. Due to the initial length of the springs, a small elastic force is present that presses the grinding wheel support against the mechanical stop.

5.3.1. Test Case 1—1-Step Profile

A 1-step burr profile with a constant height (h = 5 mm, see solid line in Figure 10 (top)) is used as an input for the dynamic simulation. The burr height is sufficiently high to make the projection of the cutting force along the sliding direction exceed the elastic force threshold of the mechanism (dashed line in Figure 10). The simulation results, i.e., the position and velocity of the grinding wheel support in function of time, are presented in Figure 10 (middle and bottom). From the analysis of Figure 10, it can be noticed that the grinding wheel support moves backwards (positive values of x) thanks to the compliance of the mechanism; moreover, an equilibrium condition is achieved (after a transient of about 0.25 s with a small overshoot) between the cutting force projection and the elastic reaction force and, therefore, a cut profile with constant height is obtained. Therefore, the simulation results for this test case are in line with what was expected in this case (Condition 2 of Section 5.3). The burr height, initially equal to 5 mm, after processing is about 2.5 mm. This means that a second cutting cycle is necessary to completely remove the burr. The maximum (normal) deburring force reduces to 148 N from the nominal value of 300 N (corresponding to the nominal height of 5 mm), i.e., roughly a 50% (automatic) force reduction is obtained, which is related to the 50% reduction of the cutting height (2.5 instead of 5 mm).

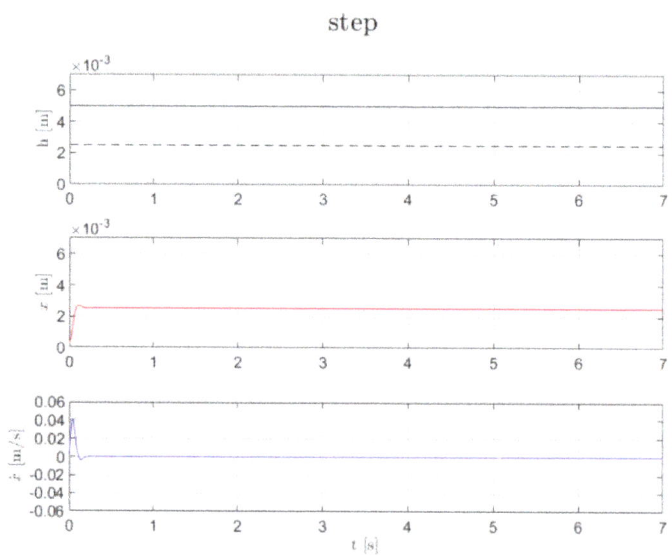

Figure 10. Step burr profile (h = 5 mm, **top**); position (**middle**) and velocity (**bottom**) of the grinding wheel support.

5.3.2. Test Case 2—3-Steps Burr Profile "Low–Low–High"

A 3-steps burr profile "low–low–high" (see Figure 11 (top)) is used as an input for the dynamic simulation. "Low"/"high" means that the burr height is not/is sufficiently high to make the projection of the cutting force along the sliding direction exceed the elastic force threshold of the mechanism. In this test case, steps with increasing height are investigated. In the first two steps the cutting force is below the elastic force threshold, whereas in the third step the cutting force is sufficiently high (above the threshold) to make the mechanism work in the compliant mode of operation. The simulation results are presented in Figure 11 (middle and bottom). It can be noticed that the grinding wheel support does not move while the first two steps are processed (precision cut), and it moves backwards thanks to the compliance of the mechanism while the third step is processed. Similarly to the previous test case, an equilibrium condition is achieved (after a transient of about 0.3 s after the beginning of step 3) and a cut profile with constant height is obtained for step 3, as it can be noticed in Figure 11. Additionally, in this test case the simulation results confirm the expected dynamic behavior of the system: the grinding wheel support moves backward only when the cutting force projection exceeds the elastic force threshold, and when this happens a constant cut profile is obtained, after an initial transient (Condition 1 and Condition 2 of Section 5.3). From another point of view, for small burr heights the mechanism is (ideally) infinitely stiff, whereas it becomes compliant for high burr heights. Some small oscillations can be noticed when the burr height changes value but is not sufficiently high to allow the mechanism to work in the compliant mode (i.e., the grinding wheel support does not move and the cutting precision is ensured). This is due to the presence of the mechanical stop, which is not infinitely stiff. Nevertheless, this effect causes a negligible effect on the grinding wheel position, as it can be noticed in Figure 11 (middle). The maximum (normal) deburring force reduces again to 148 N from the nominal value of 240 N (corresponding to the nominal height of 4 mm), i.e., roughly a 38% (automatic) force reduction is obtained, which is related to the 38% reduction of the cutting height (2.5 instead of 4 mm).

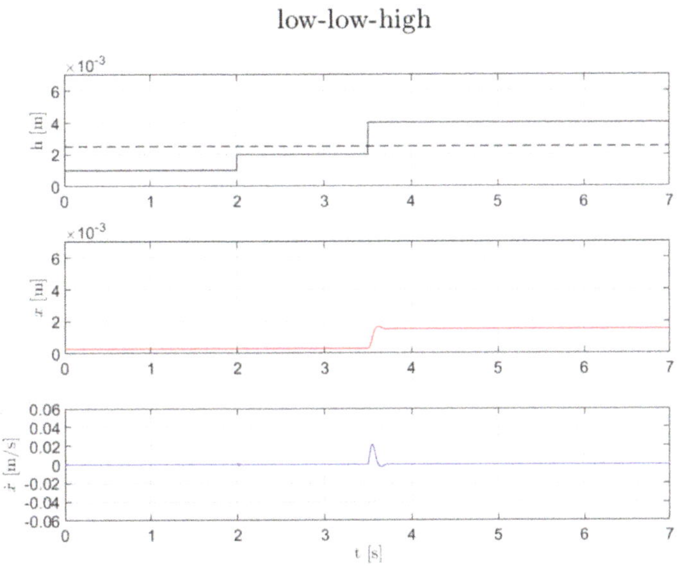

Figure 11. Three-steps burr profile ("low–low–high" (**top**)); position (**middle**) and velocity (**bottom**) of the grinding wheel support.

5.3.3. Test Case 3—3-Steps Burr Profile "High–High–High"

A 3-steps burr profile "high–high–high" (see Figure 12 (top)) is used as an input for the dynamic simulation. Similarly to the previous test case, 3 steps with increasing height are investigated. Differently with respect to the previous test case, in this case, for all the 3 steps the cutting force is sufficiently high to make the mechanism work in the compliant mode of operation. The aim of this test case is to verify that, when it passes from one step to another, the system moves from an equilibrium condition to a new equilibrium condition. The simulation results are presented in Figure 12 (middle and bottom). It can be noticed that, at each step, the grinding wheel moves backwards thanks to the compliance of the mechanism. Three different equilibrium conditions are achieved (after a transient of about 0.25 s each time) with increasing height of the residual burr and, inside each step, a cut profile with constant height is obtained (after the transient), as it can be noticed in Figure 12. The simulation results of this test case confirm the expected dynamic behavior of the system: each time that the system encounters a higher step, it moves to a new equilibrium condition with a higher height of residual burr (in particular, Condition 2 of Section 5.3 is satisfied for all the 3 steps). The maximum (normal) deburring force reduces to 147.3 N from the nominal value of 360 N (corresponding to the nominal height of 6 mm), i.e., roughly a 60% (automatic) force reduction is obtained, which is related to the 60% reduction of the cutting height (2.5 instead of 6 mm).

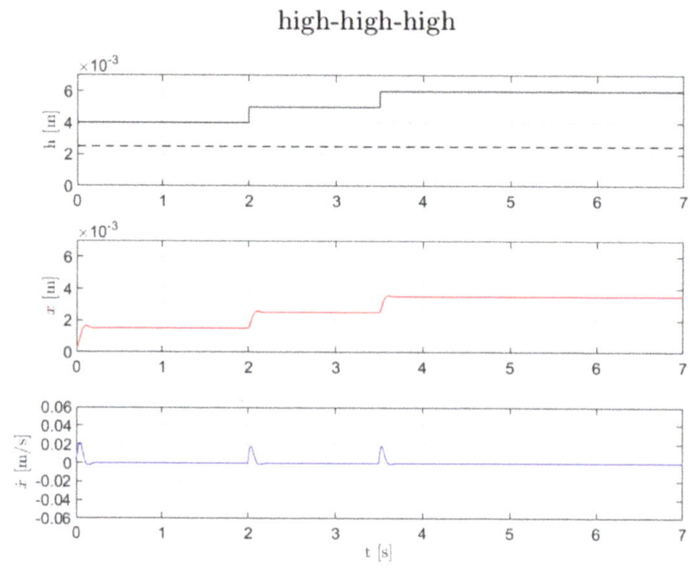

Figure 12. Three-steps burr profile ("high–high–high" (**top**)); position (**middle**) and velocity (**bottom**) of the grinding wheel support.

5.3.4. Test Case 4—3-Steps Burr Profile "High–Highest–High"

A 3-steps burr profile "high–highest–high" (see Figure 13 (top)) is used as an input for the dynamic simulation. The burr height is h = 4 mm in the first step, then it becomes h = 6 mm in the second step, and finally it returns to h = 4 mm in the third step. In all the three steps the burr height is sufficiently high to make the mechanism work in the compliant mode of operation. The main aim of this test case is to verify that, if the burr height decreases, the compliant mechanism moves forward and returns towards the TDC, as specified in Condition 3 of Section 5.3. The simulation results are presented in Figure 13 (middle and bottom). It can be noticed that, during the first two steps, the grinding wheel moves backwards thanks to the compliance of the mechanism, and two different equilibrium conditions are achieved with increasing height of the residual burr. Moreover, in the third step, the compliant mechanism moves forward and rapidly returns towards the TDC. In this step, a new equilibrium condition is obtained, with the same height of residual burr as in the step 1 (as expected, since the burr height is equal in steps 1 and 3). Therefore, the mechanism works properly and, in particular, Condition 3 of Section 5.3 is satisfied.

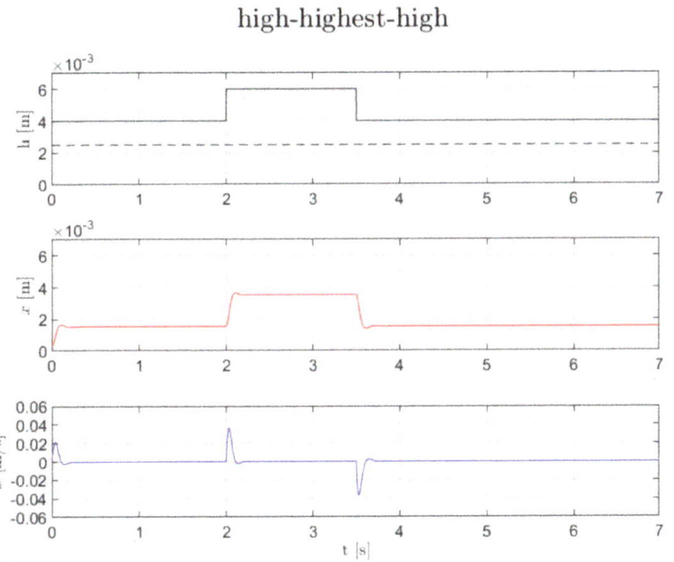

Figure 13. Three-steps burr profile ("high–highest–high" (**top**)); position (**middle**) and velocity (**bottom**) of the grinding wheel support.

5.3.5. Test Case 5—3-Steps Burr Profile "High–Low–High"

A 3-steps burr profile "high–low–high" (see Figure 14 (top)) is used as an input for the dynamic simulation. In the first and third step the burr height is sufficiently high to make the mechanism work in the compliant mode of operation, whereas in the second step the burr height is sufficiently low to generate a cutting force projection below the elastic force threshold. The main aim of this test case is to verify that, when the burr height decreases and generates a cutting force projection below the elastic force threshold, the mechanism returns in the initial configuration, i.e., with the grinding wheel support in contact with the mechanical stop. The simulation results are presented in Figure 14 (middle and bottom). It can be noticed that, during the first step, the grinding wheel moves backwards thanks to the compliance of the mechanism, and an equilibrium condition is achieved with a certain height of the residual burr. In the second step, the compliant mechanism moves forward and rapidly returns towards the TDC (Condition 3 of Section 5.3 is satisfied). Since the cutting force projection is below the elastic force threshold, the mechanism would tend to go to the TDC, but due the mechanical stop this is not possible. Indeed, an impact between the grinding wheel support and the mechanical stop takes place, as it can be noticed in Figure 14, in which small bounces are reported after $t = 1.5$ s. Of course, the amplitude and damping of these bounces depend on the dynamic parameters of the mechanism (masses, inertias, springs stiffness, and damping) and on the characteristics (stiffness and damping) of the mechanical stop. After the bounces, the mechanism remains in the initial configuration (x is about 0.3 mm, as it can be noticed in Figure 14), up to the beginning of the third step. Finally, in the third step, a new equilibrium condition is obtained, with a height of residual burr higher than in the first step (as expected, since the burr height is higher).

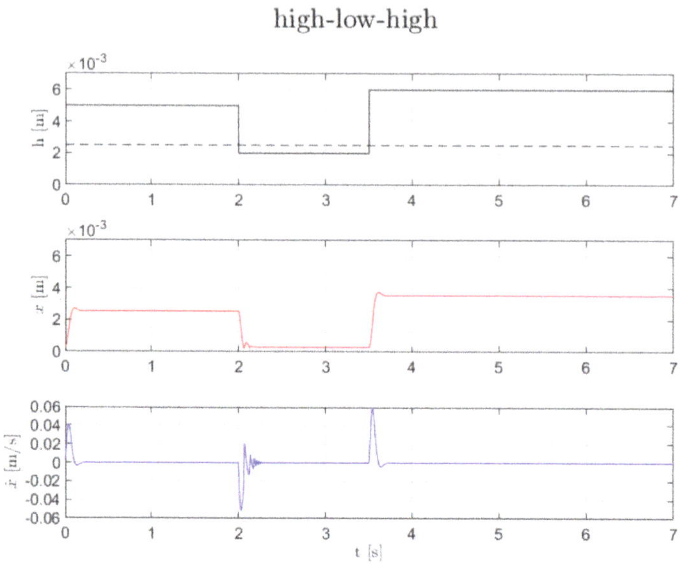

Figure 14. Three-steps burr profile ("high–low–high" (**top**)); position (**middle**) and velocity (**bottom**) of the grinding wheel support.

5.3.6. Test Case 6—Generic Burr Profile

A set of generic burr profiles have been generated using a random variation of burr height in order to test the mechanism in more realistic scenarios. The random burr profiles are then used as an input for the dynamic simulations. Random profiles have been generated starting from the power spectral density (PSD) of the random process [33]. The three most common shapes for the noise have been assumed, namely white noise (flat PSD), flicker-noise (PSD with slope -1 in a log-log diagram), and random-walk noise (PSD with slope-2). The latter is usually adopted when it come to the roughness of road surfaces (ISO 8608:2016). The shape of the PSD is related to the frequency distribution of the profile, while the area under the PSD is related to the RMS (squared) of the random profile. It has been assumed that the profiles, which are generated with a frequency of 1000 Hz, have RMS between 1 and 4 mm.

A random burr profile using a PSD slope of −2 and a RMS of 2.5 mm and the related simulation results are presented in Figure 15. It can be noticed that the proposed mechanism performs well also for a generic burr profile. In particular, the compliance of the mechanism is exploited to reduce the cutting forces and, as a consequence, a second cutting cycle is necessary to completely remove the burr (see Figure 16). Similarly to the Test case 5 (Section 5.3.5) small bounces on the mechanical stop are visible in the middle plot of both Figure 15; Figure 16 (for example after t = 0.55 s, after t = 1.9 s, etc.). From the analysis of Figure 16, it can be noticed that after the second cutting cycle a very high precision is ensured: most of the burr profile is at 2.85 * 10^{-4} mm that is the equilibrium position of the mechanical stop, and very small variations (<1 * 10^{-5} m) with respect to this value are present. Indeed, in the second cutting cycle the grinding wheel support is always in contact with the mechanical stop, and the small variations in the burr profile are due to the (low) compliance of the mechanical stop. In this case the compliance of the mechanism is not exploited, since the maximum burr height is below 2.5 mm, which is the minimum burr height that is necessary to make the mechanism work in the compliant mode, as explained in Section 5.3. Very similar conclusions can be drawn when other values of PSD slope and RMS are used (as above specified).

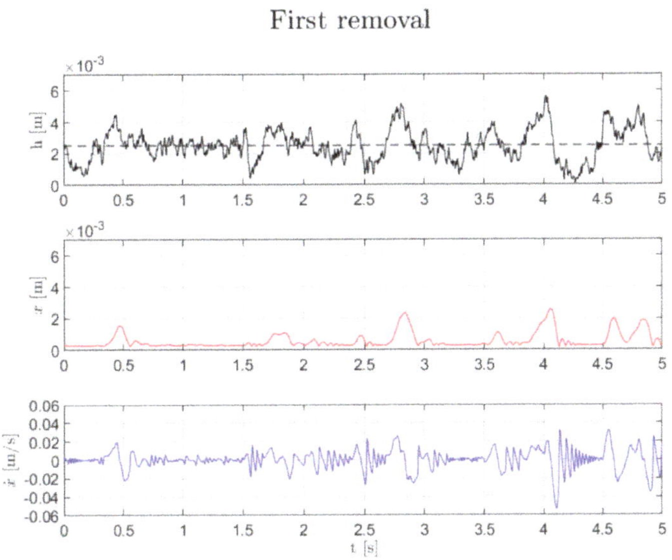

Figure 15. Generic burr profile, first cutting cycle (**top**); position (**middle**) and velocity (**bottom**) of the grinding wheel support.

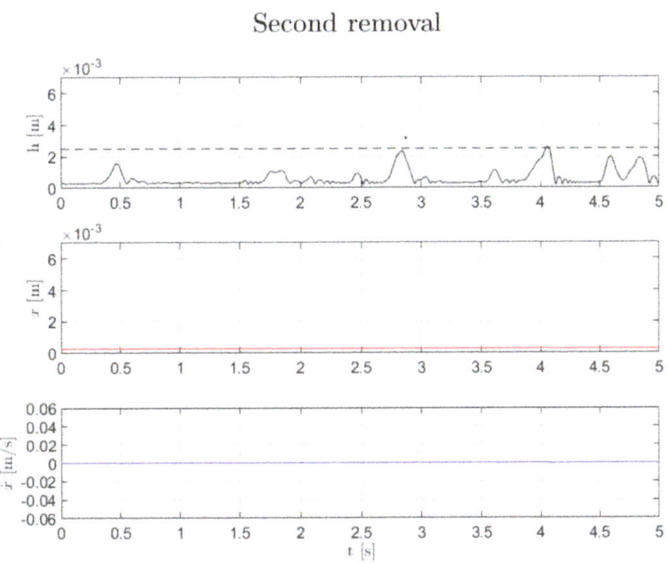

Figure 16. Generic burr profile, second cutting cycle (**top**); position (**middle**) and velocity (**bottom**) of the grinding wheel support. Please notice that the scale of these graphs are the same of Figure 15 for comparison purposes.

6. Conclusions

A mechanism has been designed to be used in robotic deburring to ensure a precision cut and to reduce cutting forces and avoid breaks of the robot/spindle in the case of high and irregular burr profiles. The mechanism is based on a classic slider-crank, which is provided with springs in order to give the desired force-displacement response. In the event that the burr is too high, it exploits its compliance to reduce the cutting forces automatically, and is able to return to the baseline configuration when the burr thickness is acceptable again. After the specification of the desired stiffness curve, the main design parameters of the system have been derived thanks to an optimization method. The effectiveness of the proposed mechanism is verified by means of dynamic simulations using selected test cases. The simulations showed that, when the compliance of the mechanism is exploited to reduce the cutting forces, a second cutting cycle is necessary to completely remove the burr and ensure a high cutting precision. With the dataset considered, the (normal) deburring forces remain always below 150 N thanks to the design of the mechanism, even if the burr height would generate nominal forces much larger. This has been shown both with basic profiles, such as steps, and with more realistic random profiles, thus ensuring cutting without the risk of damaging the cutting tool. A limitation of the present work is the fact that the proposed system needs to be validated experimentally, paying particular attention to the vibrations induced due to the cutting operation, which could make it necessary to improve the design. The experimental validation of the proposed mechanism and the development of a dynamic model that includes the manipulator model will be part of future work.

Author Contributions: Conceptualization, M.B., S.C. and M.M.; methodology, M.B., S.C. and M.M.; software, M.B.; formal analysis, M.B.; investigation, M.B., S.C. and M.M.; writing—original draft preparation, M.B. and S.C.; writing—review and editing, S.C. and M.M.; supervision, S.C. and M.M. All authors have read and agreed to the published version of the manuscript.

Funding: This research received no external funding.

Institutional Review Board Statement: Not applicable.

Informed Consent Statement: Not applicable.

Data Availability Statement: Not applicable.

Conflicts of Interest: The authors declare no conflict of interest.

References

1. Onstein, I.F.; Semeniuta, O.; Bjerkeng, M. Deburring Using Robot Manipulators: A Review. In Proceedings of the 3rd International Symposium on Small-scale Intelligent Manufacturing Systems (SIMS), Gjøvik, Norway, 10–12 June 2020.
2. Brunete, A.; Gambao, E.; Koskinen, J.; Heikkilä, T.; Kaldestad, K.B.; Tyapin, I.; Hovland, G.; Surdilovic, D.; Hernando, M.; Bottero, A.; et al. Hard material small-batch industrial machining robot. *Robot. Comput. Integr. Manuf.* **2018**, *54*, 185–199. [CrossRef]
3. Pan, Z.; Zhang, H.; Zhu, Z.; Wang, J. Chatter analysis of robotic machining process. *J. Mater. Process. Technol.* **2006**, *173*, 301–309. [CrossRef]
4. Cordes, M.; Hintze, W.; Altintas, Y. Chatter stability in robotic milling. *Robot. Comput. Integr. Manuf.* **2019**, *55*, 11–18. [CrossRef]
5. Gillespie, L.K. *Deburring and Edge Finishing Handbook*; Society of Manufacturing Engineers: Southfiled, MI, USA, 1999.
6. Li, W.; Wang, Y.; Fan, S.; Xu, J. Wear of diamond grinding wheels and material removal rate of silicon nitrides under different machining conditions. *Mater. Lett.* **2007**, *61*, 54–58. [CrossRef]
7. Gillespie, L.K. Deburring precision miniature parts. *Precis. Eng.* **1979**, *1*, 189–198. [CrossRef]
8. Babbar, A.; Jain, V.; Gupta, D. In vivo evaluation of machining forces, torque, and bone quality during skull bone grinding. Proceedings of the Institution of Mechanical Engineers, Part H. *J. Eng. Med.* **2020**, *234*, 626–638. [CrossRef] [PubMed]
9. Babbar, A.; Jain, V.; Gupta, D.; Agrawal, D.; Prakash, C.; Singh, S.; Wu, L.Y.; Zheng, H.Y.; Królczyk, G.; Bogdan-Chudy, M. Experimental analysis of Wear and Multi-Shape Burr Loading during Neurosurgical Bone Grinding. *J. Mater. Res. Technol.* **2021**, *12*, 15–28. [CrossRef]
10. Li, C.; Li, X.; Wu, Y.; Zhang, F.; Huang, H. Deformation mechanism and force modelling of the grinding of YAG single crystals. *Int. J. Mach. Tools Manuf.* **2019**, *143*, 23–37. [CrossRef]
11. Tao, Y.; Zheng, J.; Lin, Y. A sliding mode control-based on a RBF neural network for deburring industry robotic systems. *Int. J. Adv. Robot. Syst.* **2016**, *13*, 8. [CrossRef]
12. Qian, Y.; Yuan, J.; Bao, S.; Gao, L. Sensorless Hybrid Normal-Force Controller with Surface Prediction. In Proceedings of the IEEE International Conference on Robotics and Biomimetics (ROBIO), Yunnan, China, 6–8 December 2019.
13. Pappachan, B.K.; Caesarendra, W.; Tjahjowidodo, T.; Wijaya, T. Frequency domain analysis of sensor data for event classification in real-time robot assisted deburring. *Sensors* **2017**, *17*, 1247. [CrossRef]
14. Kakoi, H.; Yanagihara, K.; Akashi, K.; Tsuchiya, K. Development of Vertical Articulated Robot Deburring System by Using Sensor Feedback. In *IOP Conference Series: Materials Science and Engineering*; IOP: Bristol, UK, 2020.

15. Hu, J.; Kabir, A.M.; Hartford, S.M.; Gupta, S.K.; Pagilla, P.R. Robotic deburring and chamfering of complex geometries in high-mix/low-volume production applications. In Proceedings of the IEEE 16th International Conference on Automation Science and Engineering (CASE), Virtual meeting, 20–21 August 2020.
16. El Naser, Y.H.; Atali, G.; Karayel, D.; Özkan, S.S. Prototyping an Industrial Robot Arm for Deburring in Machining. *Akad. Platf. Mühendislik Fen Bilimleri Derg.* **2020**, *8*, 304–309.
17. Guo, W.; Li, R.; Zhu, Y.; Yang, T.; Qin, R.; Hu, Z. A Robotic Deburring Methodology for Tool Path Planning and Process Parameter Control of a Five-Degree-of-Freedom Robot Manipulator. *Appl. Sci.* **2019**, *9*, 2033. [CrossRef]
18. Lai, C.Y.; Chavez, D.E.V.; Ding, S. Transformable parallel-serial manipulator for robotic machining. *Int. J. Adv. Manuf. Technol.* **2018**, *97*, 2987–2996. [CrossRef]
19. Vuong, N.-D.; Li, R.; Chew, C.-M.; Jafari, A.; Polden, J. A novel variable stiffness mechanism with linear spring characteristic for machining operations. *Robotica* **2017**, *35*, 1627. [CrossRef]
20. Wang, W.; Fu, X.; Li, Y.; Yun, C. Design and implementation of a variable stiffness actuator based on flexible gear rack mechanism. *Robotica* **2018**, *36*, 448–462. [CrossRef]
21. Barrett, E.; Reiling, M.; Barbieri, G.; Fumagalli, M.; Carloni, R. Mechatronic design of a variable stiffness robotic arm. In Proceedings of the IEEE/RSJ International Conference on Intelligent Robots and Systems (IROS), Vancouver, BC, Canada, 24–28 September 2017.
22. Petit, F.; Daasch, A.; Albu-Schäffer, A. Backstepping Control of Variable Stiffness Robots. *IEEE Trans. Control Syst. Technol.* **2015**, *23*, 2195–2202. [CrossRef]
23. Sciavicco, L.; Siciliano, B. *Modelling and Control of Robot Manipulators*; Springer Science & Business Media: Berlin/Heidelberg, Germany, 2012.
24. Doria, A.; Cocuzza, S.; Comand, N.; Bottin, M.; Rossi, A. Analysis of the compliance properties of an industrial robot with the Mozzi axis approach. *Robotics* **2019**, *8*, 80. [CrossRef]
25. Bottin, M.; Cocuzza, S.; Comand, N.; Doria, A. Modeling and identification of an industrial robot with a selective modal approach. *Appl. Sci.* **2020**, *10*, 4619. [CrossRef]
26. Nemec, B.; Yasuda, K.; Ude, A. A Virtual Mechanism Approach for Exploiting Functional Redundancy in Finishing Operations. *IEEE Trans. Autom. Sci. Eng.* **2020**. [CrossRef]
27. Esquivel, E.; Carbone, G.; Ceccarelli, M.; Jáuregui, J.C. A Dynamic Compensation for Roll Hemming Process. *IEEE Access* **2018**, *6*, 18264–18275. [CrossRef]
28. Görgülü, İ.; Carbone, G.; Dede, M.İ.C. Time efficient stiffness model computation for a parallel haptic mechanism via the virtual joint method. *Mech. Mach. Theory* **2020**, *143*, 103614. [CrossRef]
29. Kalpakjian, S. *Manufacturing Processes for Engineering Materials*; Pearson Education India: Tamil Nadu, India, 1984.
30. Malkin, S.; Guo, C. *Grinding Technology: Theory and Application of Machining with Abrasives*; Industrial Press Inc.: New York, NY, USA, 2008.
31. Marinescu, I.D.; Hitchiner, M.P.; Uhlmann, E.; Rowe, W.B.; Inasaki, I. *Handbook of Machining with Grinding Wheels*, 2nd ed.; CRC Press: Boca Raton, FL, USA, 2016.
32. Her, M.G.; Kazerooni, H. Automated robotic deburring of parts using compliance control. *J. Dyn. Sys. Meas. Control.* **1991**, *113*, 60–66. [CrossRef]
33. Shinozuka, M.; Deodatis, G. Simulation of stochastic processes by spectral representation. *Appl. Mech. Rev.* **1991**, *44*, 191–204. [CrossRef]

applied sciences

MDPI

Article

Initial Estimation of Kinematic Structure of a Robotic Manipulator as an Input for Its Synthesis

Daniel Huczala [1,*], Tomáš Kot [1], Martin Pfurner [2], Dominik Heczko [1] and Petr Oščádal [1] and Vladimír Mostýn [1]

[1] Department of Robotics, Faculty of Mechanical Engineering, VSB-TU of Ostrava, 17. Listopadu 2172/15, 708 00 Ostrava-Poruba, Czech Republic; tomas.kot@vsb.cz (T.K.); dominik.heczko@vsb.cz (D.H.); petr.oscadal@vsb.cz (P.O.); vladimir.mostyn@vsb.cz (V.M.)

[2] Unit of Geometry and Surveying, University of Innsbruck, Technikerstr. 13, A-6020 Innsbruck, Austria; martin.pfurner@uibk.ac.at

* Correspondence: daniel.huczala@vsb.cz

Abstract: Researchers often deal with the synthesis of the kinematic structure of a robotic manipulator to determine the optimal manipulator for a given task. This approach can lower the cost of the manipulator and allow it to achieve poses that might be unreachable by universal manipulators in an existing constrained environment. Numerical methods are broadly used to find the optimum design but they often require an estimated initial kinematic structure as input, especially if local-optimum-search algorithms are used. This paper presents four different algorithms for such an estimation using the standard Denavit–Hartenberg convention. Two of the algorithms are able to reach a given position and the other two can reach both position and orientation using Bézier splines approximation and vector algebra. The results are demonstrated with three chosen example poses and are evaluated by measuring manipulability and the total link length of the final kinematic structures.

Keywords: manipulator design; robot kinematics; synthesis of kinematic structure

Citation: Huczala, D.; Kot, T.; Pfurner, M.; Heczko, D.; Oščádal, P.; Mostýn, V. Initial Estimation of Kinematic Structure of a Robotic Manipulator as an Input for Its Synthesis. *Appl. Sci.* **2021**, *11*, 3548. https://doi.org/10.3390/app11083548

Academic Editor: Giuseppe Carbone

Received: 19 March 2021
Accepted: 13 April 2021
Published: 15 April 2021

1. Introduction

Nowadays, manufacturing industry, the most common type of robotic manipulator has six serial axes (degrees of freedom—DoFs) that are arranged in a so-called universal kinematic structure, e.g., the robots of ABB, KUKA, Fanuc, Yaskawa, and many others. In a very simplified way, the typical process for the deployment of a robot is to analyse the manufacturing process and workplace area first, followed by the choosing of a universal robotic manipulator and simulating the process. If the manipulator can reach the desired poses and fulfil the given task, further deployment can be considered.

However, using this universal kinematic structure is not always necessary, when a manipulator with fewer axes is suitable to perform the given task, or even possible, if the universal manipulator can face unavoidable collisions in an already existing environment. Additionally, they might not fulfil the desireed advanced operation conditions, such as manipulability [1] and kinematic reliability [2]. Therefore, researchers are focused on the topic of the synthesis of the kinematic structure of manipulators, which means finding such a kinematic structure that is optimal for a given task. This approach of deployment of highly customised manipulators may lead to benefits like lowered energy consumption, accelerated manufacturing process cycles, or deploying manipulators in highly dense-built workplaces. An example of such a general structure is presented by Brandstötter et al., who delivered the so-called curved manipulator (CuMa) [3] with possible modifications of its structure during the operational process [4]. This is achieved by changing the temperature in the links so they become flexible. A different approach to a deformable manipulator was taken by Xu et al. [5], where the links are composed of a few components and it is possible to change the orientation between them. Clark et al. [6] uses air pressure to change the kinematic structure of the presented malleable robot. The custom design of manipulators is desired not only in the manufacturing industry, but for example, in healthcare for helping with human upper limb rehabilitation [7] and shoulder joint rehabilitation [8].

There are two approaches to the synthesis of the kinematic structure of a manipulator. Analytically, it was solved by Hauenstein et al. [9] in the synthesis of three-revolute spacial chain for five poses. However, once a path becomes more complex and requires more degrees of freedom (more manipulator axes), the numerical approach is applied utilising optimisation algorithms. Among them, evolutionary robotics with genetic algorithms (GA) are probably the most common. Chocron et al. [10] used an adaptive multi-chromosome evolutionary algorithm to build a modular manipulator. Furthermore, GA can be used to build a manipulator from adaptive modules to perform a desired task [11]. Pastor et al. [12] compared straight, rounded, and curved mechanism links synthesised using GA. Valsamos et al. created so-called pseudo-joints (links which can be modified) and proposed a GA that tries to find the kinematic structure with the best manipulability [13]. It was later verified in an experiment with a real manipulator [14]. The synthesis of a parallel manipulator is addressed in [15]. As an example of non-industrial application, the work by Zeiaee et al. [16] deals with the optimisation of an eight-DoFs upper-limb exoskeleton.

In addition to GAs, the global optimum of an objective function can be searched with Simulated Annealing algorithm [17] or by a heuristic-guided tree search algorithm [18]. Another numerical method for finding the optimal manipulator for a given task is to search for a local minimum of an objective function. To solve this, a nonlinear programming (NLP) can be used, as it was implemented by Dogra et al. [19] for the design of a modular manipulator. In the paper, an optimal kinematic structure was proposed based on the minimisation of the joint torques. Another usage of NLP is trying to find an optimal design minimizing the path length in joint space [20].

The results of the already described papers [19,20] seem promising in their application of nonlinear programming to solve task-based custom manipulator design, however, there is one unanswered question in their work. In general, nonlinear programming traditionally requires that a starting point is given as part of the problem data, and comparative numerical testing is done using these traditional starting points [21]. In the case of robotic manipulators, if there is a given random path for a robotic manipulator, how should the starting point (initial values, initial estimation) of its kinematic structure look like?

The work [19] mentions a set of input values without any detailed explanation of how those values were determined. In [20], eight initial seeds are applied, but they are random values, which might be a cause of why no solution was found at all in some cases. In [17], they searched for a global optimum; however, the input values are also random, which may extend the time of solving the objective function. Even in a book by Ghafil et al. [22] which serves as an introduction to the optimisation of kinematic structures, the initial values for all described methods are obtained randomly without any detailed discussion. Therefore, a possible answer to the previously stated question will be addressed in this paper using multiple approaches.

In this paper, a geometric analysis to estimate kinematic structures and related calculations are proposed and discussed. The outcomes may avoid relying on randomness, which in the case of local-optimum-search algorithms may more frequently lead to convergence, making them reliable but still much faster than global-optimum-search methods. Moreover, the results can also be used in the previously mentioned algorithms using GAs, where they can serve as an optional first generation input, and in global-optimum-search algorithms, where they can reduce the time needed for the optimisation. In addition to that, the procedure may serve as an input for other custom manipulator design challenges such as collision avoidance [23].

A Denavit–Hartenberg notation [24] (standard DH parameters) was applied in the presented algorithms to obtain a general kinematic structure of a robot. The DH parameters were widely used, however, they also bring some disadvantages in the case of general structures. These are discussed later. Some algorithms behind the automatic placement of DH parameters have already been presented. In [25], a vector-algebra is applied to extract the parameters. Corke [26] creates a string of elementary translations and rotations from the user-defined base coordinate to the end-effector and factorises the string afterwards. An approach by Rajeevlochana et al. [27] is using line geometry to obtain the parameters. In addition to that, some researchers proposed their algorithms and they also identified (and verified) the DH parameters with an external sensor device in simulation [28] or experiments [29]. However, all of these algorithms were applied to already existing robots or similar devices. On the other hand, the algorithms presented in this paper are here to determine and "create" a non-existing manipulator (its kinematic structure) that can achieve a freely given pose.

2. Materials and Methods

When an algorithm is searching for a local minima of a function, the results may differ upon the choice of initial values. To find the optimal solution for two initial values with different outcomes, the cost function has to be compared afterwards. More accurate initial values can lead to fewer iterations that the algorithm needs, which can also save computing time. In this section, four algorithms of automatic assignment of DH parameters are presented, so they can serve as initial values for the synthesis of kinematic structure. The inputs are the position of the base of a robot, its tool-center point (TCP) pose, and the number of joints.

In this paper, we denote the transformation matrix between two frames as **J** with axis vectors and position coordinates as shown in Equation (1). The $\vec{n}$ (normal) is the X axis vector, the $\vec{o}$ (orientation) is Y axis vector, the $\vec{a}$ (approach) is Z axis vector, and the $\vec{t}$ (translation) is the position coordinate vector of a pose. **J** is a special Euclidean group of rigid body displacements in three dimensions (SE3) representing 3D rigid-body motion:

$$\mathbf{J} = \begin{bmatrix} \vec{n} & \vec{o} & \vec{a} & \vec{t} \\ 0 & 0 & 0 & 1 \end{bmatrix} \tag{1}$$

We also use unit vectors of the X, Y, and Z axes. They are denoted as $\vec{i}$, $\vec{j}$, and $\vec{k}$:

$$\vec{i} = \begin{bmatrix} 1 \\ 0 \\ 0 \end{bmatrix}; \quad \vec{j} = \begin{bmatrix} 0 \\ 1 \\ 0 \end{bmatrix}; \quad \vec{k} = \begin{bmatrix} 0 \\ 0 \\ 1 \end{bmatrix} \tag{2}$$

For visualisation and work with SE3 groups, we used the MATLAB®, and Robotics Toolbox that was made by P. Corke. It is described in his book [30] and accessible as open source Github repository [31].

DH parameters are the most suitable and easily applicable technique for kinematic structures that have parallel or orthogonal axes. The typical procedure is that one has a robot and places the coordinate frames following the convention [24]. However, what to do when there is a given pose that is needed to be reached while no robot is chosen yet? This is the problem for the synthesis of kinematic structure. There is one important issue related to DH parameters. Between two general poses (right-hand rule following coordinate frames), it is uncertain if the transformation matrix $\mathbf{J}_{i-1,i}$ from $(i-1)$th pose to ith pose can be achieved following the typical procedure as the multiplication of a rotation matrix of θ_i around the z_{i-1} axis, the translation matrix of d_i along the z_{i-1} axis, the translation matrix of a_i along hte x_i axis, and the rotation matrix of α_i around the x_i axis, as shown in the following equations:

$$\mathrm{Rot}(z_{i-1}, \theta_i) = \begin{bmatrix} \cos\theta & -\sin\theta & 0 & 0 \\ \sin\theta & \cos\theta & 0 & 0 \\ 0 & 0 & 1 & 0 \\ 0 & 0 & 0 & 1 \end{bmatrix} \tag{3}$$

$$\mathrm{Trans}(z_{i-1}, d_i) = \begin{bmatrix} 1 & 0 & 0 & 0 \\ 0 & 1 & 0 & 0 \\ 0 & 0 & 1 & d_i \\ 0 & 0 & 0 & 1 \end{bmatrix} \tag{4}$$

$$\mathrm{Trans}(x_i, a_i) = \begin{bmatrix} 1 & 0 & 0 & a_i \\ 0 & 1 & 0 & 0 \\ 0 & 0 & 1 & 0 \\ 0 & 0 & 0 & 1 \end{bmatrix} \tag{5}$$

$$\mathrm{Rot}(x_i, \alpha_i) = \begin{bmatrix} 1 & 0 & 0 & 0 \\ 0 & \cos\alpha & -\sin\alpha & 0 \\ 0 & \sin\alpha & \cos\alpha & 0 \\ 0 & 0 & 0 & 1 \end{bmatrix} \tag{6}$$

$$\mathbf{J}_{i-1,i} = \mathrm{Rot}(z_{i-1}, \theta_i)\,\mathrm{Trans}(z_{i-1}, d_i)\,\mathrm{Trans}(x_i, a_i)\,\mathrm{Rot}(x_i, \alpha) \tag{7}$$

It is clear that rotation and translation around and along the y axis are missing to achieve all possible poses. In DH convention, it is mitigated by placing the joint coordinate frames in a specific way and applying Equation (7), as explained in [24]. However, if there are two general poses $(i-1)$th and ith which do not follow DH convention, it is possible to find a common perpendicular between their two z_{i-1} and z_i axes. Please see Figure 1. Rotation around z_{i-1} to the direction of the perpendicular is θ_i. d_i is the distance from the x_{i-1} axis to the intersection point P of the perpendicular and z_{i-1} axis. The distance along the perpendicular is equal to the a_i distance between these two frames. Finally, the rotation around the x_i axis to the direction of z_i is α_i. If another displacement of d_{i+1} is added and a rotation θ_{i+1} is applied, the previously unreachable (by four DH parameters) general pose ith becomes achievable by four DH parameters of $(i-1)$th joint and two DH parameters d_{i+1} and θ_{i+1} of the ith joint. It can also represent an end-effector coordinate frame if the $(i-1)$th joint was the last joint. This approach is presented in detail in [27,29]. For the following calculations, we enhanced a script made by Brodsky [32] to find a common perpendicular and intersection points P and Q. This can be found in the Supplementary Material of this paper.

We used 3 poses to demonstrate the strong and weak points of the four presented algorithms to synthesise manipulators guiding their end-effector through the given position or pose, so everyone can choose the right solution for its implementation. They are also compared in Section 3 by manipulability and arm length. The first pose is a general one. The second pose is also general, but with a small offset between its Z axis and the Z axis of the base frame. The third pose has the parallel Z axis with the base Z axis. The poses are visualised in Figure 2:

$$Pose(1) = \begin{bmatrix} -0.50 & -0.18 & -0.84 & -1.0 \\ -0.06 & 0.98 & -0.17 & 0.9 \\ 0.86 & -0.02 & -0.50 & 0.8 \\ 0 & 0 & 0 & 1 \end{bmatrix} \tag{8}$$

$$Pose(2) = \begin{bmatrix} 0.81 & -0.34 & -0.46 & -0.4 \\ 0.48 & -0.02 & 0.87 & 0.6 \\ -0.30 & -0.93 & 0.14 & 0.7 \\ 0 & 0 & 0 & 1 \end{bmatrix} \tag{9}$$

$$Pose(3) = \begin{bmatrix} 0 & 1 & 0 & 0.4 \\ 1 & 0 & 0 & 0.6 \\ 0 & 0 & -1 & 0.7 \\ 0 & 0 & 0 & 1 \end{bmatrix} \tag{10}$$

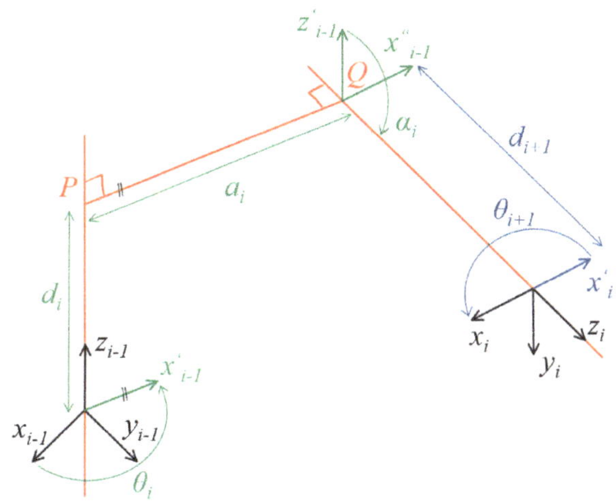

Figure 1. The principle of obtaining DH parameters between two general poses. Standard DH parameters are green; additional parameters are marked with blue colour.

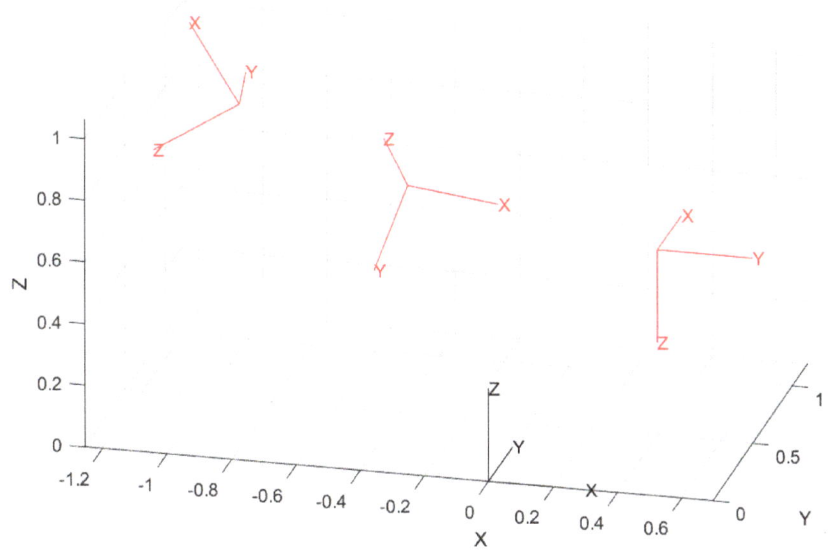

Figure 2. The poses (red) chosen for the demonstration of the working principle. Base coordinate frame has a black colour. $Pose(1)$ is on the left, $Pose(2)$ is in the middle, and $Pose(3)$ is on the right.

Two of the four presented algorithms utilised Bézier curves (splines) which are easy to implement between two given coordinate frames. For the presented calculations, only four control points are required to define a Bézier curve. We used a script by Bai [33] to calculate the curve. The control points P_{1-4} were calculated using the following equations, where $\vec{t}_b$ is the base point coordinate, $\vec{t}_p$ is the pose point coordinate, $\vec{a}_b$ is the rotational vector of the Z axis of the base, $\vec{a}_p$ is the rotational vector of the Z axis of the pose, and p is the parameter related to the Euclidean distance between the base and the pose. The Bézier splines are visualised in Figure 3 for the chosen poses.

$$P_1 = \vec{t}_b \tag{11}$$

$$P_2 = \vec{t}_b + p\vec{a}_b \tag{12}$$

$$P_3 = \vec{t}_p - p\vec{a}_p \tag{13}$$

$$P_4 = \vec{t}_p \tag{14}$$

$$p = \frac{||\vec{t}_p - \vec{t}_b||}{2} \tag{15}$$

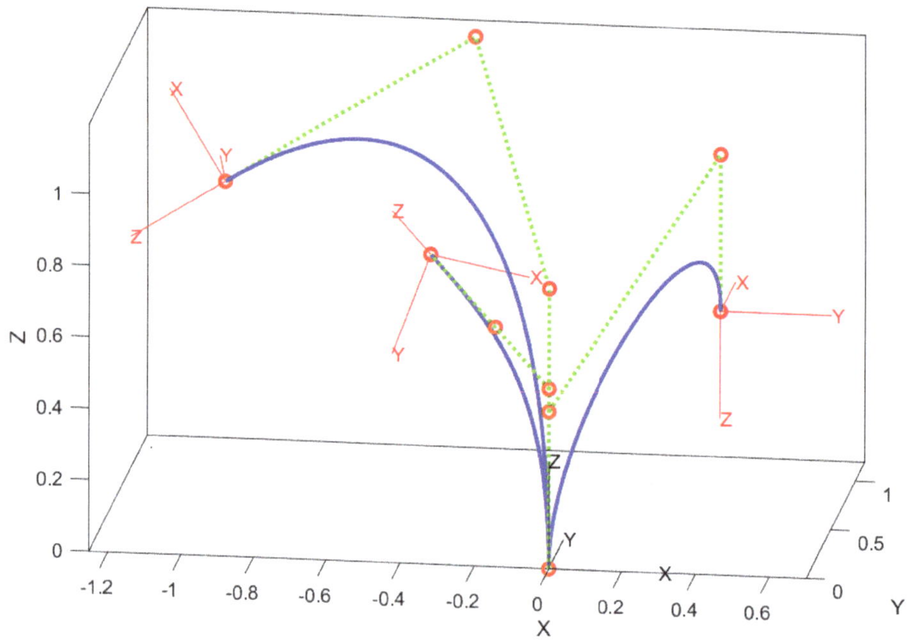

Figure 3. Bézier splines (blue curves) between the poses and base; control points are shown as red circles.

The generated kinematic structures that have served as examples in this paper have 4 joints; however, the presented algorithms are general and can provide a solution from 3 to an unlimited number of joints. In the following subsections, all 4 procedures are presented. The types A and B only deal with the position (translational part) of a given pose, so they do not fulfil the given orientation. However, this might be enough in some cases. The other two types C and D are able to achieve a pose including orientation using the common perpendicular approach, but in some specific poses it generates structures with joints in collision. The A and C types are obtained using vector algebra only, and the B and D types use Bézier's curve approximation.

Three variables are input for all presented algorithms. It is the transformation of the robot base, the transformation of the TCP pose, both with respect to the world coordinate frame, and the desired number of joints. In our case, the base is an identity matrix. We used 3 transformations of the poses presented before, and the number of joints is four, as already said.

2.1. Type A—The Nearest Distance to Achieve a Position

This simple structure is obtained by finding the distance between the poses projected into the XY base plane. Only a positional vector of a pose is reached while the orientation is not taken into account. The implementation of such a structure is easy and straightforward. The idea is presented in Figure 4:

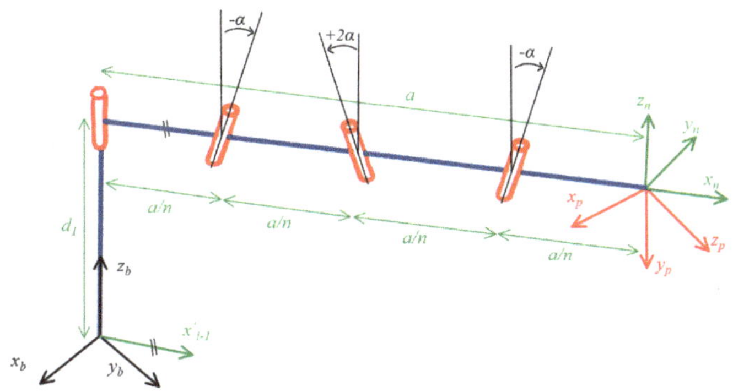

Figure 4. The schematic of type A estimation; the base frame is black, the pose frame is red, and the end-effector frame is green and does not fulfil the orientation of the pose.

The first step is to find the length of $a_{0..n}$—the distance between joints in the X axis direction. The length is the projection of $\vec{t}_p$, the pose position vector, in the XY plane of the base, so only the X and Y coordinates are applied in Equation (16). n is the number of joints:

$$a_{1..n} = \frac{\sqrt{(\vec{t}_{p,x} - \vec{t}_{b,x})^2 + (\vec{t}_{p,y} - \vec{t}_{b,y})^2}}{n} \tag{16}$$

Then, find the length of d_1—the offset of the joint along the Z axis. Only the Z axis coordinates of the two position vectors are applied:

$$d_1 = \vec{t}_{p,z} - \vec{t}_{b,z} \tag{17}$$

The $\theta_{1..n}$ are joint variables, and their offset is set to 0 degrees. The other parameters, $d_{2..n}$ and $\alpha_{1..n}$ can be set either to zero or they can be freely defined as $\pm$values, for example. We chose $\alpha_1 = \pi/4$, $\alpha_2 = -\pi/4$, etc. One must be careful in the case of an even/odd number of joints—the sum of such tweaks needs to be equal to zero.

2.2. Type B—Joints on Bézier Curve to Achieve a Position

This method places joints between the base and the pose on a Bézier curve. To be able to obtain DH parameters, the proposed algorithm is orienting the $(i-1)$th joint (its rotational matrix) in a way that the ith joint lies in the XZ plane of the previous joint. The schematic is shown in Figure 5:

Using Equations (11)–(15), the Bézier spline is approximated between a given base and a pose. The number of approximated points is equal to the number of joints n. $Q_{1..n}$ is the set of these points–coordinates of each point with respect to the base frame.

Let us define a set $\mathbf{J}_{1..n}$ of SE3 objects representing the translation and orientation of the joints in the manipulator's default position. As a first step, all $\mathbf{J}_{1..n}$ are set to be equal to the given base (in our case, an identity matrix). We also define an SE3 object $\mathbf{J}_{n+1}$ representing the given end-effector pose. Now, for joints $\mathbf{J}_{2..n}$, the following procedure is done.

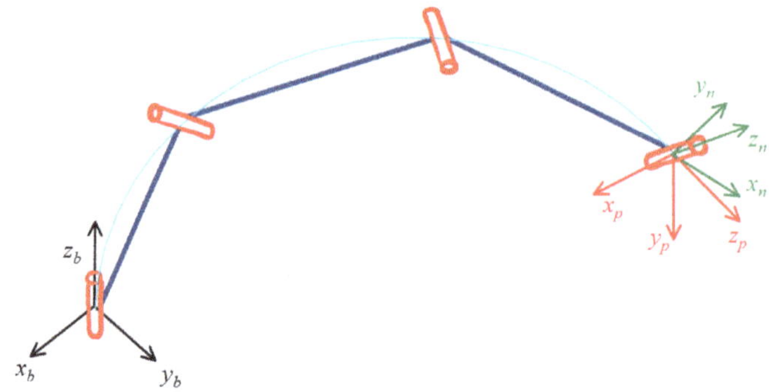

Figure 5. The schematic of type B estimation; the base frame is black, the pose frame is red, and the end-effector frame is green and does not fulfil the orientation of the pose. Bézier spline is shown as a light blue curve.

The ith joint is equal to the previous $(i-1)$th joint:

$$\mathbf{J}_i = \mathbf{J}_{i-1} \tag{18}$$

The frame $\mathbf{J}_i$ is translated on the Bézier curve changing its translation vector $\vec{t}_i$:

$$\vec{t}_i = Q_i \tag{19}$$

The position vector $\vec{t}_i$ of $\mathbf{J}_i$ is expressed in the coordinate frame of the previous $\mathbf{J}_{i-1}$ using its inverse matrix:

$$\vec{t'}_i = \mathbf{J}_{i-1}^{-1}\vec{t}_i \tag{20}$$

A projection of the $\vec{t}_i$ vector in the XY plane of the $\mathbf{J}_{i-1}$ frame is determined:

$$\vec{t''}_i = \vec{t'}_i - \begin{bmatrix} 0 \\ 0 \\ \vec{t'}_{i,z} \end{bmatrix} \tag{21}$$

The angle θ_i (DH parameter) between the unit vector $\vec{i}$ of the $\mathbf{J}_{i-1}$ frame and the projection of the $\vec{t}_i$ vector is calculated as

$$\theta_i = \tan^{-1}\left(\frac{||\vec{i} \times \vec{t''}_i||}{\vec{i} \cdot \vec{t''}_i}\right) \tag{22}$$

While using a right-handed coordinate frame, it is necessary to check if an angle is rotating around an axis in the positive (counter clockwise) or negative (clockwise) direction. To determine this, we used the projection property of the dot product between the two vectors. In this case, if the dot product of the X_i axis is in the negative direction of the Y_{i-1} axis, the angle θ_i has to be multiplied by -1:

$$\theta_i = \begin{cases} -\theta_i, & \textbf{if } \vec{j} \cdot \vec{t''}_i < 0 \\ \theta_i, & \textbf{otherwise} \end{cases} \tag{23}$$

Now, $\mathbf{J}_i$ can be updated using the matrix multiplication:

$$\mathbf{J}_i = \mathbf{J}_i \text{Rot}(z_{i-1}, \theta_i) \tag{24}$$

Thanks to the known translations, $_i$ and a_i, between the frames (from the Bézier curve approximation), and the calculated θ_i, only the angle α_i is missing among the DH parameters. The steps to determine it are similar as in the case of θ. Following the DH convention, the ith and $(i+1)$th frames are involved.

The position vector of $\mathbf{J}_{i+1}$ on the Bézier curve is given:

$$\vec{t}_{i+1} = Q_{i+1} \tag{25}$$

Position vector $\vec{t}_{i+1}$ of $\mathbf{J}_{i+1}$ is expressed in the coordinate frame of the currently determining frame $\mathbf{J}_i$ using its inverse matrix:

$$\vec{t'}_{i+1} = \mathbf{J}_i^{-1}\vec{t}_{i+1} \tag{26}$$

A projection of the $\vec{t}_{i+1}$ vector in the YZ plane of the $\mathbf{J}_i$ frame is calculated:

$$\vec{t''}_{i+1} = \vec{t'}_{i+1} - \begin{bmatrix} \vec{t'}_{i+1,x} \\ 0 \\ 0 \end{bmatrix} \tag{27}$$

The angle α_i is between the unit vector $\vec{k}$ of the $\mathbf{J}_i$ frame and the projection of the $\vec{t}_{i+1}$ vector, calculated as the inverse tangent fraction of the cross and dot products of those two vectors:

$$\alpha_i = \tan^{-1}\left(\frac{||\vec{k} \times \vec{t''}_{i+1}||}{\vec{k} \cdot \vec{t''}_{i+1}}\right) \tag{28}$$

Using a right-hand rule for coordinate frames, if the dot product of the Z_i axis is in the negative direction of the Y_{i+1} axis, the angle θ_i has to be multiplied by -1. Y_{i+1} is calculated as a cross product of $\vec{t''}_{i+1}$ and $\vec{i}$ vectors:

$$\alpha_i = \begin{cases} -\alpha_i, & \textbf{if } (\vec{t''}_{i+1} \times \vec{i}) \cdot \vec{k} < 0 \\ \alpha_i, & \textbf{otherwise} \end{cases} \tag{29}$$

Now, the final form of $\mathbf{J}_i$ that fulfils the DH convention between $(i-1)$th and ith can be obtained:

$$\mathbf{J}_i = \mathbf{J}_i \mathrm{Rot}(x_i, \alpha_i) \tag{30}$$

This procedure works smoothly for all joints. However, it will probably not be possible to obtain such DH parameters between the last joint $\mathbf{J}_n$ and the given pose $\mathbf{J}_{n+1}$ to reach the pose with the right orientation. Therefore, this algorithm is extended as the type D estimation in Section 2.4.

2.3. *Type C—Achieving a Pose with Common Perpendicular*

This algorithm finds a common perpendicular between the Z axis of the base and the Z axis of the pose. The joints are placed on this perpendicular line, and the last joint is oriented in the direction of the Z axis. Both the position and orientation can be achieved using this approach, as Figure 6 shows.

At first, a common perpendicular and intersection points P and Q are determined between the Z_b and Z_p axes using the script made by Brodsky [32]. However, his algorithm was not providing good results if the 2 lines were parallel, so we enhanced it and added some functionality to mitigate this issue.

The next step is to calculate the angle α_{sum} between those two axes:

$$\alpha_{sum} = \tan^{-1}\left(\frac{||\vec{a_b} \times \vec{a_p}||}{\vec{a_b} \cdot \vec{a_p}}\right) \tag{31}$$

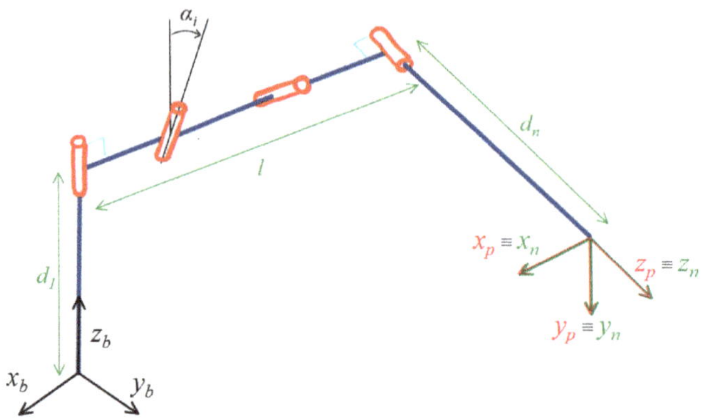

Figure 6. The schematic of the type C estimation; the base frame is black, the pose frame is red, and the end-effector frame is green and coincident with the pose.

Again, when using a right-handed coordinate system, it is necessary to determine whether the angle is positive or negative. We check the pose Z_{i+1} axis as a projection in the base Y_i axis:

$$\alpha_{sum} = \begin{cases} -\alpha_{sum}, & \textbf{if } \vec{o_b} \cdot \vec{a_p} < 0 \\ \alpha_{sum}, & \textbf{otherwise} \end{cases} \tag{32}$$

$\alpha_{1..n}$ is the angle between the Z axes of particular joints:

$$\alpha_{1..n-1} = \frac{\alpha_{sum}}{n-1} \tag{33}$$

$$\alpha_n = 0 \tag{34}$$

Now, we can calculate the rest of the DH parameters. $a_{1..n}$ is the distance between the joints. l is the length of a common perpendicular, and n is the number of joints:

$$a_{1..n-1} = \frac{l}{n-1} \tag{35}$$

$$a_n = 0 \tag{36}$$

d_0 is the distance from the base coordinate frame to the P-intersection point of the $\vec{a_b}$ and common perpendicular. d_n is the distance from the Q, the intersection point of the $\vec{a_p}$ and a common perpendicular, to the pose coordinate frame. d_n is also the translation of the end-effector from the last joint:

$$d_0 = (P - \vec{t_b}) \cdot \vec{a_b} \tag{37}$$

$$d_{1..n-1} = 0 \tag{38}$$

$$d_n = (\vec{t_p} - Q) \cdot \vec{a_p} \tag{39}$$

2.4. Type D—Joints on Bézier Curve while the Last Lies on Common Perpendicular to Achieve a Pose

This method extends type B estimation by adding a common perpendicular approach, used in type C, between the two last joints. This assures reaching the pose including orientation, as shown in Figure 7. The beginning steps are the same as in type C, but only from $\mathbf{J}_1$ to $\mathbf{J}_{n-1}$. The transformation of the last joint is determined using the following procedure.

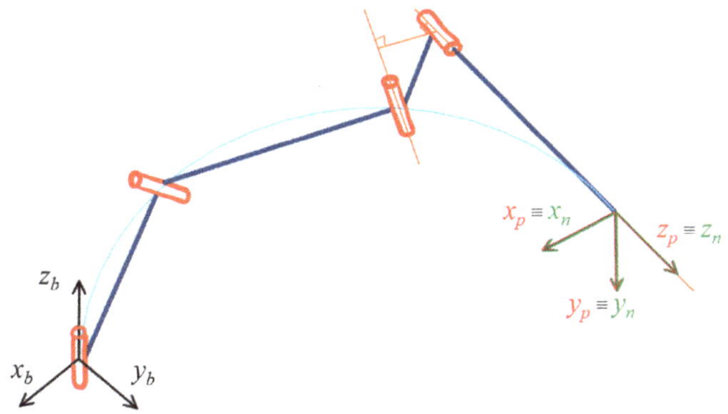

Figure 7. The schematic of the type D estimation; the base frame is black, the pose frame is red, the end-effector frame is green and coincident with the pose. Bézier spline is shown as a light blue curve.

Again, a common perpendicular and the intersection points P and Q are found between the joint $\mathbf{J}_{n-1}$ and the TCP pose $\mathbf{J}_{n+1}$ using the already presented ways. $\mathbf{J}_n$ is set equal to $\mathbf{J}_{n-1}$:

$$\mathbf{J}_n = \mathbf{J}_{n-1} \tag{40}$$

Angle θ_n between X_{n-1} and X_n axes is obtained:

$$\theta_n = \tan^{-1}\left(\frac{||\vec{n}_{n-1} \times \vec{PQ}||}{\vec{n}_{n-1} \cdot \vec{PQ}}\right) \tag{41}$$

Right-hand rule check is performed:

$$\theta_n = \begin{cases} -\theta_n, & \textbf{if } \vec{o}_{n-1} \cdot \vec{PQ} < 0 \\ \theta_n, & \textbf{otherwise} \end{cases} \tag{42}$$

$\mathbf{J}_n$ is rotated around Z_{n-1} axis afterwards:

$$\mathbf{J}_n = \mathbf{J}_n \mathrm{Rot}(z_{n-1}, \theta_n) \tag{43}$$

Translation of the $\mathbf{J}_n$ along the Z_n and X_n is obtained by changing its translational vector $\vec{t_n}$, and it is equal to the coordinates of point Q:

$$\vec{t_n} = Q \tag{44}$$

Angle α_n between Z_n and Z_{n+1} axes is calculated:

$$\alpha_n = \tan^{-1}\left(\frac{||\vec{a_n} \times \vec{a}_{n+1}||}{\vec{a_n} \cdot \vec{a}_{n+1}}\right) \tag{45}$$

Right-hand rule check, if the dot product of Z_{i+1} axis is in the positive direction of the Y_i axis, the angle α_n has to be multiplied by -1:

$$\alpha_n = \begin{cases} -\alpha_n, & \textbf{if } \vec{o_n} \cdot \vec{a}_{n+1} > 0 \\ \alpha_n, & \textbf{otherwise} \end{cases} \tag{46}$$

The final transformation matrix of the last joint $\mathbf{J}_n$ is obtained:

$$\mathbf{J}_n = \mathbf{J}_n \mathrm{Rot}(x_n, \alpha_n) \tag{47}$$

From this point, when all frames $\mathbf{J}_{1..n}$ representing the joints are known, it is possible to derive the DH parameters between them.

3. Results

This section presents the kinematic structures generated by the presented algorithms for the three poses defined earlier (Figure 2). The results are discussed and later compared by manipulability measure and manipulator length. A table with calculated DH parameters is also included. The visualisation of the final kinematic structures is shown in Figures 8–10.

The types of estimation A and B are not able to reach a pose in terms of orientation in general; however, in some cases (as shown in Figure 10b) the real solution was found for the type B. In addition, if compared with a similar D result (Figure 10d) for $Pose(3)$, solution B provides shorter links. Furthermore, the A and B types are generated in a way where no collision of joints should occur.

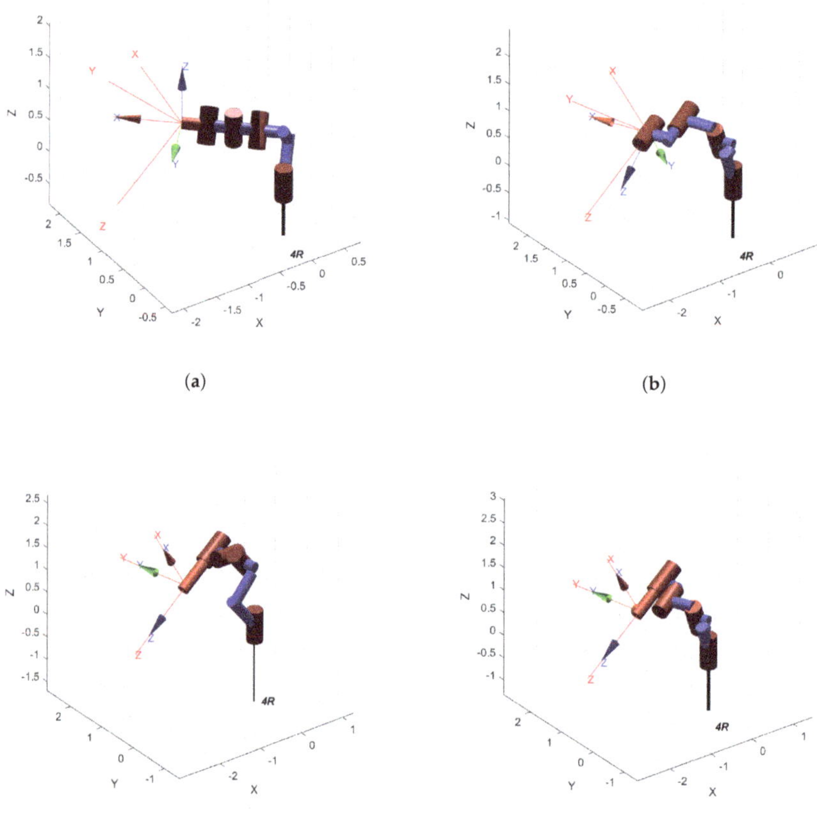

Figure 8. Initial estimation results for Pose (1): (**a**) type A estimation; (**b**) type B estimation; (**c**) type C estimation; and (**d**) type D estimation.

The type C may perform very well if the Z axes of the base and pose are parallel, as shown in Figure 10c; on the other hand, if the axes are very close to each other (the perpendicular distance is short), the joints are in collision, as shown in Figure 9c.

Placing joints on an approximated spline provides the most general result of the provided algorithms, as shown in Figures 8d and 9d, but it is struggling with parallel axes—see Figure 10d. This could be mitigated by tuning the Bézier curve driven point related to the pose and placing the joints not in a plane that is defined by the two parallel Z axes.

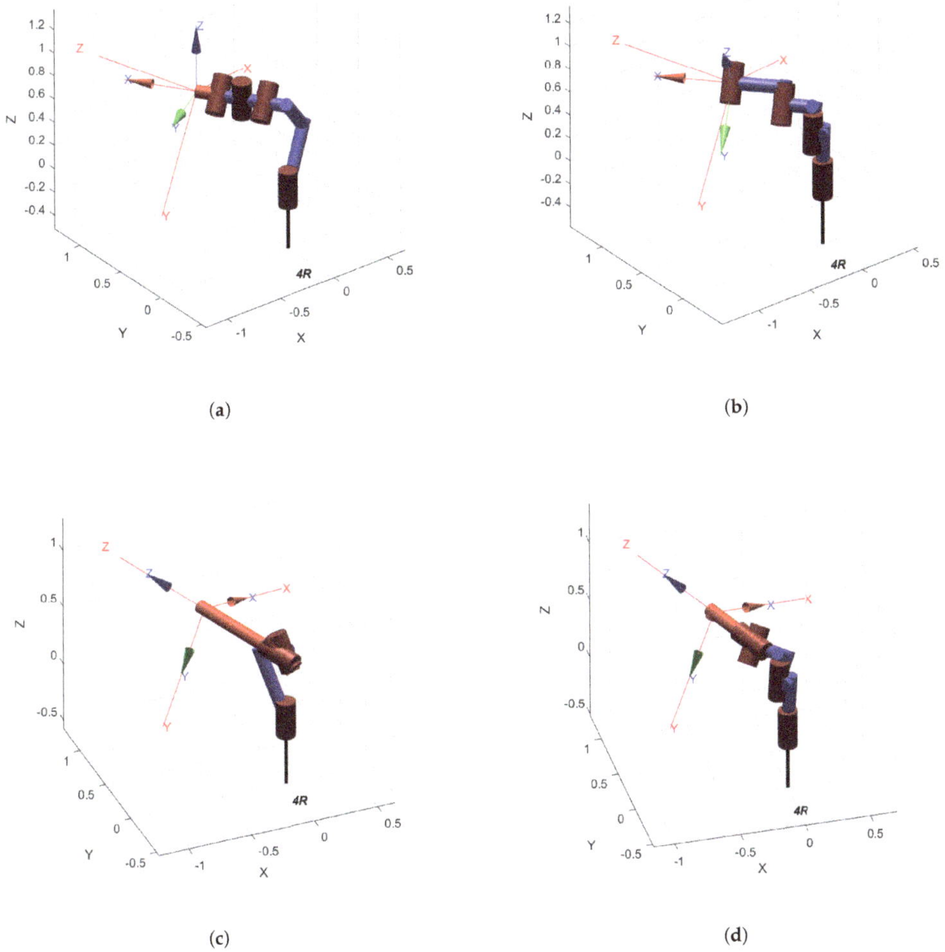

Figure 9. Initial estimation results for Pose (2): (**a**) type A estimation; (**b**) type B estimation; (**c**) type C estimation; and (**d**) type D estimation.

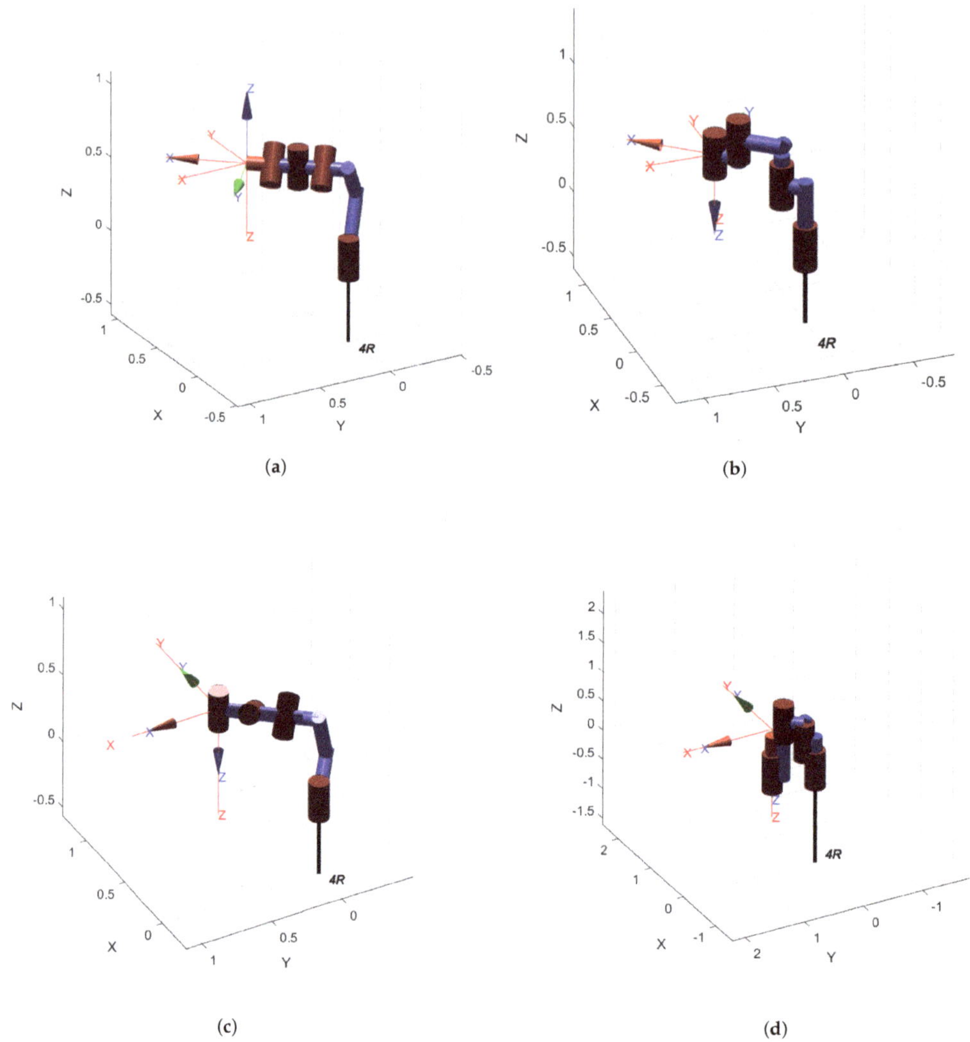

Figure 10. Initial estimation results for Pose (3): (**a**) type A estimation; (**b**) type B estimation; (**c**) type C estimation; and (**d**) type D estimation.

3.1. Resulting DH Parameters

For a better evaluation, we include Table 1 with the generated DH parameters for $Pose(1)$. The angles θ are considered as variables with zero offset.

Table 1. The DH parameters of manipulators for $Pose(1)$.

Joint	d_i (m)	a_i (m)	α_i (rad)	Joint	d_i (m)	a_i (m)	α_i (rad)
1	0.8	0.34	0.79	1	0.64	0.29	−0.39
2	0	0.34	−0.79	2	0.32	0.56	−1.48
3	0	0.34	0.79	3	0.46	0.38	0
4	0	0.34	−0.79	4	0	0	−0.28
	a Type A estimation				**b** Type B estimation		
Joint	**d_i (m)**	**a_i (m)**	**α_i (rad)**	**Joint**	**d_i (m)**	**a_i (m)**	**α_i (rad)**
1	1.27	0.36	−0.7	1	0.64	0.29	−0.40
2	0	0.36	−0.7	2	0.32	0.55	−1.48
3	0	0.36	−0.7	3	−0.33	0.30	−0.29
4	0.93	0	0	4	0.83	0	0
	c Type C estimation				**d** Type D estimation		

3.2. Manipulator Length Comparison

In general, longer links of a manipulator demand more powerful motors because of higher torques. In Figure 11, there is a bar plot comparing the lengths of the resulting manipulators. The length was determined using Equation (48), where a_i and d_i are DH parameters of i_{th} link. n is the number of joints:

$$L = \sum_{i=1}^{n} \sqrt{a_i^2 + d_i^2} \tag{48}$$

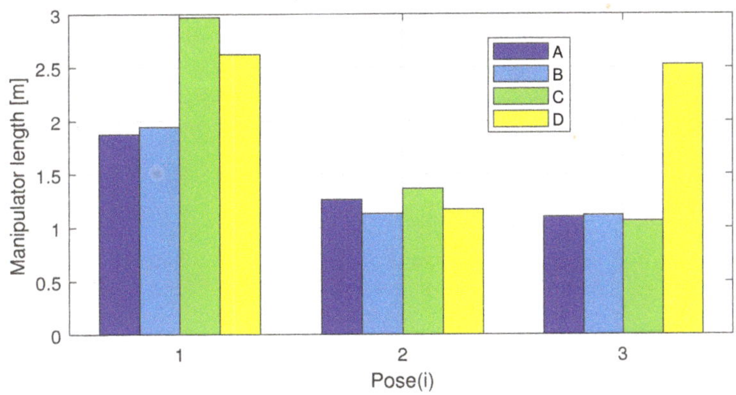

Figure 11. Comparison of the length of generated manipulators for every pose.

3.3. Manipulability Comparison

To compare the results, we decided to evaluate the manipulability of the calculated kinematic structures. This scalar measure was obtained using the Yoshikawa algorithm [1], which describes how spherical the end-effector velocity ellipsoid is. It differs between 0 and 1, where the value 1 shows the best manipulability in all axes. If the value is close to 0, the mechanism might be dealing with singularities.

The results are shown for both translational and rotational motions in Figure 12 as a logarithmic graph, because the measure differs significantly between particular kinematic structures. It should be kept in mind that the presented manipulators have less than six DoFs, which is one of the reasons why the manipulability measure by Yoshikawa evaluates them with low numbers.

As expected, the type A kinematic structure has to deal with singularities and the manipulability tends to be the lowest for the given poses. Type B has almost the same translational manipulability as type D, but the last two joints are in a singular position, so the rotational manipulability drops in the case of type B kinematic structure. Type C performs better than types A and B. In the case of general poses $Pose(1,2)$, type D provides the highest manipulability measures. However, for $Pose(3)$ when Z axes are parallel, the manoeuvrability of the type D algorithm drops under the values of type C.

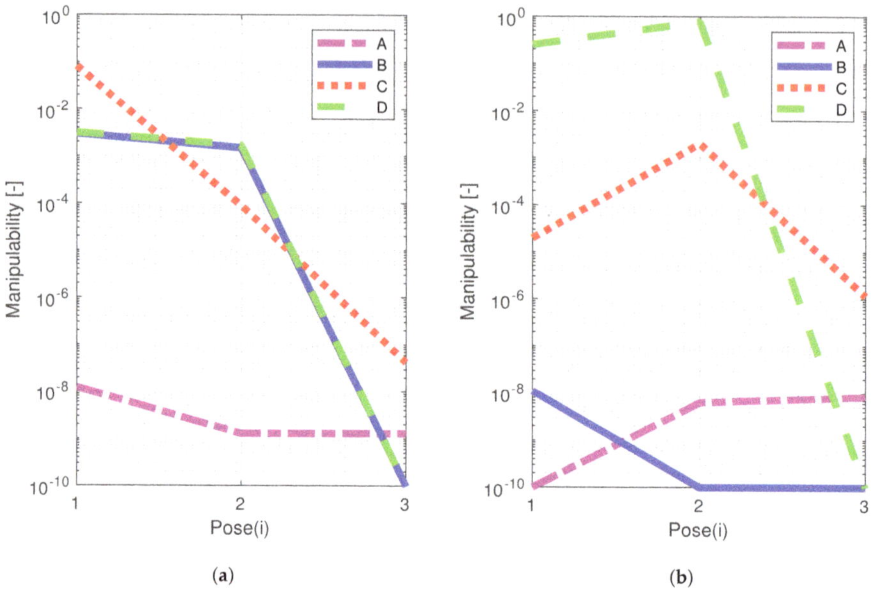

Figure 12. Logaritmic graph comparing the manipulability measures for given poses (1–3): (**a**) for translational motion; and (**b**) for rotational motion.

4. Discussion

The outcomes of this paper are aimed to be utilised in the synthesis of kinematic structures of robotic manipulators. Other applications are also possible, for example, in rapid kinematic analysis and simulation to evaluate kinematic options if only one position or pose on a trajectory is desired to be reached. However, only topics related to the synthesis problem are going to be discussed here.

This paper provided four algorithms to estimate the initial value of a kinematic structure for later implementation in optimisation algorithms. As mentioned previously, especially in algorithms searching the local minimum of an objective function, every initial guess can provide different results. Therefore, we decided to present the types (A and B) of estimation even though they do not fulfil the given pose in terms of orientation. The reason is that an optimisation algorithm may overcome this issue later and the solution could achieve the pose on a given trajectory with a better final value of the objective function than with the other presented types (C and D). This always depends on the type and properties of the trajectory. Therefore, we suggest implementing all four estimation types into an optimisation algorithm and compare the results afterwards.

If one would like to know which one of the four structures is the best, this question is not easy to answer. We chose three poses to demonstrate the advantages and disadvantages of particular algorithms and compared the final structures for these poses by manoeuvrability and length. The type D provides the most general structure that can reach any pose, but one must be careful in cases when some axes of the base and the pose are parallel, for example. We suggest to always visualise all initial structures for better evaluation.

The presented algorithms are based on the Denavit–Hartenberg convention, generating DH parameters. During this work, a question arose, of how suitable the DH convention is for general structures and especially for their synthesis for given trajectory. As mentioned previously, it is possible to find DH parameters between two coordinate frames using the common perpendicular approach, but it is required to "borrow" another two DH parameters from the next joint and to add them to the already existing four parameters. This is obvious since we need six parameters (three translations and three rotations) in general to describe the motion of a frame in Euclidean space. Therefore, using an optimisation algorithm to synthesise a manipulator that fulfils the DH convention may be limiting. The frames representing joints cannot freely rotate and translate wherever the algorithm tends to, but they have to follow the XZ planes in which the common perpendiculars between neighbouring joints lie. On top of that, the algorithm may not find the global or local minimum because of this limit at all. The reason is that traditionally DH parameters serve for the description of existing robots and once they are obtained, some local coordinates of the joints may be located outside of the rigid body along the Z axis, and although they are still closely tied to the particular joint and representing its kinematics, they are not representing the real (physical) position of the joint. On the other hand, in terms of synthesis when a rigid body does not exist yet, the location (transformation) of the coordinate frames of joints is the only known and crucial parameter and its variability should not be limited during the synthesis process anyhow.

The comparison of the synthesis of kinematic structure using DH convention and other standard approaches, such as screw theory [34] for instance, will be an interesting topic for future research. However, this matter has no impact on the work presented in this paper. The kinematic structures obtained using the four algorithms may be translated into any other standard description of the structure of a manipulator.

5. Conclusions

This paper presents four mathematical algorithms to find an initial estimation of a kinematic structure of a serial robotic manipulator. The input values are the number of joints, the position of the base of the manipulator, and the target pose of its TCP. The outputs are the standard DH parameters of a kinematic structure that can reach the given pose either by position or by position and orientation. The presented methods are applicable in the topic of synthesis of the kinematic structure of robotic manipulators, where they can serve as an initial guess of a structure.

The examples of three poses for all four algorithms are shown and compared using the manipulability measure and the arm length. However, it is not easy to determine which method out of these four is the best as it is always depending on the input, especially the TCP pose. The manipulability measure indicates that the D type algorithm, which can fully satisfy any given pose, provides the best manipulability in general, because it avoids placing joints on a single line and places them on a Bézier curve instead. In addition to that, the joints also tend to prevent collisions. On the other hand, if the given pose is representing a specific case, as a parallel axis or axes with a base frame axis or axes, one should be always watchful and comparison with the other presented algorithms is suggested. As expected, algorithm types A and B achieved the best results in the case of the arm length measure, when only a position is desired. However, in some cases, the C and D algorithms may fully achieve a pose with a manipulator with similarly long links.

Future steps are to implement this method in an optimisation algorithm and to observe if the convergence is faster or if the obtained result and the values of objective functions are better.

The algorithms were tested in MATLAB and the scripts are available as Supplementary Material for this paper or with eventual updates as an open source package on a public repository [35].

Supplementary Materials: The source code of algorithms presented in this paper is available as open source repository on the Github page of the Department of Robotics, VSB-Technical University of Ostrava, Czech Republic: github.com/robot-vsb-cz/initial-estimation accessed on 15 April 2021.

Author Contributions: Conceptualisation, D.H. (Daniel Huczala) and T.K.; methodology, D.H. (Daniel Huczala); software, D.H. (Daniel Huczala) and P.O.; validation, D.H. (Daniel Huczala), M.P. and D.H. (Dominik Heczko); formal analysis, V.M.; investigation, D.H. (Dominik Heczko); resources, P.O.; data curation, T.K.; writing—original draft preparation, D.H. (Daniel Huczala); writing—review and editing, T.K. and M.P.; visualisation, D.H. (Dominik Heczko); supervision, M.P.; project administration, V.M.; funding acquisition, V.M. All authors have read and agreed to the published version of the manuscript.

Funding: This article has been elaborated under support of the project Research Centre of Advanced Mechatronic Systems, reg. no. *CZ*.02.1.01/0.0/0.0/16_019/0000867 in the frame of the Operational Program Research, Development and Education. This article has been also supported by specific research project *SP*2021/47 and financed by the state budget of the Czech Republic.

Institutional Review Board Statement: Not applicable.

Informed Consent Statement: Not applicable.

Data Availability Statement: Source code available on: github.com/robot-vsb-cz/initial-estimation accessed on 15 April 2021.

Conflicts of Interest: The authors declare no conflict of interest.

Abbreviations

Abbreviations

The following abbreviations are used in this manuscript:

CuMa	Curved Manipulator
DoF	Degree of Freedom
DH	Denavit–Hartenberg
GA	Genetic Algorithm
NLP	Nonlinear Programming
SE3	Euclidean Group of Rigid Body Displacements in Three Dimensions
TCP	Tool-Center Point

References

1. Yoshikawa, T. Translational and Rotational Manipulability of Robotic Manipulators. In Proceedings of the 1990 American Control Conference, San Diego, CA, USA, 23–25 May 1990; doi:10.23919/acc.1990.4790733. [CrossRef]
2. Lara-Molina, F.A.; Dumur, D. A Fuzzy Approach for the Kinematic Reliability Assessment of Robotic Manipulators. *Robotica* **2021**, 1–15. [CrossRef]
3. Brandstötter, M.; Angerer, A.; Hofbaur, M. The curved manipulator (cuma-type arm): Realization of a serial manipulator with general structure in modular design. In Proceedings of the 14th IFToMM World Congress, Taipei, Taiwan, 25–30 October 2015; pp. 403–409. [CrossRef]
4. Brandstötter, M.; Gallina, P.; Seriani, S.; Hofbaur, M. Task-Dependent Structural Modifications on Reconfigurable General Serial Manipulators. In *Advances in Service and Industrial Robotics*; Springer International Publishing: New York, NY, USA, 2018; pp. 316–324._33. [CrossRef]
5. Xu, S.; Li, G.; Song, D.; Sun, L.; Liu, J. Real-time Shape Recognition of a Deformable Link by Using Self-Organizing Map. In Proceedings of the 2018 IEEE 14th International Conference on Automation Science and Engineering (CASE), Munich, Germany, 20–24 August 2018; doi:10.1109/coase.2018.8560514. [CrossRef]
6. Clark, A.B.; Rojas, N. Design and Workspace Characterisation of Malleable Robots. In Proceedings of the 2020 IEEE International Conference on Robotics and Automation (ICRA), Paris, France, 31 May–31 August 2020; doi:10.1109/icra40945.2020.9197439. [CrossRef]
7. Pang, Z.; Wang, T.; Wang, Z.; Yu, J.; Sun, Z.; Liu, S. Design and Analysis of a Wearable Upper Limb Rehabilitation Robot with Characteristics of Tension Mechanism. *Appl. Sci.* **2020**, *10*, 2101. [CrossRef]
8. Yan, H.; Wang, H.; Chen, P.; Niu, J.; Ning, Y.; Li, S.; Wang, X. Configuration Design of an Upper Limb Rehabilitation Robot with a Generalized Shoulder Joint. *Appl. Sci.* **2021**, *11*, 2080. [CrossRef]
9. Hauenstein, J.D.; Wampler, C.W.; Pfurner, M. Synthesis of three-revolute spatial chains for body guidance. *Mech. Mach. Theory* **2017**, *110*, 61–72. [CrossRef]
10. Chocron, O.; Bidaud, P. Evolutionary algorithms in kinematic design of robotic systems. In Proceedings of the 1997 IEEE/RSJ International Conference on Intelligent Robot and Systems, Innovative Robotics for Real-World Applications, IROS'97, Grenoble, France, 11 September 1997; doi:10.1109/iros.1997.655148. [CrossRef]

11. Chung, W.; Han, J.; Youm, Y.; Kim, S. Task based design of modular robot manipulator using efficient genetic algorithm. In Proceedings of the International Conference on Robotics and Automation, Albuquerque, NM, USA, 25 April 1997; doi:10.1109/robot.1997.620087. [CrossRef]
12. Pastor, R.; Huczala, D.; Bobovský, Z.; Grushko, S. Genetic Optimization of a Manipulator: Comparison between Straight, Rounded, and Curved Mechanism Links. *Appl. Sci.* **2021**, *11*, 2471. [CrossRef]
13. Valsamos, C.; Moulianitis, V.C.; Aspragathos, N. Metamorphic Structure Representation: Designing and Evaluating Anatomies of Metamorphic Manipulators. In *Advances in Reconfigurable Mechanisms and Robots I*; Springer: London, UK, 2012; pp. 3–11._1. [CrossRef]
14. Katrantzis, E.F.; Aspragathos, N.A.; Valsamos, C.D.; Moulianitis, V.C. Anatomy Optimization and Experimental Verification of a Metamorphic Manipulator. In Proceedings of the 2018 International Conference on Reconfigurable Mechanisms and Robots (ReMAR), Delft, The Netherlands, 20–22 June 2018; doi:10.1109/remar.2018.8449880. [CrossRef]
15. Hamida, I.B.; Laribi, M.A.; Mlika, A.; Romdhane, L.; Zeghloul, S. Dimensional Synthesis and Performance Evaluation of Four Translational Parallel Manipulators. *Robotica* **2020**, *39*, 233–249. [CrossRef]
16. Zeiaee, A.; Soltani-Zarrin, R.; Langari, R.; Tafreshi, R. Kinematic Design Optimization of an Eight Degree-of-Freedom Upper-Limb Exoskeleton. *Robotica* **2019**, *37*, 2073–2086. [CrossRef]
17. Patel, S.; Sobh, T. Task based synthesis of serial manipulators. *J. Adv. Res.* **2015**, *6*, 479–492. [CrossRef] [PubMed]
18. Ha, S.; Coros, S.; Alspach, A.; Bern, J.M.; Kim, J.; Yamane, K. Computational Design of Robotic Devices From High-Level Motion Specifications. *IEEE Trans. Robot.* **2018**, *34*, 1240–1251. [CrossRef]
19. Dogra, A.; Padhee, S.S.; Singla, E. An Optimal Architectural Design for Unconventional Modular Reconfigurable Manipulation System. *J. Mech. Des.* **2020**, 1–29. [CrossRef]
20. Whitman, J.; Choset, H. Task-Specific Manipulator Design and Trajectory Synthesis. *IEEE Robot. Autom. Lett.* **2019**, *4*, 301–308. [CrossRef]
21. Vanderbei, R.J.; Shanno, D.F. An interior-point algorithm for nonconvex nonlinear programming. *Comput. Optim. Appl.* **1999**, *13*, 231–252. [CrossRef]
22. Ghafil, H.N.; Jármai, K. *Optimization for Robot Modelling with MATLAB*; Springer: Berlin/Heidelberg, Germany, 2020.
23. Kot, T.; Bobovský, Z.; Brandstötter, M.; Krys, V.; Virgala, I.; Novák, P. Finding Optimal Manipulator Arm Shapes to Avoid Collisions in a Static Environment. *Appl. Sci.* **2020**, *11*, 64. [CrossRef]
24. Uicker, J.J.; Denavit, J.; Hartenberg, R.S. An Iterative Method for the Displacement Analysis of Spatial Mechanisms. *J. Appl. Mech.* **1964**, *31*, 309–314. [CrossRef]
25. Barker, L.K. Vector-algebra approach to extract Denavit-Hartenberg parameters of assembled robot arms. In *NTRS-NASA Technical Reports Server*; National Aeronautics and Space Administration, Scientific and Technical Information Branch: Washington, DC, USA, 1983.
26. Corke, P. A Simple and Systematic Approach to Assigning Denavit–Hartenberg Parameters. *IEEE Trans. Robot.* **2007**, *23*, 590–594. [CrossRef]
27. Rajeevlochana, C.; Saha, S.K.; Kumar, S. Automatic extraction of DH parameters of serial manipulators using line geometry. In Proceedings of the 2nd Joint International Conference on Multibody System Dynamics, Stuttgart, Germany, 29 May–1 June 2012; pp. 1–9.
28. Klug, C.; Schmalstieg, D.; Gloor, T.; Arth, C. A Complete Workflow for Automatic Forward Kinematics Model Extraction of Robotic Total Stations Using the Denavit-Hartenberg Convention. *J. Intell. Robot. Syst.* **2018**, *95*, 311–329. [CrossRef]
29. Faria, C.; Vilaca, J.L.; Monteiro, S.; Erlhagen, W.; Bicho, E. Automatic Denavit-Hartenberg Parameter Identification for Serial Manipulators. In Proceedings of the IECON 2019-45th Annual Conference of the IEEE Industrial Electronics Society, Lisbon, Portugal, 14–17 October 2019; doi:10.1109/iecon.2019.8927455. [CrossRef]
30. Corke, P. *Robotics, Vision and Control: Fundamental Algorithms in MATLAB®*, 2nd ed.; Springer Tracts in Advanced Robotics; Springer International: New York, NY, USA, 2017.
31. Corke, P. Robotics Toolbox for Matlab. 2020. Available online: https://github.com/petercorke/robotics-toolbox-matlab (accessed on 15 April 2021).
32. Brodsky, A. Shortest Distance between Two Lines in N Dimensions. 2011. Available online: https://www.mathworks.com/matlabcentral/fileexchange/29130-shortest-distance-between-two-lines-in-n-dimensions (accessed on 15 April 2021).
33. Bai, T. 3D Bezier Curve. 2012. Available online: https://www.mathworks.com/matlabcentral/fileexchange/35154-3d-bezier-curve (accessed on 15 April 2021).
34. Lynch, K.M.; Park, F.C. *Modern Robotics*; Cambridge University Press: Cambridge, UK, 2017.
35. Huczala, D. Initial Estimation. 2021. Available online: github.com/robot-vsb-cz/initial-estimation (accessed on 15 April 2021).

applied sciences

MDPI

Article

Instant Center Identification of Single-Loop Multi-DOF Planar Linkage Using Virtual Link

Liangyi Nie [1], Huafeng Ding [1,*], Kwun-Lon Ting [2] and Andrés Kecskeméthy [3]

1 School of Mechanical Engineering and Electronic Information, China University of Geosciences, No. 388 LuMo Road, Hongshan District, Wuhan 430074, China; nieliangyi@cug.edu.cn
2 Center for Manufacturing Research, Tennessee Technological University, Cookeville, TN 38505, USA; kting@tntech.edu
3 Faculty of Engineering Sciences Institute of Mechatronics and System Dynamics, University of Duisburg-Essen, 47057 Duisburg, Germany; andres.kecskemethy@uni-due.de
* Correspondence: dhf@ysu.edu.cn

Abstract: Instant center is an important kinematic characteristic which can be used for velocity and singularity analysis, configuration synthesis and dynamics modeling of multi-degree of freedom (multi-DOF) planar linkage. The Aronhold–Kennedy theorem is famous for locating instant centers of four-bar planar linkage, but for single-loop multi-DOF linkages, it fails. Increasing with the number of the links of single-loop multi-DOF planar linkages, the lack of link relationship makes the identification of instant center become a recognized difficulty. This paper proposes a virtual link method to identify instant centers of single-loop multi-DOF planar linkage. First, three types of instant centers are redefined and the instant center identification process graph is introduced. Then, based on coupled loop chain characteristic and definition of instant center, two criteria are presented to convert single-loop multi-DOF planar linkage into a two-loop virtual linkage by adding the virtual links. Subsequently, the unchanged instant centers are identified in the virtual linkage and used to acquire all the instant centers of original single-loop multi-DOF planar linkage. As a result, the instant centers of single-loop five-bar, six-bar planar linkage with several prismatic joints are systematically researched for the first time. Finally, the validity of the proposed method is demonstrated using loop equations. It is a graphical and straightforward method and the application is wide up to single-loop multi-DOF N-bar ($N \geq 5$) planar linkage.

Keywords: loop chain; virtual link; instant center; multi-degree-of-freedom; Aronhold–Kennedy theorem

Citation: Nie, L.; Ding, H.; Ting, K.-L.; Kecskeméthy, A. Instant Center Identification of Single-Loop Multi-DOF Planar Linkage Using Virtual Link. *Appl. Sci.* **2021**, *11*, 4463. https://doi.org/10.3390/app11104463

Academic Editors: Giuseppe Carbone and Med Amine Laribi

Received: 30 March 2021
Accepted: 12 May 2021
Published: 14 May 2021

1. Introduction

The concept of instant center is proposed by Bernoulli [1], which refers to the zero velocity point for two rigid bodies in a planar motion. It is not only used to analyze kinematics of linkages such as absolute velocity and angular velocity, but also to research singular configurations [2–5], configuration synthesis [6] and dynamics modeling [7]. Generally, instant centers are obtained by Aronhold–Kennedy theorem [8], but if the linkages contain more links or link loops, the traditional method fails. The instant center which cannot be directly located by Aronhold–Kennedy theorem, is called "indeterminate instant center" [9] or indeterminate secondary instant center in Refs. [10–24]. Moreover, it exists in almost all linkages besides single-loop four-bar linkage and Stephenson linkage. Therefore, the location of indeterminate secondary instant center is a difficulty of kinematics analysis of planar linkages, which attracts many researchers' interests. Dijksman [10] proposed a graphical linkage reduction method to find coordinated centers of curvature by changing the pentagonal loop into four-bar loop. The disadvantage of this method is that the joint-joining operation may have to be carried out twice and each time in a different way. Based on singular coefficient matrix of the derived velocity equations, Yan and Hsu [11] presented an analytical method to locate instant centers of single or multiple degrees of freedom linkages. This method has tedious calculation. Klein [12] raised a graphic method of trial-and-error type to identify the right position of the center. The defect is that the process is complex and inefficient. Forster and Pennock [9,14–16] presented both analytical method and graphical technique to solve instant center problem of single-DOF, two-DOF planar indeterminate linkage and three-DOF planar six-bar linkage. The essence of this method is first to determine one indeterminate secondary instant center by combing two arbitrary possible positions, then to reverse all the indeterminate instant centers, but its implementation method is roundabout. Based on instant centers of the cam pair residing on the extension line of a primary adjacent link, Chang [17–19] introduced a virtual cam method to help locate key instant centers of the linkages up to ten-bar. This method is a graphical one and is accurate in AutoCAD, but the applicable rate is not high. In addition, in some cases, it needs to work with Pennock's method [9,14–16], then the linkages can be solved. Di Gregorio [20] presented an algorithm to calculate the positions of indeterminate secondary instant centers of the indeterminate linkages. Although the analytical method uses only pieces of information of the linkage configuration, the difficulty is how to write the equations together with the closure equations of the mechanism to allow the computation of the instant centers' positions as a function of the generalized coordinate chosen to identify the mechanism configuration. Obviously, it is not easy. Kung and Wang [21] proposed the concept of instant center walk and instant center circuit and established the recursive formula to compute the coordinates of the instant centers. However, the drawback is that the application is only for single-DOF indeterminate linkages at non-singular configurations. According to the fact that an indeterminate secondary instant pole of two-DOF spherical linkage lies somewhere on the unique great circle for a specific configuration of the linkage, Zarkandi [22] provided two techniques to convert a single-DOF spherical mechanism to a two-DOF one, and then locate all the indeterminate secondary instant centers. The main issue is how to use the techniques, and it is not universal for different linkages. Valderrama-Rodríguez [23,24] presented a screw theory approach for the computation of instant rotation axes of the spherical linkages requiring the solution and comparison of two quadratic equations. Although this method is simpler than the previous literature, the calculation is relatively complex. Diab [25] utilized the location of the instant centers to perform acceleration analysis of a four-bar mechanism. On the basis of an adequate literature analysis, Sancibrian and Sarabia [6] presented a synthesis method based on optimization in which the identification of instant center is included in the objective function for rigid-body guidance synthesis Moreover, this proved approach is robust, accurate and efficient. Depended on the fact that the positions of instant centers can determine the velocity coefficients and the virtual work of the external forces of the mechanism, Di Gregorio [7] provided a dynamic model and an algorithm to solve the dynamic problems of single-DOF planar linkage. It is simple to use and numerically effective. In addition, since the relationships between the positions of the instant centers and the absolute velocities, instant centers are possibly used in some other applications such as dynamic model of the spherical mechanisms [26,27], wrench capability analysis of the redundant mechanisms [28], dynamic balancing analysis [29], and even commercial packages. For dynamic model of the spherical parallel mechanism, based on the principle of virtual work, the dynamic model can be built with the Jacobian matrices including angular velocities. As we all know, in a certain mechanism, once the instant centers are located, the corresponding angular velocities are decided. That is, the instant centers are able to address the dynamic model. For wrench capability analysis of the redundant planar parallel manipulator, joint torques, which are decided by the forces and moments acting on the end-effector, are the kinematic condition to sustain the wrench. They can be deduced by the derivatives of absolute velocities obtained from the instant centers. For dynamic balancing analysis of a given mechanism [29], the positions of instant centers can directly be used to calculate the angular momentum which effects whether the sum of all forces and moments acting on the based are zero. For commercial packages, the identification of instant centers may be available as a plug-in of the AutoCAD tool since the instant centers have a widespread application and the proposed method is suited to be programmed. As for real devices where imperfections are unavoidable in production processes, another application of instant center is found inspired by Profs. Bucolo and Buscarino's paper [30]. In their research, the nonlinear dynamical circuits are built to help investigate the performance of a novel control strategy for imperfect systems, and the instant center may be used to calculate the dynamical equations which regulate the behavior of the circuits. Although instant center, which is a basic kinematic property, is a useful tool to analyze kinematics of the linkages [2–5,31], rigid-body synthesis [6], dynamics modeling [7], and dynamic balancing analysis [29] of planar linkages, and it can be used in spatial mechanisms [26,27], redundant mechanisms [28], imperfect systems of real devices [30], and is even available in commercial packages, the identification of instant center is difficult since the lack of relationship of the interval links in a N-bar loop ($N \geq 5$). As discussed above, the existing methods can be classified three types: (1) graphical method [10,12,17–19,22], (2) analytical method [11,20,23,24], and (3) both analytical method and graphical method [9,14–16]. The graphical method has the advantage of visualization, but the defects are complex process and low applicable rate. The analytical method has the merit of high accuracy, but the calculation is normally complicated. Both analytical method and graphical method has both advantages, but the implementation method is roundabout. The generality and simplicity of instant center identification method is still a challenge.

The motivation of this paper is to propose a universal method to identify instant centers of planar linkages, especially for single-loop multi-DOF planar linkages. Based on coupled loop chain characteristic and definition of instant center, a single-loop multi-DOF planar linkage is changed into the two-loop linkage with the added virtual links using the proposed criteria, and the unchanged instant centers are identified according to Aronhold–Kennedy theorem, then all the instant centers of the original single-loop multi-DOF planar linkage can be obtained by the instant center identification process graph. The essence of the proposed method is to cover the shortage of the lacking link relationships of the single-loop linkage using virtual link operation, then obtain the indeterminate instant centers. The virtual link operation is an auxiliary mean which does not affect the motion of the original single-loop linkage. Moreover, the instant center identification of the single-loop multi-DOF planar linkage can be the basic of the instant center identification for multi-loop multi-DOF planar linkage which is discussed in our further research. That is, the proposed method may solve the instant center problem of the planar linkages no matter how many number of the links the linkages contain.

The first contribution of this paper is to provide a virtual link method to identify instant centers of the single-loop multi-DOF planar linkage, which is a graphical method and sticks to the definition of instant center. Compared to the previous literature [10–22], the proposed method has three advantages: simplification, concision, and validity. Firstly, to build virtual link operation, a few steps are carried out and normally, only one virtual link added operation is required. There is no need to choose the specific link and to perform repetitive complex steps [9,10,12,14–16]. Secondly, the location operation is completely decided by the drawing graph, which also can be programmed and automated in AutoCAD. The closure equations are unnecessary [11,20]. Thirdly, the proposed method is only based on instantaneous configuration of the linkage and the definition of instant center, and to ensure universality, the parameters of the virtual links are arbitrary. As a result, the application of this method is universal no matter what the degrees of freedom and the components of the joints of the linkage are [17–19]. The second contribution is to redefine three types of instant centers. The new classification is more accurate and detailed compared to the existing classification, which brings convenience for further research about instant centers. The third contribution is that the instant center problem of the single-loop five-bar, six-bar planar linkages with only rotation joints and several prismatic joints is solved. Note that the instant centers of single-loop five-bar, six-bar planar linkages with several prismatic joints are the first time to be located, to our best knowledge. The proposed method provides a geometry loop insight to reveal the relationship between the formation of the instant center and the motion of the mechanism, and a new research idea for the study of instant centers.

This paper is organized as follows. In Section 2, the mathematics definition is presented and three types of instant centers are redefined. In Section 3, based on coupled loop chain characteristic and definition of instant center, two criteria to determine instant centers are provided, and the instant center identification process graph is introduced. Instant centers of the Stephenson six-bar linkage are identified to explain how the proposed method works. Then, the steps of instant center identification of single-loop multi-DOF N-bar ($N \geq 5$) planar linkages are summarized. Subsequently, the mathematical proof is shown in Section 4. Finally, in Section 5, instant centers of the single-loop five-bar and six-bar planar linkages with only rotation joints and several prismatic joints are located using the criteria, and conclusions are presented at the end of this paper.

2. Mathematic Definition and Classification of Instant Center

Instant center is defined as a point where the relative velocity of the two rigid bodies is zero in a planar motion, but the absolute velocity may be not. For a given planar linkages, the positions of instant centers are varying yielding to the input condition when the planar linkage continuously moves. In fact, the continuous motion of the linkage is formed by lots of instantaneous configurations. Each instantaneous configuration corresponds to one specified input at this instant. That is, an instantaneous configuration coming from the continuous motion linkage is a momentary configuration in which the corresponding momentary input is specified. Moreover, different instantaneous configurations correspond to different specified ratios of the inputs. If the same linkage moves from one instantaneous configuration to the other instantaneous configuration, the specified ratios of the inputs should be accordingly changed. In this paper, instant centers of the instantaneous configuration are discussed, i.e., the linkages researched in our paper are all discussed on the condition that the linkages are all the instantaneous configurations, only one specified ratio of the inputs corresponds to the corresponding instantaneous configuration. Since the momentary input condition is specified in the instantaneous configuration, the passive angles of the linkage can be decided at this instant. In another word, the link relationships (i.e., instant centers) of the instantaneous configuration can be all obtained at the exact points at this instant based on mathematic definition above. In Figure 1, the location of the instant center I_{13} can be expressed as followed.

$$V_{13} = W_1 \times r_1 = W_3 \times r_3 \quad (1)$$

where V_{13} is absolute velocity of the links 1 and 3, and W_i and r_i (i = 1 or 3) denote the angular velocity of corresponding link and distance between the binary link from the point, respectively. Equation (1) represents the fact that the absolute velocity V_{13} of the instant center I_{13} only relatives to the angular velocity W_i and the corresponding distance r_i in a instantaneous configuration where the momentary input condition is specified and the link parameters of the planar linkage are determinate. It means that the identification of instant center in the instantaneous configuration is only related to the positions and motions (i.e., distances and angular velocities discussed in above) of the participant links (links 1 and 3 in Figure 1) and is irrelevant to the other link of the instantaneous configuration. In another word, when the positions and motions of participant links are the same regardless of whether the remaining links of the linkage changed or not, the corresponding instant center is the same. This property, evolved from the definition, is the theoretical basis to research the single-loop multi-DOF planar linkage in the following sections of this paper. In addition, it is used and proved in Ref. [32] for invariant link rotatability of N-bar kinematic chains. It is worth noting that the property above is effective in instantaneous configurations, but it fails when the linkage is in a continuous motion. The reason is that the passive angular velocity of the participant links (for example links 1 and 3 in Figure 1) is timely drove by the input link (link 4) and the limits coming from sizes of the links. Although the distances and the angular velocities of other links of the moving planar linkage do not occur in Equation (1), they still influence the positions and motions of the participant links. All the locations of the instant centers are continuously varying. However, for instantaneous configurations, each instantaneous configuration corresponds to one specified input at this instant. The ratio of the inputs of an instantaneous configuration is specified, the momentary input is known and invariable.

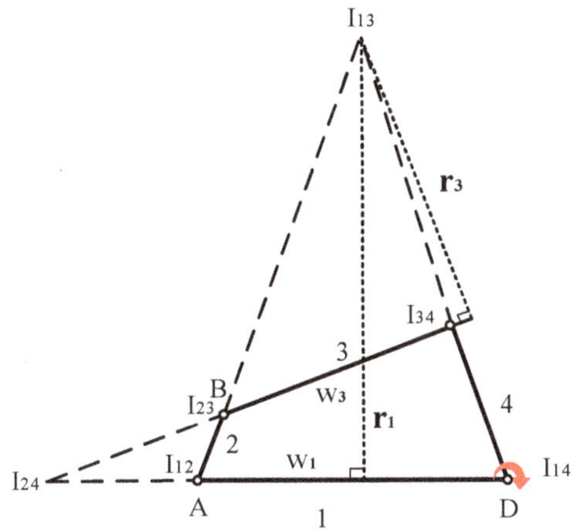

Figure 1. Planar four-bar linkage with instant centers.

In literatures above [9–24], instant centers are classified into two types: primary instant center and secondary instant center. For a given linkage, primary instant center is constantly the coincident point for a pair of rigid bodies of a planar linkage. The other instant centers are all secondary instant centers no matter whether they can be located with Aronhold–Kennedy theorem. However, in this paper, instant centers are classified into three types in view of the access: first instant center, secondary instant center and third instant center. First instant center is same to primary instant center of the existing literatures. It is the easiest to be obtained and is also the base to locate the secondary and third instant centers, such as I_{12}, I_{23}, I_{14}, and I_{34} in Figure 1. The instant centers, which can be directly received according to the established first instant centers using Aronhold–Kennedy theorem, are called as secondary instant center. It is different to the concept of secondary instant center of the existing literatures. In the new classification, all the secondary instant centers can be located with Aronhold–Kennedy theorem. They have exact positions. For Figure 1, the secondary instant centers are I_{13}, I_{24}. Apart from first and secondary instant centers, the remaining instant centers of the linkages are third instant center. Since its indeterminacy, the identification of third instant center is a difficulty for single-loop multi-DOF planar linkages. Compared to secondary instant center, third instant center can also be determined by Aronhold–Kennedy theorem, but the difference is that secondary instant center can be acquired straightforwardly using the existing first instant centers but third instant center is not. The third instant center I_{13}, formed by the link 1 and link 3 in Figure 2a, is this case. The instant center I_{13} is a point lying on the extension line of link 5, but the exact location is unknown. That is, the location of third instant center needs other addition information. According to whether third instant center exists on a known line (at least two instant centers on the line are known, normally, the extension line of the link), two types third instant center can be concluded. When a third instant center lies on a known line, the third instant center is an A type third instant center. If not, it is a B type. Taking third instant centers in Figure 2a for example, the instant center I_{13}, I_{14}, I_{24}, I_{25}, and I_{35} (also represented as L_{13}, L_{14}, L_{24}, L_{25}, and L_{35}, L means that the instant centers are on the lines, but their exact locations are unknown) are all the A type third instant centers. Obviously, there is no B type third instant center. For Figure 2b, the instant center I_{14}, I_{25} and I_{36} are all the B type third instant centers. What is noteworthy is that the shortest topological distance of the two links coming from the instant center can be used to distinguish which type the third instant center belongs to. The discrimination process is shown as follows. If the shortest topological distance is less than two units, which means that there is only one link among the two links, the corresponding instant center is an A type third instant center. For Figure 2a, the link 2 is the only link for the A type instant center I_{13} formed by the links 1 and 3. The B type third instant center I_{14} has two links (i.e., links 2 and 3) between the formed links 1 and 4 in Figure 2b. Compared to the listed literatures, the new classification is more detailed and accurate.

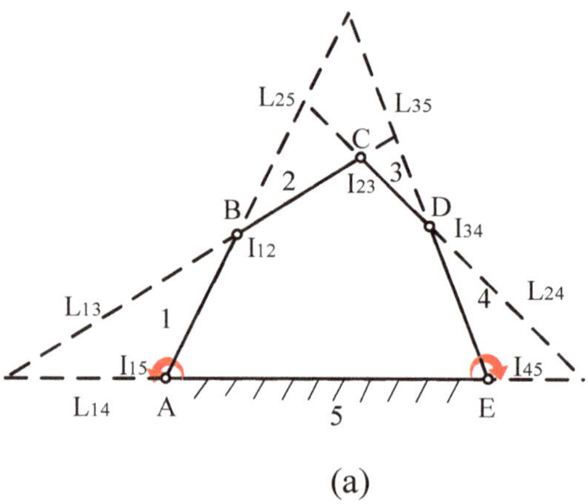

Figure 2. Two types of instant centers in (**a**) single-loop five-bar planar linkage; (**b**) single-loop six-bar planar linkage.

3. Virtual Link Method for Instant Center Identification

3.1. Criteria for Instant Center Identification

In order to explain the formation of the two proposed criteria, the instant center identification of the Stephenson linkage containing two-loop, in Figure 3, is taken as an example as follows.

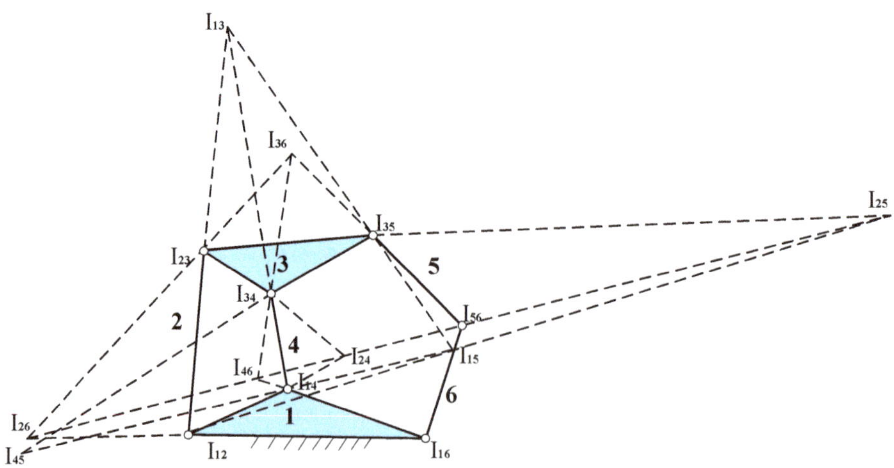

Figure 3. Instant centers of the Stephenson linkage.

The Stephenson linkage has $C_6^2 = 15$ instant centers (Figure 3). The seven instant centers I_{12}, I_{23}, I_{34}, I_{14}, I_{35}, I_{56}, and I_{16} are the first instant centers, and the secondary instant centers are the centers I_{13}, and I_{24}. The rest of the instant centers, such as I_{15}, I_{36}, I_{46}, I_{25}, I_{26}, and I_{45}, are the A type third instant centers shown in Table 1. Since no B type third instant center exists, instant centers of the Stephenson linkage are easy to be located with Aronhold–Kennedy theorem. In order to efficiently use the existing first or secondary instant centers to locate the third instant centers, the instant center identification process graph is introduced, which makes the location operation visible and efficient. The essence of the instant center identification process graph is the visualization of third instant center identification using Aronhold–Kennedy theorem. For example, the third instant center identification process graph of Stephenson linkage is shown in Figure 4. The instant center identification process graph is built by the third instant centers and the lines formed by other instant centers except for themself. The symbols "$\surd$", "$\times$", solid "$\downarrow$" and hollow "$\downarrow$" of the graph denote determinate instant center, indeterminate instant center, determinate line and indeterminate line, respectively. The steps of the identification process are shown as follows. Firstly, the third instant centers are all listed on the top. Secondly, using Aronhold–Kennedy theorem, the third instant center can be divided into four instant centers on two possible Aronhold–Kennedy lines. As for instant center I_{15}, the four centers are the instant centers I_{16}, I_{56}, I_{13}, and I_{35}. Thirdly, if the four instant centers are all determinate instant centers, i.e., four $\surd$, then the two lines can be decided (i.e., two solid $\downarrow$). For instant center I_{15}, the two lines are line $I_{16}I_{56}$ and line $I_{13}I_{35}$. Finally, the third instant center can be identified in which the two determinate lines cross. For Figure 4, all the instant centers of the Stephenson linkage can be directly obtained with Aronhold–Kennedy theorem, but for single-loop multi-DOF N-bar ($N \geq 5$) planar linkage with B type third instant centers, the situation becomes complex. In this paper, the concept of coupled loop chain characteristic and virtual link operation are proposed to solve the instant center identification of single-loop multi-DOF N-bar ($N \geq 5$) planar linkage.

Table 1. Instant centers of the Stephenson linkage.

First instant Center	I_{12} I_{23} I_{34} I_{14} I_{35} I_{56} I_{16}	
Secondary Instant Center	I_{13} I_{24}	
Third instant center	A type	I_{15} I_{36} I_{46} I_{25} I_{26} I_{45}
	B type	none

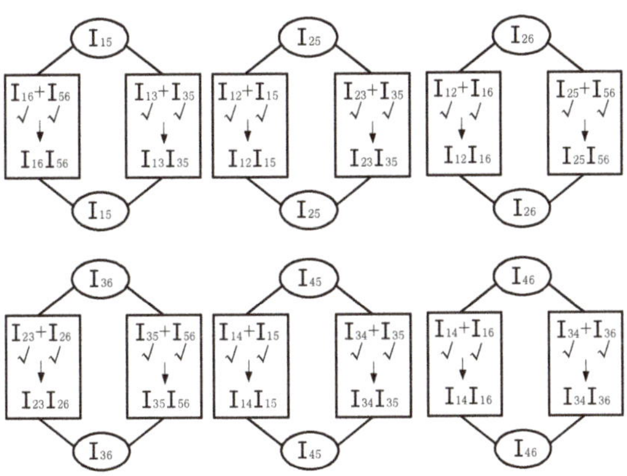

Figure 4. The instant center identification process graph of the Stephenson linkage.

3.1.1. Coupled Loop Chain Characteristic

In the view of loop chain, the instant centers of the Stephenson linkage (Figure 3) can be located as follows. Firstly, three loop chains are found: loop 1-2-3-4, loop 1-4-3-5-6, and loop 1-2-3-5-6. The coupled loop chain 3-5-6-1 is the common part of two loops if the loop 1-4-3-5-6 and loop 1-2-3-5-6 are grouped. For the loop 1-2-3-5-6, there exists one instant center ${I_{13}}^1$. However, for the other loop 1-4-3-5-6, the other instant center ${I_{13}}^2$ can also be obtained. Only when the instant centers ${I_{13}}^1$ and ${I_{13}}^2$ come cross, can the Stephenson linkage be formed. In other words, only if the common instant centers ${I_{13}}^1$ and ${I_{13}}^2$ of two loops coincide, the Stephenson linkage exists. In the loop 1-2-3-5-6 and loop 1-4-3-5-6, the instant centers ${I_{13}}^1$ and ${I_{13}}^2$ are all A type third instant centers, which lie on the extension lines of link 2 and link 4, respectively. That is, the instant center I_{13} occurs at the point of the intersection of the two line when the instant center ${I_{13}}^1$ and ${I_{13}}^2$ coincide. Similarly, if combining the two loops: loop 1-2-3-4 and loop 1-4-3-5-6, the same result can be obtained. This property, here called coupled loop chain characteristic, is discussed and demonstrated in Refs. [11–14,18].

3.1.2. Virtual Link Operation

According to discussion above, for single-loop multi-DOF N-bar ($N \geq 5$) planar linkage, two criteria and corresponding virtual link operation are presented to locate instant centers as follows.

Criterion 1: For a single-loop multi-DOF N-bar ($N \geq 5$) planar linkage, add the virtual links without changing instantaneous configuration to take shape the two-loop virtual linkage in which the instant centers are easy to be determined. Based on mathematics definition of instant center, some of the obtained instant centers in virtual linkage are equivalent to the ones in the original linkage.

Criterion 2: Based on coupled loop chain characteristic, in a certain linkage, if several common instant centers exist when combining any two loops of the linkage, once any pair of common instant centers obtained from different loop chain come cross, the rest common instant centers must coincide, accordingly.

Note that, virtual link operation is an assumption, and the virtual links are unreal. Virtual link operation is an auxiliary means to change the single-loop linkage into a virtual two-loop linkage which do not affect the motion of the original single-loop linkage. That is, the virtual two-loop linkage is actually still the single-loop linkage. The inputs and the motions of the links (not including the virtual links) in the virtual two-loop linkage are the same compared to the original single-loop linkage. The corresponding mathematical proof is discussed in detail in Section 4.

3.2. Process for Instant Center Identification

For single-loop multi-DOF N-bar ($N \geq 5$) planar linkage, as shown in Figure 5a, there are ${C_N}^2$ instant centers including no secondary instant center. According to two criteria above, the single-loop multi-DOF N-bar ($N \geq 5$) planar linkage can be transformed into the two-loop virtual planar linkage which is combined by a N-bar virtual loop and a $(N-1)$-bar virtual loop when N is odd number or two $(N-1)$-bar virtual loops when N is even number by adding the middle virtual links $N + 1'$ in Figure 5b (the number m of the middle virtual links depends on what odd number or even number N ($N \geq 5$) belongs to. For example, 5 is the first odd number for N, the number m is one for single-loop five-bar planar linkage. 7 is the second odd number for N, therefore, for seven-bar linkage, the number m is two. The number m is three in nine-bar linkage. Similarly, 6 is the first even number for N, so the number m is one for single-loop six-bar linkage, and the number m is two for eight-bar linkage. It is three in 10-bar linkage. The rule of adding virtual links is to ensure that the number of the two formed virtual loops are same or the difference is 1). If the common instant centers of the two formed virtual loops are located, all the instant centers of the two-loop virtual planar linkage can be identified, and then the instant center problem of the original N-bar planar linkage can be solved with Aronhold–Kennedy theorem. Six steps are summarized for the instant center identification of single-loop multi-DOF N-bar (N ≥ 5) planar linkage as follows. The corresponding flow chart is shown in Figure 6.

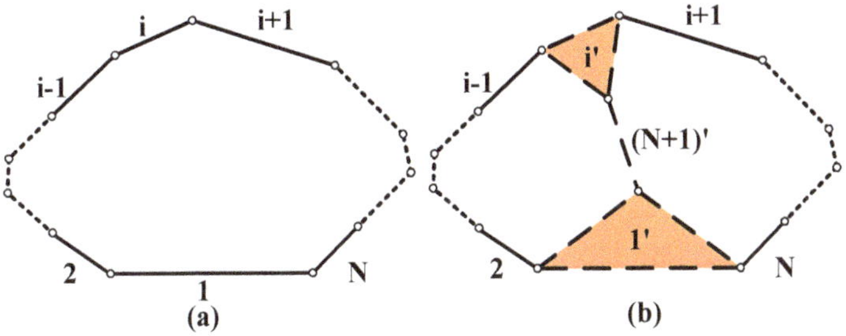

Figure 5. (**a**) Single-loop multi-DOF N-bar (N ≥ 5) planar linkage, (**b**) Two-loop virtual planar linkage with virtual links.

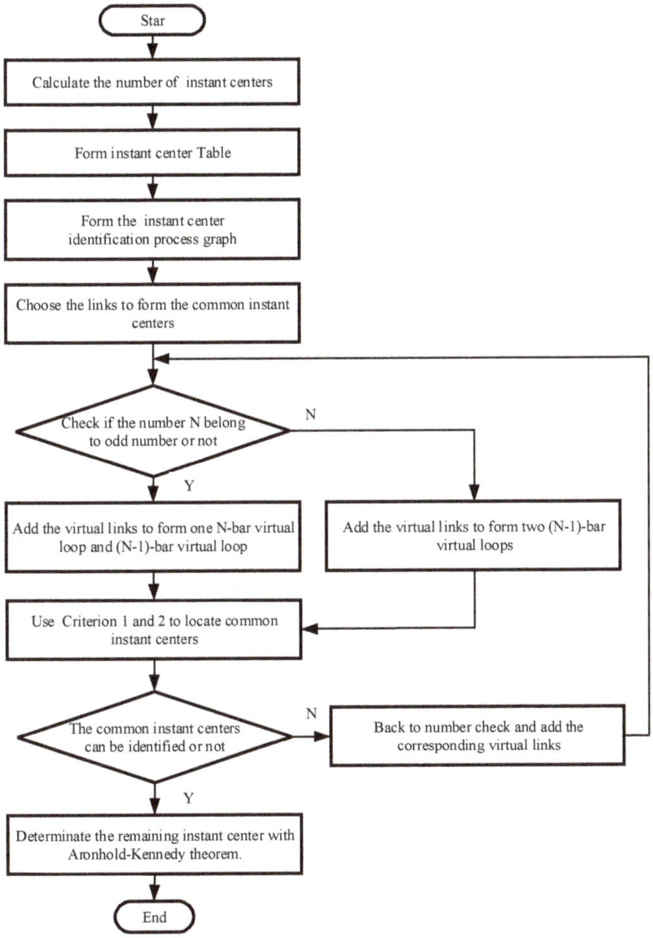

Figure 6. The flow chart of instant center identification for single-loop multi-DOF N-bar (N ≥ 5) planar linkage.

Step 1: Calculate the number of instant centers and classify the instant centers according to the proposed classification, then form the instant center Table.

Step 2: Form the instant center identification process graph and choose the right links to build the common instant centers. The common instant centers can be formed by the unchanged links after virtual link operation or directly come from the virtual links. The common instant centers and virtual link operation are interdependent. In another word, the rule of adding virtual links discussed above is the rule to choose the common instant centers, to some extent.

Step 3: Check if the number N ($N \geq 5$) belongs to odd number or not.

Step 4: Based on the results from Steps 2 and 3, add the corresponding virtual links to form two-loop virtual planar linkage.

Step 5: Use Criterions 1 and 2 to locate the common instant centers of the two-loop virtual planar linkage. If not, back to Step 3 to add new virtual links in the existing virtual loop until the common instant centers can be obtained. Step 3-Step 4-Step 5 is a iteration loop. The proposed method is suitable to be programmed.

Step 6: Determinate the remaining instant centers of the original N-bar planar linkage which fails to be located in Step 5 using Aronhold–Kennedy theorem.

4. Mathematical Proof

Loop equation is a common mathematical tool to analyze the kinematics of planar linkages. It is employed here to verify the validity of virtual link operation in the instantaneous configuration. Taking the single-loop five-bar planar linkage (red part in Figure 7a, i.e., loop1-2-3-4-5), for example, the known variables are the input angles θ_2, θ_5 and the link parameters a_1, a_2, a_3, a_4, and a_5, i.e., the passive angles θ_3, and θ_4 are unknown. Choosing the angle θ_4 as the output, the loop equation of the single-loop five-bar planar linkage can be expressed as

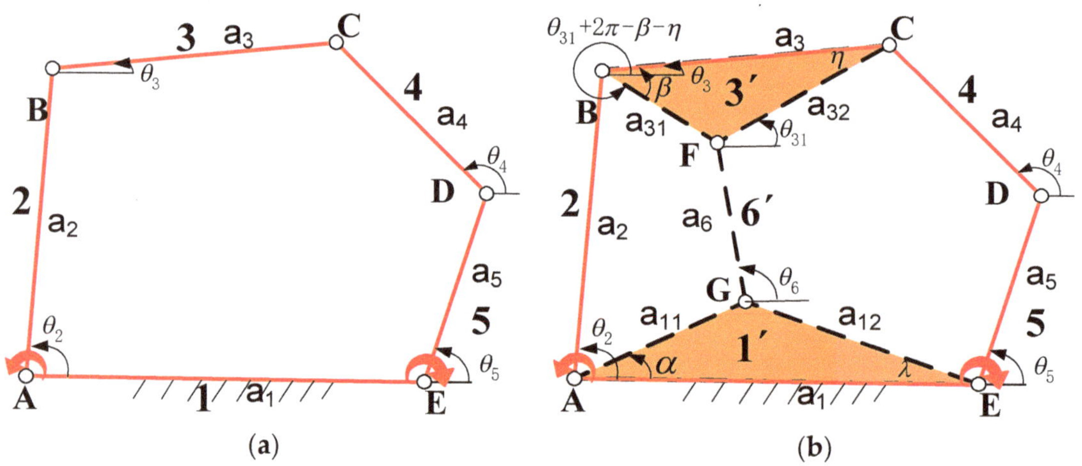

Figure 7. (**a**) Single-loop two-DOF five-bar planar linkage, (**b**) Two-loop virtual Stephenson planar linkage.

Mathematical proof for single-loop five-bar planar linkage.

Loop ABCDE:

$$a_1 + a_2 e^{i\theta_2} + a_3 e^{i\theta_3} - a_4 e^{i\theta_4} - a_5 e^{i\theta_5} = 0 \tag{2}$$

Using Euler formula, Equation (2) can be written as the following two equations:

$$a_1 + a_2 \cos\theta_2 + a_3 \cos\theta_3 - a_4 \cos\theta_4 - a_5 \cos\theta_5 = 0 \tag{3}$$

$$a_2 \sin\theta_2 + a_3 \sin\theta_3 - a_4 \sin\theta_4 - a_5 \sin\theta_5 \doteq 0 \tag{4}$$

Eliminating the unknown angle θ_3 in Equations (3) and (4) according to the trigonometric function $(\sin\theta_3)^2 + (\cos\theta_3)^2 = 1$,

$$\theta_4 = M_1(\theta_2, \theta_5) \tag{5}$$

Differentiating Equation (5) with respect to time, yields,

$$W_4 = \dot{\theta}_4 = M_2(\dot{\theta}_2, \dot{\theta}_5) \tag{6}$$

where M_i (i = 1, 2) are the coefficients in the light of θ_2 and θ_5, which can be obtained with mathematical software Maple. Note that M_i is constant. Equation (6) shows that the angular velocity W_4 of the single-loop five-bar planar linkage is only related to the input angle θ_2, θ_5.

After adding the virtual links, the single-loop five-bar linkage (Figure 7a) is changed into a virtual Stephenson linkage (Figure 7b). According to discussion above, the sizes and positions of the virtual links 1′, 3′, and 6′ are arbitrary, hence, there are no known parameters added. Note that the links 1 and 3 (Figure 7a) and the links 1′and 3′ in Figure 7b are different. Seeing from Figure 7a, the link 3 has only one angular velocity, the derivative of angle θ_3 in loop ABCDE. However, the link 3′ (Figure 7b) has three angular velocities: the derivative of angle θ_3 in loop ABCDE, the derivative of angle θ_{31} in loop GFCDE, the derivative of angle ($\theta_{31} + 2\pi - \beta - \eta$) in loop ABFG. As for the links 1 and 1′, the similar result can be concluded. Therefore, the instant center $I_{1'3'}$ in the virtual Stephenson linkage is not same to the instant center I_{13} in the single-loop five-bar planar linkage. In fact, the virtual link operation is an assumption. The virtual links are unreal. The virtual Stephenson linkage is actually still the single-loop five-bar linkage. The inputs and the motions of the links (not including the virtual links) in the virtual Stephenson linkage are the same compared to the original single-loop five-bar linkage. The instant centers which are formed by the virtual links 1′, 3′, and 6′, such as $I_{1'2}$, $I_{1'3'}$, $I_{1'4}$, $I_{1'5}$, $I_{1'6'}$, $I_{23'}$, $I_{26'}$, $I_{3'4}$, $I_{3'5}$, $I_{3'6'}$, $I_{46'}$, and $I_{56'}$ in Figure 7b, do not change the motion of the original five-bar linkage. That is, the instant centers $I_{1'2}$, $I_{1'3'}$, $I_{1'4}$, $I_{1'5}$, $I_{23'}$, $I_{3'4}$, and $I_{3'5'}$ in the virtual Stephenson linkage are all different to the instant center I_{12}, I_{13}, I_{14}, I_{15}, I_{23}, I_{34}, and I_{35} in the original single-loop five-bar linkage.

The loop equation of the virtual Stephenson linkage (Figure 7b) can be expressed as Loop ABFG:

$$a_2e^{i\theta_2} + a_{31}e^{i(2\pi+\theta_3-\beta)} - a_{11}e^{i\alpha} - a_6e^{i\theta_6} = 0 \tag{7}$$

Loop GFCDE:

$$a_{12}e^{i(\pi-\lambda)} + a_6e^{i\theta_6} + a_{32}e^{i\theta_{31}} - a_4e^{i\theta_4} - a_5e^{i\theta_5} = 0 \tag{8}$$

Adding Equation (7) to Equation (8) to eliminate the passive angle θ_6, Equation (9) can be obtained as follows:

$$a_2e^{i\theta_2} + a_{31}e^{i(2\pi+\theta_3-\beta)} - a_{11}e^{i\alpha} + a_{12}e^{i(\pi-\lambda)} + a_{32}e^{i\theta_{31}} - a_4e^{i\theta_4} - a_5e^{i\theta_5} = 0 \tag{9}$$

In fact, the elimination operation above is to form the loop ABCDE (red part in Figure 7b), i.e., Equation (9) is the loop equation of loop ABCDE, as a result, Equation (9) should equal to Equation (2).

Using Euler formula, Equation (9) can be written as the following two equations:

$$\begin{aligned} &a_2\cos\theta_2 + a_{31}\cos(\theta_3 - \beta) - a_{11}\cos\alpha - a_{12}\cos\lambda \\ &+a_{32}\cos\theta_{31} - a_4\cos\theta_4 - a_5\cos\theta_5 = 0 \end{aligned} \tag{10}$$

$$\begin{aligned} &a_2\sin\theta_2 + a_{31}\sin(\theta_3 - \beta) - a_{11}\sin\alpha + a_{12}\sin\lambda \\ &+a_{32}\sin\theta_{31} - a_4\sin\theta_4 - a_5\sin\theta_5 = 0 \end{aligned} \tag{11}$$

Since the angle $\theta_3 = \theta_{31} - \eta$, combining Equations (10) and (11) to eliminate the passive angle θ_3 with tangent-half-angle formula (firstly, $\cos\theta_3 = 1 - m_2/(1 + m^2)$, $\sin\theta_3 = 2m/(1 + m^2)$, where $m = \tan(\theta_3/2)$, are substituted into Equations (10) and (11), then eliminate the common term m of the two equations), Equation (12) can be obtained.

$$\theta_4 = N_1(\theta_2, \theta_5) \tag{12}$$

Differentiating Equation (12) with respect to time, yields,

$$W_4 = \dot{\theta}_4 = N_2(\dot{\theta}_2, \dot{\theta}_5) \tag{13}$$

where N_i (i = 1, 2) are the coefficients in the light of θ_2 and θ_5. Note that N_i are composed of the unknown angles β, η, α, λ (β, η are the interior angles of the link 3′, α, λ are the interior angles of the link 1′) and the arbitrary link parameters a_{11}, a_{12}, a_6, a_{31}, and a_{32} (a_{11}, a_{12} are the link parameters of the links 1′, a_{31}, and a_{32} are the link parameters of the links 3′). Contrasted the single-loop five-bar linkage in Figure 7a with the virtual Stephenson linkage in Figure 7b, since the parameters α, λ, β, η, a_{11}, a_{12}, a_{31}, and a_{32} are random, that is, a lot of virtual Stephenson linkages can be obtained, which correspond to the same single-loop five-bar linkage. However, according to Equations (6) and (13), it is obvious that the angular velocities W_4 of Equations (6) and (13) are only related to the input angles θ_2, θ_5 no matter whether the virtual links are added or not. When the single-loop five-bar linkage is in the instantaneous configuration, i.e., the ratio of the inputs (W_2/W_5) is specified, the two angular velocities W_4 all coming from the loop ABCDE in which the links 4 correspond the same derivative of angle θ_4, are the same. Then the instant centers I_{24}, which is decided by (W_4/W_2), are the same. That is, the virtual links of the virtual Stephenson linkage do not change the motion of the original single-loop five-bar linkage. The virtual Stephenson linkage is actually still the original single-loop five-bar linkage. Virtual link operation is an auxiliary means to change the single-loop linkage into a virtual two-loop linkage which do not affect the motion of the original single-loop linkage. Therefore, the unchanged instant centers I_{24} and I_{25} in Figure 7a,b, formed by the common part of the original single-loop five-bar linkage and the virtual Stephenson linkage, are the same. In other words, the instant centers I_{24} and I_{25} obtained from the virtual Stephenson linkage can be used in the instant center identification of the original single-loop five-bar linkage, and in the virtual Stephenson linkage, the instant centers I_{24} and I_{25} are easy to be obtained. The validity of virtual link operation is demonstrated.

5. Instant Center Identification of the Single-Loop Planar Linkages

5.1. Single-Loop Five-Bar Planar Linkage

5.1.1. Single-Loop Five-Bar Planar Linkage with Only Rotation Joints

The five-bar planar linkage (Figure 2a) is the simplest single-loop planar linkage beside the four-bar planar linkage. It has ten instant centers which are shown in Table 2. The instant centers I_{12}, I_{23}, I_{34}, I_{45}, and I_{15} are the first instant centers, and the A type third instant centers include the instant centers I_{13}, I_{14}, I_{24}, I_{25}, and I_{35}. Lacking the secondary instant centers, the third instant centers cannot be determined as shown in Figure 8.

Table 2. Instant centers of five-bar planar linkage.

First Instant Center		I_{12} I_{23} I_{34} I_{45} I_{15}
Secondary Instant Center		None
Third instant center	A type	I_{13} I_{14} I_{24} I_{25} I_{35}
	B type	none

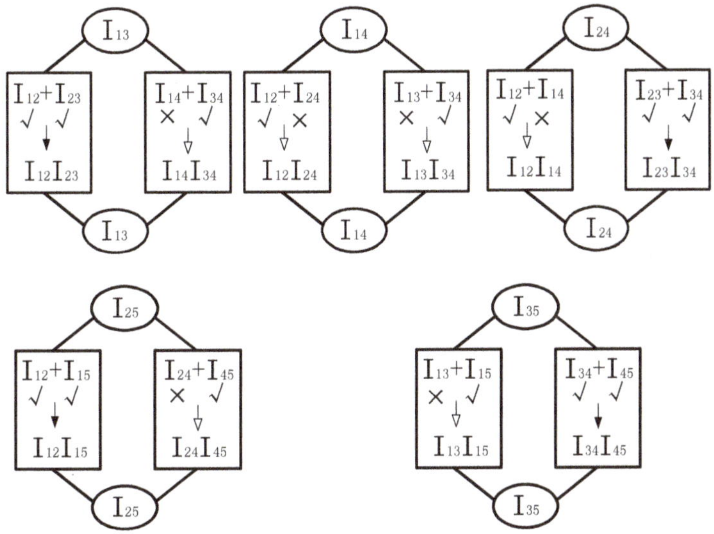

Figure 8. The instant center identification process graph of five-bar planar linkage without virtual links.

Using Criterions 1 and 2, the instant centers of five-bar planar linkage (Figure 2a) can be solved. After adding the virtual links, the virtual loops are formed and the secondary instant centers occur. The created secondary instant center can be used to get the A type third instant centers. Since the parameters and positions of inserted virtual links are arbitrary, the transformation of the configuration is not out of generality. In Figure 9, the original five-bar planar linkage (Figure 2a) is converted into the virtual Stephenson linkage with the virtual links 2′, 5′, and 6′. Note that the virtual links 2′ and 5′ are different to the links 2 and 5 (Figure 2a). The instant centers of the virtual Stephenson linkage are shown in Table 3. With Aronhold–Kennedy theorem, the secondary instant centers $I_{16'}$ and $I_{2'5'}$ (which is also the common instant center of the loop 1-2′-6′-5′ and loop 2′-3-4-5′-6′) and the A type third instant centers I_{13}, I_{14}, $I_{2'4}$, $I_{35'}$, $I_{36'}$, and I_{45}' are easy to be located in Figure 10. Compared the links 1, 3, and 4 (red segment in Figure 9) in the five-bar planar linkage and the virtual Stephenson linkage, the locations and motions of links are invariant, that is, the unchanged instant centers I_{13}, and I_{14} of two planar linkage are equivalent. Seeing from Figure 8, as a result, the remaining A type third instant centers (i.e., I_{24}, I_{25}, and I_{35}) of the single-loop five-bar planar linkage are solvable using the instant center identification process graph. Note that the instant center $I_{2'5'}$ obtained in loop 1-2′-6′-5′ is different to the instant center I_{25} in loop 1-2-3-4-5 since the derivative of angle of the instant center $I_{2'5'}$ coming from loop 1-2′-6′-5′ and the derivative of angle of the instant center $I_{2'5'}$ coming from loop 1-2-3-4-5 are different based on the discussion in Section 4.

Table 3. Instant centers of Stephenson linkage with the virtual links.

First instant center	$I_{12'}$ $I_{2'3}$ I_{34} $I_{15'}$ $I_{45'}$ $I_{2'6'}$ $I_{5'6'}$	
Secondary instant center	$I_{16'}$ $I_{2'5'}$	
Third instant center	A type B type	I_{13} I_{14} $I_{2'4}$ $I_{35'}$ $I_{36'}$ $I_{46'}$ none

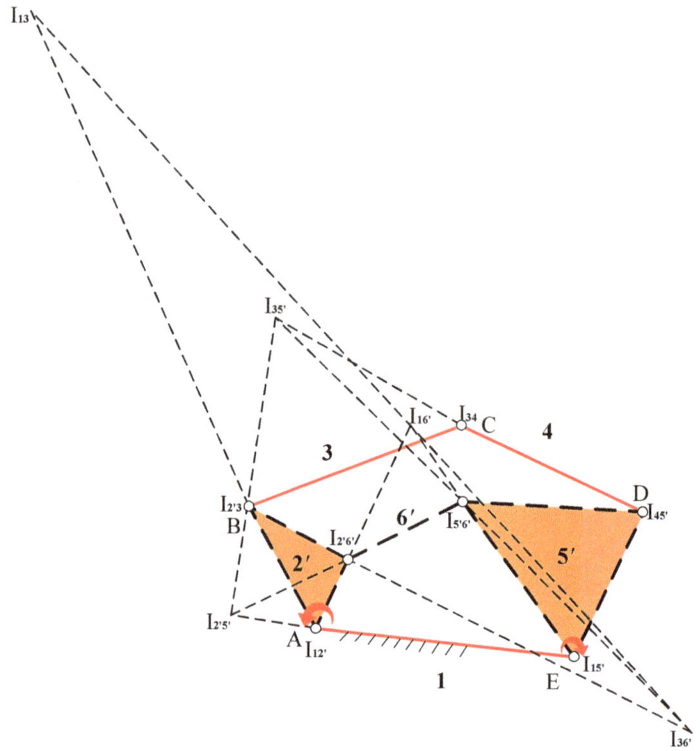

Figure 9. Single-loop five-bar planar linkage with virtual links.

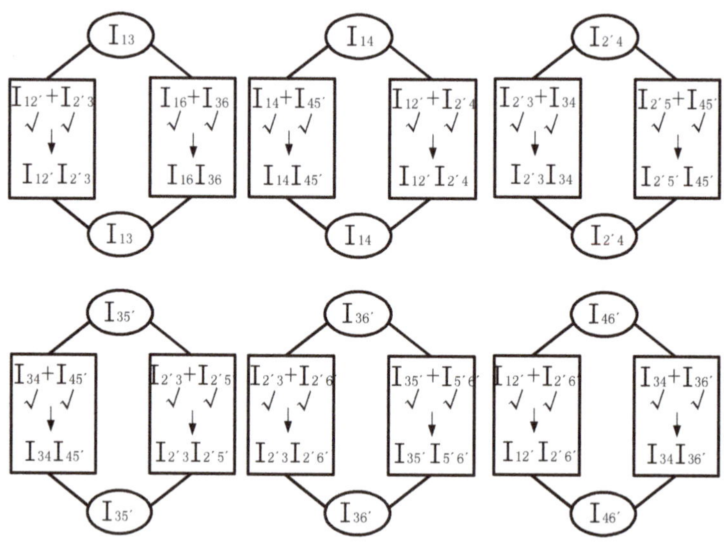

Figure 10. The instant center identification process graph of five-bar planar linkage with virtual links.

5.1.2. Single-Loop Five-Bar Planar Linkage with Prismatic Joints

The instant center of a prismatic joint lies on the line which is perpendicular to the motion pair, for example, L_{45} in Figure 11. There are three cases of single-loop five-bar planar linkages with the different number of prismatic joints. Taking the linkages in Figure 11a for example, the instant centers are shown in Table 4. Contrasted Table 2 with Table 4, the difference of the instant centers between the single-loop five-bar linkage with one prismatic joint and the single-loop five-bar linkage with only rotation joints is that one first instant center (I_{45} in Table 2) is changed into an A type third instant center (I_{45} in Table 4). Using the proposed method, the instant center problem of this kind linkages can be solved as follows. The virtual links can be added in two ways: (1) after added operation, the five-bar linkage is changed into a virtual four-bar loop with only rotation joints and a five-bar loop with one prismatic joint; (2) the five-bar linkage is changed into a virtual five-bar loop with only rotation joints and a four-bar loop with one prismatic joint, shown in the Figures 12 and 13. For Figure 12, the linkage with the prismatic joint is changed into a virtual Stephenson linkage with one prismatic joint containing loop1-2-3′-6′ with rotation joints and loop1′-6′-3′-4-5 with the prismatic joint. However, some A type third instant centers, such as $L_{1'4}$, L_{24}, $L_{46'}$, and L_{45}, in this case, cannot be obtained even under exceptional conditions that the angle between the link 4 and the sliding block is 90°. As for Figure 13a, using the corresponding virtual link operation, the linkage is transformed into a virtual Stephenson linkage including loop 1′-2-3-4′-6′ with rotation joints and loop 1′-6′-4′-5 with the prismatic joint. Although the instant centers $L_{1'4'}$, $L_{56'}$, and $L_{4'5}$ in the loop1′-6′-4′-5 cannot be located, the instant centers of loop 1′-2-3-4′-6′ with rotation joints are solvable seeing from the case in Figure 9. The corresponding virtual link operations are shown in Figure 13b. The instant center $I_{1'4'}$ is the common instant center of the loop 1′-2-3-4′-6′ and loop 1′-6′-4′-5 with the prismatic joint. In another word, the instant center $I_{1'4'}$, obtained from the loop 1′-2-3-4′-6′, can be used to identify the rest instant centers $L_{56'}$ and $L_{4'5}$ in loop 1′-6′-4′-5 with the prismatic joint. The solution for single-loop five-bar planar linkage with prismatic joint sounds like an iterative process. The instant centers of the five-bar loop with rotation joints are solved, and then used in the identification of the four-bar loop with the prismatic joint. The other cases are shown in Figure 13c,d in which the five-bar planar linkage has two prismatic joints and three prismatic joints, respectively. The five-bar planar linkage with two prismatic joints (Figure 13c) can be changed into a four-bar loop with one prismatic joint and a five-bar loop with another prismatic joint discussed in Figure 11a. The five-bar planar linkage with three prismatic joints (Figure 13d) can be changed into a four-bar loop with one prismatic joint and a five-bar loop with another two prismatic joints discussed in Figure 11b. For simplicity, this repeated discussion is omitted.

Table 4. Instant centers of five-bar planar linkage with one prismatic joint.

First instant center	I_{12} I_{23} I_{34} I_{15}	
Secondary instant center	none	
Third instant center	A type	I_{13} I_{14} I_{24} I_{25} I_{35} I_{45}
	B type	none

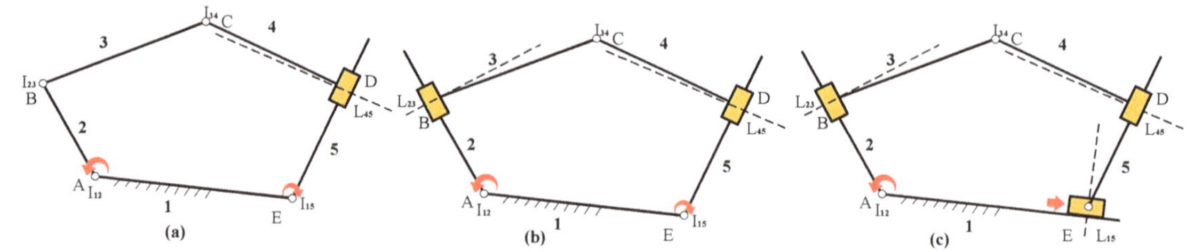

Figure 11. Single-loop five-bar planar linkage with (**a**) one prismatic joint, (**b**) two prismatic joints, (**c**) three prismatic joints.

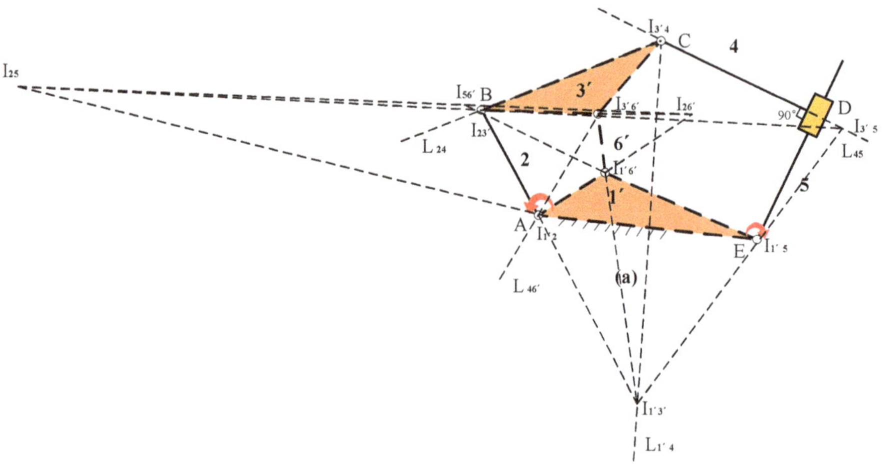

Figure 12. Instant centers of single-loop five-bar planar linkage with one prismatic joint.

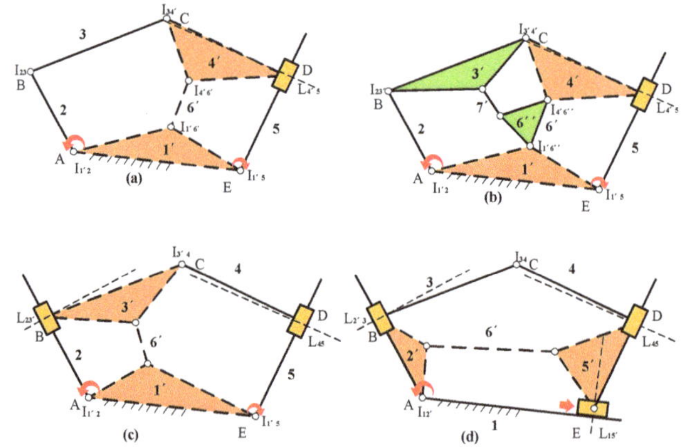

Figure 13. Single-loop five-bar planar linkage with (**a**) one prismatic joint and virtual links 1′, 4′, 6′, (**b**) one prismatic joint and virtual links 1′, 3′, 4′, 6″, 7′ using iterative configuration, (**c**) two prismatic joints and virtual links 1′, 3′, 6′, (**d**) three prismatic joints and virtual links 2′, 5′, 6′.

5.2. Single-Loop Six-Bar Planar Linkage

5.2.1. Single-Loop Six-Bar Planar Linkage with Only Rotation Joints

Single-loop six-bar planar linkage (Figure 2b) has 15 instant centers. The first instant centers I_{12}, I_{23}, I_{34}, I_{45}, I_{56}, and I_{16} and the third instant center I_{13}, I_{14}, I_{15}, I_{24}, I_{25}, I_{26}, I_{35},

I_{36}, and I_{46} are shown in Table 5, then the instant center identification process graph is established in Figure 14. Similar to the instant center identification of single-loop five-bar planar linkage, the instant center identification is shown as follows. The single-loop six-bar planar linkage with only rotation joints becomes the virtual seven-bar linkage by adding the virtual links 2′, 5′ and 7′ in Figure 15, which contains two virtual five-bar loops using Criterion 1. The instant centers are listed in Table 6 after this change. In Figure 15, the coupled loop chain 2′-7′-5′ is the common part of two virtual loop 1-2′-7′-5′-6 and 2′-3-4-5′-7′, namely, the instant center $I_{2'5'}$ formed by the common part of two virtual loop, is the common instant center. Using Criterion 2, the wanted instant centers occur if the common instant centers of the two loops coincide. It is clear that the single loop 1-2′-7′-5′-6 is just the single-loop five-bar planar linkage discussed in Section 5.1 in which the instant centers had been recognized in Figure 8. Therefore, the instant center $I_{2'5'}{}^{1}$ is known and can be used for the identification of the instant centers of the virtual loop 3-4-2′-7′-5′ when the loop is taken as a single loop. In this way, the instant centers $I_{15'}$, $I_{17'}$, $I_{2'5'}$, $I_{2'6}$, $I_{67'}$, $I_{35'}$, $I_{37'}$, $I_{2'4}$, and $I_{47'}$ are immediately obtained, and the instant center I_{13} is figured out according to the instant centers $I_{12'}$, $I_{2'3}$, $I_{17'}$, and $I_{37'}$ using Aronhold–Kennedy theorem. As the same to the instant center I_{13}, the instant centers I_{14}, I_{36}, and I_{46} are solved and brought into the original six-bar linkage in Table 7, which makes all the instant centers of the linkage determinate in Figure 14.

Table 5. Instant centers of six-bar planar linkage without virtual links.

First instant center	I_{12} I_{23} I_{34} I_{45} I_{56} I_{16}	
Secondary instant center	none	
Third instant center	A type	I_{13} I_{15} I_{24} I_{26} I_{35} I_{46}
	B type	I_{14} I_{25} I_{36}

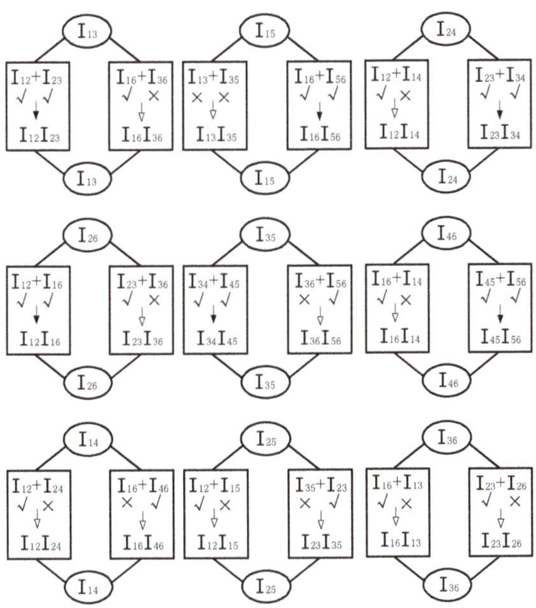

Figure 14. The instant center identification process graph of six-bar planar linkage without virtual links.

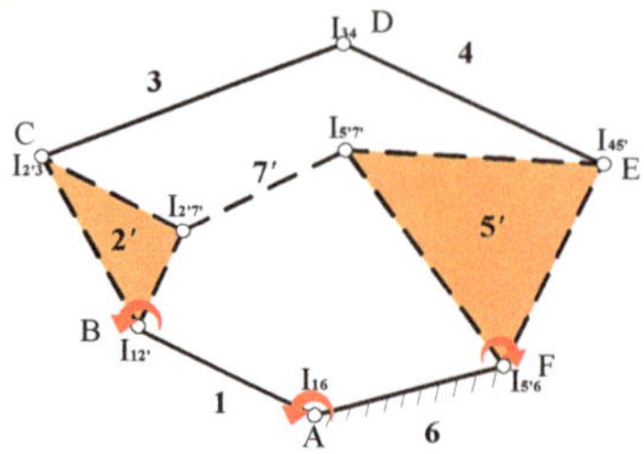

Figure 15. Single-loop six-bar planar linkage with virtual links.

Table 6. Instant centers of six-bar planar linkage with virtual links.

First instant center		$I_{12'}$ $I_{2'3}$ I_{16} $I_{27'}$ I_{34} $I_{45'}$ $I_{5'6}$ $I_{5'7'}$
Secondary instant center		none
Third instant center	A type	I_{13} $I_{15'}$ $I_{17'}$ $I_{2'4}$ $I_{2'5'}$ $I_{2'6}$ $I_{35'}$ I_{46} I_{47}
	B type	I_{14} I_{36}

Table 7. Instant centers of six-bar planar linkage with Criterions 1 and 2.

First and identified instant center	I_{12} I_{23} I_{34} I_{45} I_{56} I_{16} I_{13} I_{14} I_{36} I_{46}
Instant center by Aronhold–Kennedy theorem	I_{15} I_{24} I_{26} I_{35} I_{25}

5.2.2. Single-Loop Six-Bar Planar Linkage with Prismatic Joints

Similar to single-loop five-bar planar linkage with prismatic joints, there show three cases of single-loop six-bar linkages with several prismatic joints in Figure 16. According to discussion above, the instant center identification of the linkage is a iteration. The results coming from the linkages in Figure 15 can be used for the instant center locations in Figure 16a. The corresponding virtual link operation is shown in Figure 16d. Then, the similar steps are repeated in Figure 16b, the corresponding virtual link operation is shown in Figure 16e. In Figure 16f, the consequences of Figure 16a,b are utilized for the instant center identification of single-loop six-bar linkages with three prismatic joints. For simplicity, the detailed process is not expanded.

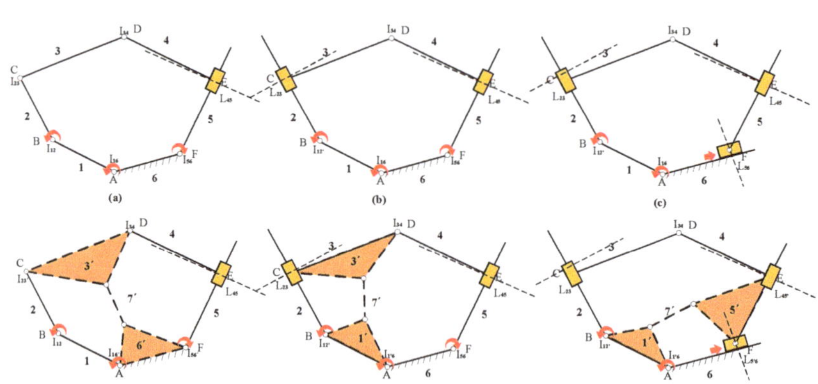

Figure 16. Single-loop six-bar planar linkage with (**a**) one prismatic joint, (**b**) two prismatic joints, (**c**) three prismatic joints, (**d**) one prismatic joint and virtual links 3′, 6′, 7′, (**e**) two prismatic joints and virtual links 1′, 3′, 7′, (**f**) three prismatic joints and virtual links 1′, 5′, 7′.

6. Conclusions

Instant center can be used in spatial mechanisms [26,27], redundant mechanisms [28], real devices [30], and is even available in commercial packages. It is a basic kinematic property which helps analyze kinematics of the linkages [2–5,31], rigid-body synthesis [6], dynamics modeling [7], dynamic balancing analysis [29], and imperfect systems of real devices [30]. Based on coupled loop chain characteristic and definition of instant center, a virtual link method to identify instant centers of the single-loop multi-DOF planar linkage is proposed. The conclusions and advantages can be summarized as follows:

(1) Three types of instant centers are redefined and the instant center identification process graph is introduced. Compared to the traditional classification of instant center, the difference is that the new classification classifies the traditional secondary instant center into three refined types: new secondary instant center, A type third instant center and B type third instant center in view of the access. That is, the new classification is more accurate and detailed, which brings conveniences for further research about instant centers.

(2) According to the instantaneous configuration of the linkage and definition of instant center, two criteria are presented and used to convert single-loop multi-DOF planar linkage into a two-loop virtual linkage by adding virtual links to acquire all the instant centers. Compared to the previous graphical methods, the proposed method has less operation than Dijksman's method and Klein's method facing the instant center identification of the simple planar linkage, such as single-loop five-bar planar linkage, and it has higher applicable rate than Chang's method. As for Pennock's method, the proposed method does not need the analytical calculation. However, its disadvantage is that it must be programmed facing the instant center identification of the single-loop N-bar (N > 6) planar linkage. Moreover, the proposed method only works in the instantaneous configuration, for the continuous motion of planar linkage, it fails.

(3) The instant centers of the single-loop five-bar, six-bar linkages with several prismatic joints are the first time to be located.

(4) The proposed method provides a geometry loop insight to reveal the relationship between the formation of the instant center and the motion of single-loop multi-DOF planar linkage, and a new research idea for the study of instant centers.

Author Contributions: L.N., H.D. and A.K. conducted the numerical analyses and wrote the majority of the paper. H.D. supervised the findings and organized and structured the paper. K.-L.T. reviewed the paper and given the constructive suggestion to improve the quality of the paper. All authors have read and agreed to the published version of the manuscript.

Funding: This paper has been supported by the Natural Science Foundation of China (No. 51975544) and the Natural Science Foundation of Hubei Province (No.2019AAA071).

Data Availability Statement: The data in this study can be requested from the corresponding author.

Acknowledgments: The first author would like to thank Academic Editors, Giuseppe Carbone, Med Amine Laribi, Assistant Editor, Erica Li, and two anonymous referees, whose hard work improved the quality of the paper.

Conflicts of Interest: The authors declare that they have no conflict of interest.

References

1. Hartenberg, R.; Danavit, J. *Kinematic Synthesis of Linkages*; McGraw-Hill: New York, NY, USA, 1964.
2. Hunt, K.H. *Kinematic Geometry of Mechanisms*; Oxford University Press: New York, NY, USA, 1978.
3. Wang, J.; Ting, K.L.; Zhao, D.X. Equivalent Linkages and Dead Center Positions of Planar Single-DOF Complex Linkages. In Proceedings of the ASME 2013 International Mechanical Engineering Congress and Exposition, San Diego, CA, USA, 13–21 November 2013.

4. Wang, J.; Nie, L.Y.; Wang, Q.; Sun, J.; You, Y.; Zhao, D.; Ting, K.-L. Singularity Analysis of Planar Multiple-DOF Linkages. In Proceedings of the ASME 2014 International Design Engineering Technical Conferences and Computers and Information in Engineering Conference, Buffalo, NY, USA, 17–20 August 2014.
5. Wang, J.; Nie, L.Y.; Zhao, D. Equivalent Five-Bar Linkages for the Singularity Analysis of Two-DOF Seven-Bar Linkages. In Proceedings of the ASME 2017 International Design Engineering Technical Conferences and Computers and Information in Engineering Conference, Cleveland, OH, USA, 6–9 August 2017.
6. Sancibrian, R.; Sarabia, E.G.; Sedano, A.; Blanco, J.M. A general method for the optimal synthesis of mechanisms using prescribed instant center positions. *Appl. Math. Model* **2016**, *40*, 2206–2222. [CrossRef]
7. Di Gregorio, R. A novel dynamic model for single degree-of-freedom planar mechanisms based on instant centers. *J. Mech. Robot.* **2016**, *8*, 011013. [CrossRef]
8. Kumar Mallik, A.; Ghosh, A.; Dittirich, G. *Kinematic Analysis and Synthesis of Mechanisms*; CRC: Boca Raton, FL, USA, 1994.
9. Foster, D.E.; Pennock, G.R. A Graphical Method to Find the Secondary Instantaneous Centers of Zero Velocity for the Double Butterfly Linkage. *J. Mech. Des.* **2003**, *125*, 268–274. [CrossRef]
10. Dijksman, E.A. Geometric determination of coordinated centers of curvature in network mechanisms through linkage reduction. *Mech. Mach. Theory* **1984**, *19*, 289–295. [CrossRef]
11. Yan, H.S.; Hsu, M.H. An Analytical Method for Locating Instantaneous Velocity Centers. In Proceedings of the 22nd ASME Biennial Mechanisms Conference, Scottsdale, AZ, USA, 13–16 September 1992; pp. 47, 353–359.
12. Klein, A.W. *Kinematics of Machinery*; McGraw-Hill: New York, NY, USA, 1917.
13. Foster, D.E.; Pennock, G.R. Graphical Methods to Locate the Secondary Instantaneous Centers of Single-Degree-of-Freedom Indeterminate Linkages. *J. Mech. Des.* **2005**, *127*, 249–256. [CrossRef]
14. Penne, R.; Crapo, H. A general graphical procedure for finding motion centers of planar mechanisms. *Adv. Appl. Math.* **2007**, *38*, 419–444. [CrossRef]
15. Foster, D.E.; Pennock, G.R. A study of the instantaneous centers of velocity for two-degree-of-freedom planar linkages. *Mech. Mach. Theory* **2010**, *45*, 641–657. [CrossRef]
16. Foster, D.E.; Pennock, G.R. A study of the instantaneous centers of velocity for the 3-dof planar six-bar linkage. *Mech. Mach. Theory* **2011**, *46*, 1276–1300. [CrossRef]
17. Chang, Y.P.; Her, I. A virtual cam method for locating instant centers of kinematic indeterminate linkages. *J. Mech. Des.* **2008**, *130*, 062304. [CrossRef]
18. Liu, Z.; Chang, Y. The Virtual Cam Method Application and Verification on Locating Key Instant Centers of Ten-Bar 1-DOF Indeterminate Linkages. In Proceedings of the ASME International Mechanical Engineering Congress and Exposition. American Society of Mechanical Engineers, Tampa, FL, USA, 3–9 November 2017. V04AT05A014.
19. Liu, Z.; Chang, Y. The Virtual Cam–Hexagon Method Authentication on Locating Key Instant Centers of All Planar Single Degree of Freedom Kinematically Indeterminate Linkages up to Ten-Bar. In Proceedings of the ASME 2018 International Mechanical Engineering Congress and Exposition. Volume 4A: Dynamics, Vibration, and Control, Pittsburgh, PA, USA, 9–15 November 2018. V04AT06A027.
20. Di Gregorio, R. An algorithm for analytically calculating the positions of the secondary instant centers of indeterminate linkages. *J. Mech. Des.* **2008**, *130*, 042303. [CrossRef]
21. Kung, C.M.; Wang, L.C.T. Analytical method for locating the secondary instant centers of indeterminate planar linkages. *J. Mech. Eng. Sci.* **2009**, *223*, 491–502. [CrossRef]
22. Zarkandi, S. Geometrical methods to locate secondary instantaneous poles of single-DOF indeterminate spherical mechanisms. *J. Mech. Eng.* **2010**, *41*, 80–88. [CrossRef]
23. Valderrama-Rodríguez, J.I.; Rico, J.M.; Cervantes-Sánchez, J.J. A screw theory approach to compute instantaneous rotation axes of indeterminate spherical linkages. *Mech. Des. Struct. Mach.* **2020**, 1–41. [CrossRef]
24. Valderrama-Rodríguez, J.I.; Rico, J.M.; Cervantes-Sánchez, J.J. A general method for the determination of the instantaneous screw axes of one-degree-of-freedom spatial mechanisms. *Mech. Sci.* **2020**, *11*, 91–99. [CrossRef]
25. Diab, N. A new graphical technique for acceleration analysis of four bar mechanisms using the instantaneous center of zero acceleration. *Sn Appl. Sci.* **2021**, *3*, 1–12. [CrossRef]
26. Danaei, B.; Arian, A.; Masouleh, M.T.; Kalhor, A. Dynamic modeling and base inertial parameters determination of a 2-DOF spherical parallel mechanism. *Multibody Syst. Dyn.* **2017**, *41*, 367–390. [CrossRef]
27. Chen, X.; Guo, J. Effects of Spherical Clearance Joint on Dynamics of Redundant Driving Spatial Parallel Mechanism. *Robotica* **2020**, *39*, 1064–1080. [CrossRef]
28. Boudreau, R.; Nokleby, S.; Gallant, M. Wrench Capabilities of a Kinematically Redundant Planar Parallel Manipulator. *Robotica* **2021**, *First View*, 1–16. [CrossRef]
29. Wei, B.; Zhang, D. A review of dynamic balancing for robotic mechanisms. *Robotica* **2021**, *39*, 55–71. [CrossRef]
30. Bucolo, M.; Buscarino, A.; Famoso, C.; Fortuna, L.; Frasca, M. Control of imperfect dynamical systems. *Nonlinear Dyn.* **2019**, *98*, 2989–2999. [CrossRef]
31. Rojas, N.; Thomas, F. The Forward Kinematics of 3-RPR Planar Robots: A Review and a Distance-Based Formulation. *IEEE Trans. Robot.* **2010**, *27*, 143–150. [CrossRef]
32. Shyu, J.H.; Ting, K.L. Invariant link rotatability of N-bar kinematic chains. *J. Mech. Des.* **1994**, *116*, 343–347. [CrossRef]

applied sciences

MDPI

Article

ResQbot 2.0: An Improved Design of a Mobile Rescue Robot with an Inflatable Neck Securing Device for Safe Casualty Extraction

Roni Permana Saputra [1,2,*,†], Nemanja Rakicevic [1], Isabelle Kuder [3], Joel Bilsdorfer [3], Alexander Gough [3], Alexandra Dakin [3], Emma de Cocker [3], Shaun Rock [4], Richard Harpin [4] and Petar Kormushev [1]

1 Robot Intelligence Lab, Dyson School of Design Engineering, Imperial College London, London SW7 2AZ, UK; n.rakicevic@imperial.ac.uk (N.R.); p.kormushev@imperial.ac.uk (P.K.)
2 Research Center for Electrical Power and Mechatronics, Indonesian Institute of Sciences—LIPI, Bandung 40135, Indonesia
3 Department of Mechanical Engineering, Imperial College London, London SW7 2AZ, UK; isabelle.kuder16@imperial.ac.uk (I.K.); marius.bilsdorfer16@imperial.ac.uk (J.B.); alexander.gough16@imperial.ac.uk (A.G.); alexandra.dakin16@imperial.ac.uk (A.D.); emma.de-cocker16@imperial.ac.uk (E.d.C.)
4 London Ambulance Service NHS Trust, London SE1 8SD, UK; shaun.rock2@nhs.net (S.R.); richard.harpin1@nhs.net (R.H.)
* Correspondence: r.saputra16@imperial.ac.uk
† Website: https://www.imperial.ac.uk/robot-intelligence/robots/resqbot/ (accessed on 15 May 2021).

Citation: Saputra, R.P.; Rakicevic, N.; Kuder, I.; Bilsdorfer, J.; Gough, A.; Dakin, A.; de Cocker, E.; Rock, S.; Harpin, R.; Kormushev, P. ResQbot 2.0: An Improved Design of a Mobile Rescue Robot with an Inflatable Neck Securing Device for Safe Casualty Extraction. *Appl. Sci.* **2021**, *11*, 5414. https://doi.org/10.3390/app11125414

Academic Editors: Giuseppe Carbone and Med Amine Laribi

Received: 18 May 2021
Accepted: 7 June 2021
Published: 10 June 2021

Abstract: Despite the fact that a large number of research studies have been conducted in the field of search and rescue robotics, significantly little attention has been given to the development of rescue robots capable of performing physical rescue interventions, including loading and transporting victims to a safe zone—i.e., casualty extraction tasks. The aim of this study is to develop a mobile rescue robot that could assist first responders when saving casualties from a dangerous area by performing a casualty extraction procedure whilst ensuring that no additional injury is caused by the operation and no additional lives are put at risk. In this paper, we present a novel design of ResQbot 2.0—a mobile rescue robot designed for performing the casualty extraction task. This robot is a stretcher-type casualty extraction robot, which is a significantly improved version of the initial proof-of-concept prototype, ResQbot (retrospectively referred to as ResQbot 1.0), that has been developed in our previous work. The proposed designs and development of the mechanical system of ResQbot 2.0, as well as the method for safely loading a full-body casualty onto the robot's 'stretcher bed', are described in detail based on the conducted literature review, evaluation of our previous work, and feedback provided by medical professionals. We perform simulation experiments in the Gazebo physics engine simulator to verify the proposed design and the casualty extraction procedure. The simulation results demonstrate the capability of ResQbot 2.0 to carry out safe casualty extractions successfully.

Keywords: rescue robotics; search and rescue; robot design; mobile robot; patient transfer; casualty extraction

1. Introduction

Responses to natural or human-made disasters—such as chemical, biological, radiological, and nuclear (CBRN) incidents—are always a race against time. Extracting casualties from a hazardous scene is such an example of an emergency case in which a significant amount of pressure and risk is placed on the people working as the first responders. Efficient and timely action is crucial since it is known that the mortality rate increases and reaches a peak after 48 h, implying that the chance of survival drops significantly after this period [1–5]. While it is crucial to act fast, ensuring the rescue operation is safely performed is also critical. Apart from ensuring the safety of the first responders, minimising the possibility of creating further damage or additional risk to the casualties is a high priority [5].

Currently, research and development in robotics and its applications is an actively growing field, including the development of robotics applications in assisting first responders in rescue missions, in which the main objective is to improve the search and rescue (SAR) operation to become faster and safer. A wide range of robotics design has been introduced and deployed for specifically performing SAR missions [4–11]. Most of these robots are designed to assist with some specific tasks that can be grouped into four general categories according to their purpose in the SAR response process, namely search, extraction, evacuation, and treatment [5].

The field of search robotics has received the most attention so far, particularly unmanned aerial vehicles (UAVs), which have been used in many rescue operations [12–18], collecting evidence about the position of a missing person, but not interacting with the casualty. However, the field of casualty extraction using rescue robots is significantly less mature, with fewer applicatFions and proposed designs. The main reason for this is that these robots are often significantly larger and more complex due to the medical casualty handling requirements [9]. Other than academic institutions, most of the research and many technical implementation proposals were carried out by military organisations [6].

This paper presents a novel design of a mobile rescue robot, called ResQbot 2.0, capable of safely rescuing a casualty lying on the ground (i.e., casualty extraction procedure). This robot design is a significantly improved version of the initial proof-of-concept prototype—ResQbot (retrospectively referred to as ResQbot 1.0)—that has been developed in our previous work, and has been presented in [19–23]. The main contributions of this work can be summarised as follows:

(1) The novel ResQbot 2.0 design comprises seven main novel components: A differential-drive mobile base; a stretcher bed tilting mechanism; a stretcher bed sliding mechanism; a motorised stretcher bed conveyor module, a pair of motorised stretcher strap modules; and a neck securing device module.
(2) The proposed methodology of casualty extraction procedure using ResQbot 2.0 for safely loading a full-body casualty onto the robot's 'stretcher bed'.
(3) Validation of the proposed design and the casualty extraction procedure via simulation experiments in the Gazebo physics engine simulator.

The remaining of this paper is organised as follows: In Section 2, we review the state of the art on the major work related to the existing mobile rescue robots designed for casualty extraction. In Section 3, we describe the design specification and development of the robot, as well as the medical and safety considerations. The proposed robot design and method of casualty extraction procedure using the proposed robot design are described in Section 4. Furthermore, the results and discussion of the proposed design are presented in Section 5. Finally, we conclude our findings in Section 6.

2. Related Work

In this section, we present an overview of the existing mobile rescue robots designed for casualty extraction. The summary of major contributions to the field of casualty extraction is depicted in Figure 1 [23–31].

One of the very first casualty extraction proposal incorporating a mobile robot is the iRobot Valkyrie project, as presented in [26,32]. In this proposal, the idea is implementing a flexible stretcher that is tethered to a multi-purpose mobile robot (see Figure 1, iRobot Valkyrie [24]). First, an operator or medic is remotely operating the robot to find the casualty. Then, the casualty needs to roll with its own strength onto the stretcher. After the casualty rolls safely on the stretcher, the robot then pulls away the stretcher to the safe place for further treatment [26].

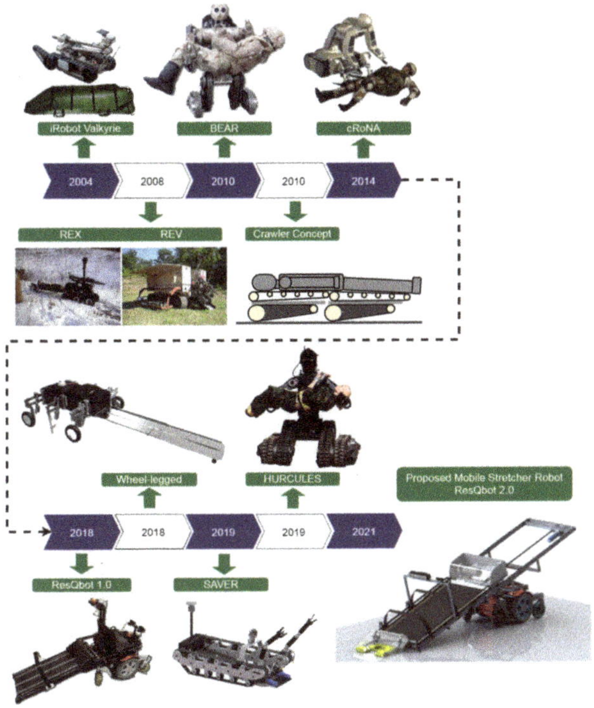

Figure 1. Timeline of the major contributions to the field of casualty extraction (photos on the diagram are adapted from [23–31]).

The US Department of Defence then introduced the robotic extraction (REX) and robotic evacuation (REV) rescue robot system for casualty extraction missions as presented in [25]. This system combines several differently-sized, unmanned ground vehicles (UGV) to perform a rescue mission. A small, mobile manipulator (REX) is used for short-range extraction from the site of injury to a larger and faster vehicle (REV), which transports the casualty to a medical facility (see Figure 1, REX and REV). The system is part of a more extensive tactical amphibious ground support system (TAGS). This project was proposed to extract battlefield casualties from hostile environments and from under fire, designed mainly for outdoor battlefield terrain [6].

One of the most sophisticated robot platforms designed and developed specifically for casualty extraction procedures is a robot with a semi-humanoid design. The humanoid robot is roughly the size of a human male. Its upper torso is equipped with a heavy-duty dual-arm manipulator built on top of a mobile base with tank tracks on its thighs and calves. The battlefield extraction assist robot (BEAR) developed by Vecna Technologies [6,7], the combat robotic nursing assistant (cRoNA) [8] from Hstar Technologies [33], and the humanoid rescue robot for calamity response (HURCULES) robot [29] developed by the Agency for Defense Development, South Korea [34,35] are examples of a semi-humanoid form mobile robot platform, designed and developed specifically for casualty-extraction procedures (see Figure 1, BEAR, cRoNA, HURCULES). Compared to all previous designs, these robots are designed to be highly agile while performing casualty extraction using their arms by scooping, lifting up, and carrying the casualty (see Figure 2a). While this casualty extraction procedure seems to be flexible, feasible, and mimics how a normal person handles a casualty, medical safety remains an unsolved issue as this design does not stabilise the sensitive areas, including the casualty's spine, neck, and head.

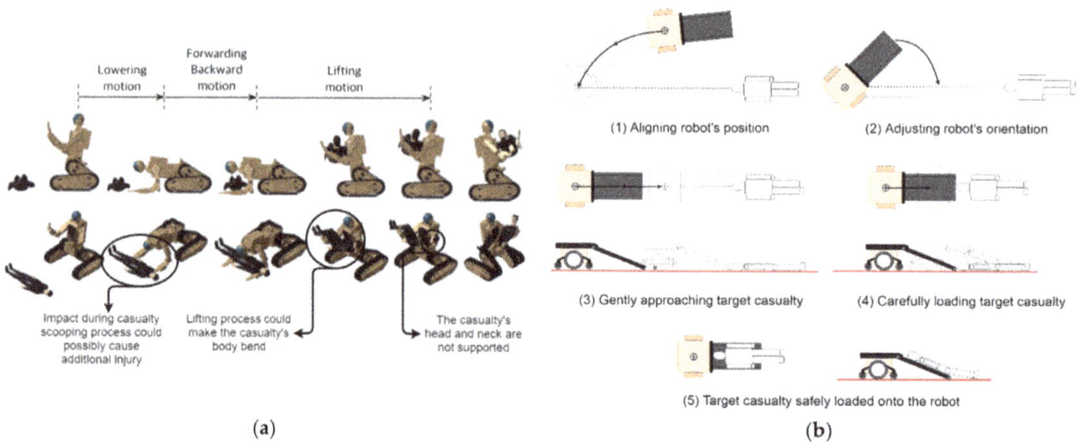

Figure 2. Casualty extraction procedures using two different types of casualty extraction robots. (**a**) Humanoid-type robot construction. The casualty extraction procedure has been presented in [29]. We found that there are at least three safety concerns in the typical humanoid-type robot construction procedure, the impact during the casualty scooping process could possibly bend the casualty's body sideways, the lifting process could make the casualty's body bend forward, and provide minimal support for the casualty's head and neck during the transportation phase. (**b**) Stretcher-type robot construction. The presented casualty extraction procedure introduced as the loco-manipulation approach has been presented in [23].

In [29], the authors highlight that one of the noticeable features in the mechanical design of the HURCULES robot is to use the worm gear in the joint to maintain the safety of the casualty even with the power off and to reduce the energy through a selected operating mode. Moreover, unlike the upper body of a conventional humanoid robot, a chest plate is installed and used to properly distribute the casualty's weight to the dual-arm manipulator and the chest plate when carrying the casualty. Nevertheless, the lack of body support (especially spine, neck, and head) during the procedure, in comparison to the conventional stretcher, remains a critical concern. In addition, in terms of operating the robots, controlling such complex robots performing intricate and sensitive tasks is a significant challenge. Teleoperating such a complex system most likely requires more than one highly skilled and experienced operator with complex teleoperation devices.

Alternative designs using stretcher-type constructions or litters that aim for a smoother pick-up and transportation process of casualty extraction procedure have also been investigated. Examples of proposals for such systems are the robot presented by Iwano et al. in [36–38] and the Tokyo Fire Department [39]. These robots are designed to perform casualty extraction using a conveyor belt mechanism to pick up a casualty without having to move the body significantly in the process. Once the casualty is properly loaded on top of the robot, the conveyor belt base then properly supports the casualty's body so it can additionally serve as a stretcher bed. Thus, this robot design and method for casualty extraction is expected to be safer and would minimise the possibility of causing additional injuries to the casualty during the casualty extraction process.

Ning in [28] has also presented a similar design of a stretcher-type casualty ex-traction robot. A unique feature of this design is that it incorporates a wheel-legged structure that can be raised or lowered using linear actuators. The purpose of the design is to improve the robot's adaptability in a complex disaster scene. On the other hand, a very recently published work in [30,40] presents a casualty extraction robot design called semi-autonomous victim extraction robot (SAVER) [41,42]. This robot is equipped with two manipulator arms and a head support system. The arms are designed for the telemanipulation process to gently adjust the pose of the casualty prior to the loading process. The head support system is designed to grasp the injured person and stabilise the head and neck while gently pulling the casualty on board.

In our previous works presented in [19–23], we have developed a proof-of-concept mobile rescue robot for casualty extraction, called ResQbot (see Figure 1, ResQbot 1.0). This robot is a stretcher-type mobile rescue robot designed to safely load a casualty using the loco-manipulation approach [23] that uses a combination of the robot's wheeled locomotion and the belt conveyor mechanism to load a casualty (i.e., manipulation task), as illustrated in Figure 2b. The loco-manipulation approach allows the robot to load the casualty while ensuring that key safety thresholds (based on [43,44]) are adhered to and avoiding potential causes of additional injury to the casualty, such as head, neck or spinal cord injuries [23]. The stretcher bed conveyor of this robot is also equipped with a stretcher strap mechanism to safely secure the casualty during transportation.

Despite the promising results obtained from the evaluation experiments on the loco-manipulation approach using ResQbot 1.0 [23], ResQbot 1.0 is only capable of securing half the body of the casualty (i.e., upper body) onto the robot's stretcher bed module (see Figure 3a). During the casualty extraction mission, the robot will drag the casualty's legs during transportation (Videos of ResQbot 1.0 are available at https://www.imperial.ac.uk/robot-intelligence/robots/resqbot/ (accessed on 15 May 2021)). Even though we believe that in an emergency, this procedure is still highly acceptable—in fact, first responders also frequently perform the same procedure [45–47]—this procedure could potentially cause severe damage to the casualty.

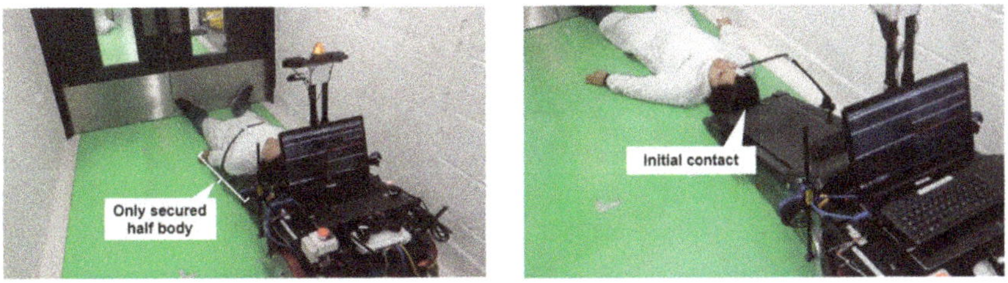

(a) (b)

Figure 3. Two critical concerns of the ResQbot 1.0 design: (**a**) ResQbot 1.0 can only load and secure half of the casualty's body (i.e., upper body). (**b**) The initial contact between the robot and the casualty's head during the casualty loading process remains a critical safety concern (the potential cause of head or neck injury).

Another critical concern on the ResQbot 1.0 design is the fact that the loading process, as part of the casualty extraction routine, is initiated from the head of the casualty. Based on the experimental evaluations, the process can satisfy the safety metric (we refer readers to [19,23] for more details about the experiments). Nevertheless, the process (i.e., initiating loading the casualty from the head) is still raising a safety concern in terms of potential damage to the casualty's head and neck during the initial contact between the robot and the casualty's head (see Figure 3b).

3. Design Specification

3.1. Design Objectives

The design specification is based on the literature review, evaluation of our previous work, and feedback provided by medical professionals. Based on our previous work, there are at least two main concerns: ResQbot's inability to safely load the casualty's entire body and safety regarding the robot's initial contact with the casualty's head during casualty extraction procedures. Therefore, in this study, we focus on three main design objectives:

- Optimising the design mechanism to safely load a casualty's entire body onto the robot's stretcher bed.

- Designing a mechanism that provides more protection to the casualty's head and neck during the extraction process.
- Optimising the robot's compact design and manoeuvrability in narrow environments.

3.2. Design Assumptions

In order to limit the scope of this work, the following assumptions about the robot's deployment environment were made at the start of the design process:

- The robot would be working in an urban environment with flat surfaces. Some possible scenarios include areas of gas leaks and radiation/chemical contamination.
- The robot would not have to deal with stairs.
- The casualty would be lying flat with hands at the sides (in readiness for loading the casualty onto the robot).
- The methodology for controlling the ResQbot 2.0 platform (e.g., teleoperation or autonomous modes) lies beyond the scope of this paper.

3.3. Medical Considerations and Research

As mentioned in the introduction, one of the main priorities in casualty extraction procedures is to ensure minimal harm or risk. One of the main design considerations is to minimise any possible traumatic injuries during the casualty extraction procedure. Any unfavourable handling might aggravate injuries, particularly to the neck, head, and spinal cord (which includes the cervical spine). For this reason, medical professionals were consulted during all design stages of ResQbot 2.0, and a survey of the literature was carried out [48–57].

According to our literature review, it was found that spinal cord injuries and unstable fractures are some of the major concerns. Such injuries can occur either at the tetraplegic (the neck region) or paraplegic level (lower back region) [48–57]. Injuries at the tetraplegic level can cause impairment or loss of motor or sensory functions in the cervical segments of the spinal cord, affecting arms and legs [48]. To avoid this type of injury, any additional backward (hyperextension) or forward (hyperflexion) bending of the neck, as well as compression or rotation must be avoided [48]. To prevent potential paraplegic traumas, any forward or backward bending of the lower back must also be avoided, although this is less critical.

Possible existing injuries must be taken into account when performing a casualty extraction, especially the possibility of secondary spinal cord injuries, including neurogenic shocks, post-traumatic ischemia or failure to stabilise and immobilise an unstable fracture, which might cause bone fragments to move towards, put pressure on or cut the spinal cord [48]. One of the standard operating procedures is to place the cervical spine in a neutral position and attach a stabilisation unit to the patient, as proposed in [30]. The main objective of this procedure is to minimise translation and rotation of the head in order to avoid the aggravation of spinal injuries during lifting or transportation. In classical ambulant care, post-trauma stabilisation usually includes fitting a cervical collar to the patient's neck as a frame to immobilise the head. However, recent work has illustrated that a cervical collar is not indispensable, and might even restrict a patient's airway [50,51]. Therefore, cranio-thoracic stabilisation methods, such as sandbags or stabilisation blocks that maintain the cervical spine straight without a cervical collar [52,53], have received increased attention.

Another safety consideration related to casualty extraction procedures, as shown in Figures 2b and 3b, is the possibility of head injury during contact between the robot and the casualty's head. A number of research studies have focused on evaluating and developing methods and technical devices that could protect against both spinal cord and head injuries [43,44]. In [43], Engsberg presents an investigation into the possible spinal cord and head injuries caused by an impacting force. This study evaluates several impact-testing methods as well as a selection of injury threshold limits. In a separate study, EURailSafe [44] presented a report on the evaluation of bot head and neck injuries, including injury mechanisms, criteria, and tolerance levels. We used the data from [43,44] to obtain the key safety thresholds and validate our design results.

4. Proposed Design and Method

4.1. Proposed Method and Design for Casualty Loading

In this study, we adopted the loco-manipulation approach presented in our previous work [23] for the casualty extraction method. This method was evaluated in the study presented in [19,23]; it has shown potential for the safe loading of casualties and satisfies several safety measures [43,58]. One of the objectives of this study is to propose a robot design to safely load an average-size person's entire body onto the robot's stretcher bed and secure it safely.

In order to achieve this objective, the size of the robot's stretcher bed and conveyor must be increased. This increase in length would lead to an increase in the overall size of the robot. Figure 4a,b illustrates the different robot sizes. The size of the stretcher bed can be increased without a significant change to the robot's design. There are at least two major problems with this design: (1) The robot's overall length increases by 1.5 times the length of the original ResQbot 1.0. This will cause the robot to struggle while manoeuvring in typical indoor environments, and (2) the more extended stretcher bed module (potentially loaded with a casualty) towed behind the differential-drive mobile base would make it difficult to manoeuvre the robot and keep the stretcher bed stable, since the mobile base would require much more effort to turn with an asymmetrical load towed behind the mobile base. Moreover, slight turns by the mobile base would cause much more movement on the stretcher bed, making it challenging to keep the stretcher bed stable during transportation.

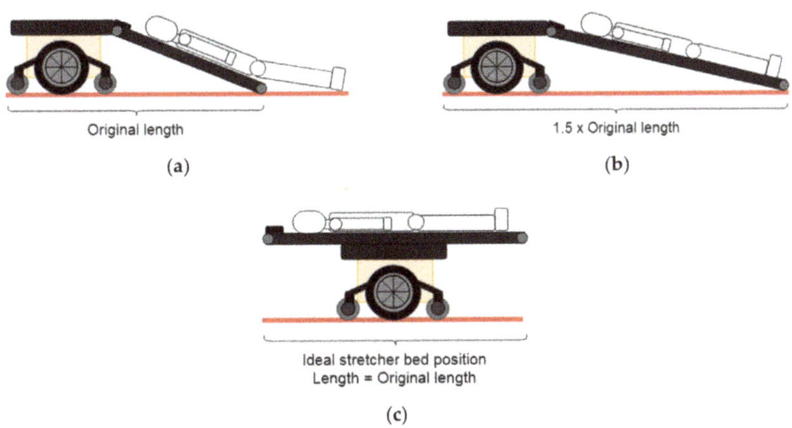

Figure 4. Illustration showing the ResQbot 1.0 stretcher-bed conveyor in comparison to the longer stretcher-bed conveyor designs that enable safe loading of the casualty's entire body onto the stretcher bed. (**a**) ResQbot 1.0 original size. (**b**) ResQbot 1.0 with an expanded stretcher bed. (**c**) The ideal position of the stretcher bed on the differential-drive mobile base (i.e., proposed design).

Figure 4c illustrates the ideal position of the stretcher bed on the mobile base. In this position, the load would be uniformly distributed to the robot's mobile base. In comparison with the original ResQbot 1.0 configuration (Figure 4a), upgrading the robot's stretcher bed size would result in a similar overall length when configured as shown in Figure 4c. Moreover, since the robot's configuration in Figure 4c is symmetrical, the mobile base's manoeuvrability will be improved seeing as it is a mobile base with a differential-drive wheel mechanism with the driving wheel placed in the middle. Consequently, the stretcher bed would be more stable while manoeuvring the robot during transportation—even when loaded with a casualty.

To accommodate the configuration shown in Figure 4c and retain the robot's loco-manipulation capability to load the casualty, we propose adding tilting and sliding mechanisms to the stretcher module on the mobile base platform. Figure 5 illustrates the stretcher bed sliding and tilting mechanisms proposed for ResQbot 2.0.

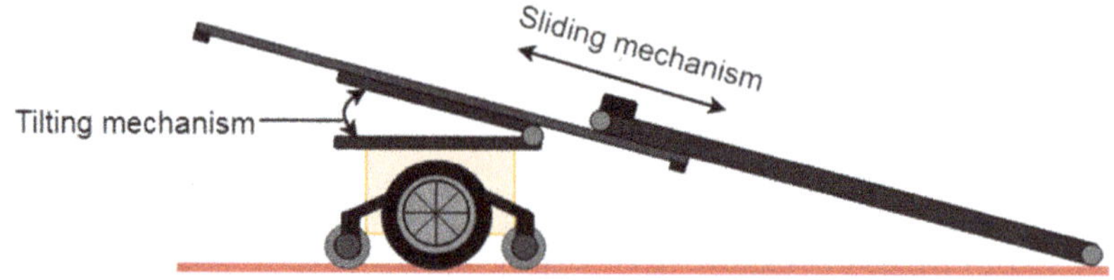

Figure 5. Illustration of proposed tilting mechanism and sliding mechanism adapted in the ResQbot 2.0 proposed design. These mechanisms were adapted in order to accommodate the upgraded size of the stretcher bed—capable of loading the entire body of an average-size casualty—while maintaining the overall compact design of the robot.

With these additional mechanisms, ResQbot 2.0 could perform casualty extraction in a similar manner to ResQbot 1.0, but with a slightly modified procedure. Figure 6 illustrates the proposed new casualty extraction procedure using ResQbot 2.0.

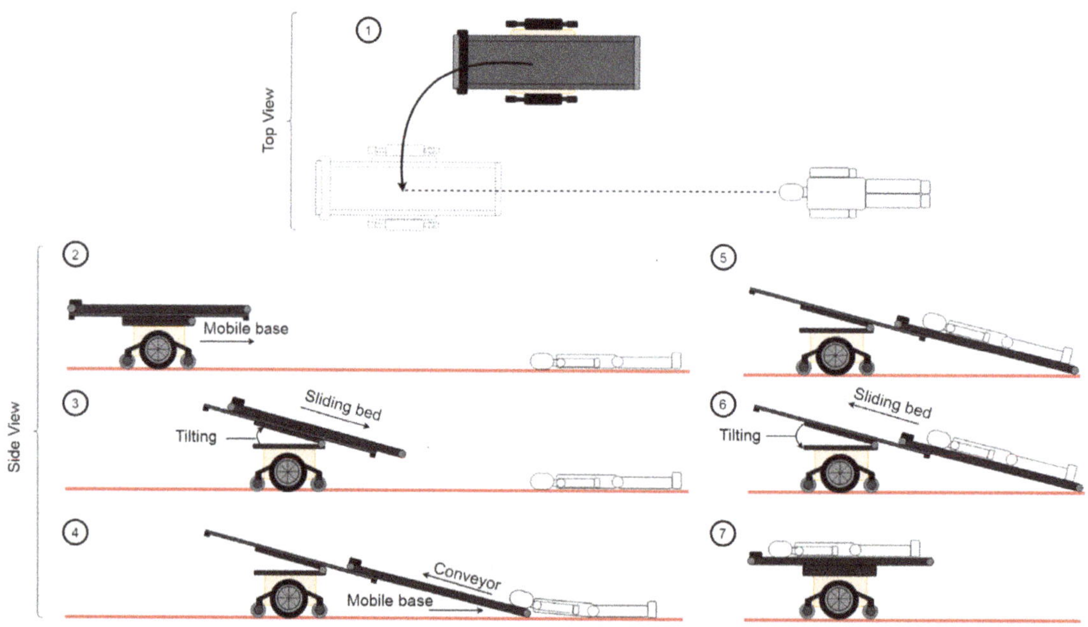

Figure 6. Illustration of the proposed casualty extraction procedure using ResQbot 2.0 comprising of seven main sequential phases: (**1**) Aligning the robot's pose with respect to the casualty's orientation. (**2**) Approaching the casualty in the desired target position in readiness to load the casualty. (**3**) Changing to the loading configuration by tilting and sliding the robot's stretcher bed to the desired configuration. (**4**) Synchronising the robot's mobile base and the conveyor belt movements enables the robot to gently load the casualty onto the stretcher bed. (**5**) Once the casualty is correctly positioned on the robot, the stretcher's strap mechanism securely fastens, stabilises, and safely immobilises the casualty on the stretcher bed. (**6**) Changing back to the compact configuration by sliding up and tilting down the robot's stretcher bed. (**7**) The robot is ready to transport the casualty to the medical area for further treatment.

This procedure involves several major phases:

(1) Relative pose adjustment: The robot aligns its relative pose with respect to the victim in preparation for the loco-manipulation routine.
(2) Approaching the target casualty: The robot gently and safely approaches the casualty and makes contact with the casualty's head to initiate loading.
(3) Changing to the loading configuration (i.e., sliding and tilting the stretcher bed): The robot's stretcher bed frame tilts up to the desired loading angle and then slides down until it touches the ground in readiness to load the casualty.
(4) Loading the target casualty: By balancing the movement of the base and the motion of the belt conveyor, the robot smoothly loads the casualty onboard.
(5) Fastening the stretcher strap: Once the casualty is fully onboard, the strapping mechanism secures, stabilises, and immobilises the casualty, minimising the risk of additional harm due to undesired movements.
(6) Changing to the compact configuration: Once the casualty is properly secured, the stretcher bed slides up and tilts down to its original compact configuration.
(7) Transportation: The robot is now ready for the transportation phase. The casualty is brought to the medical area for further treatment.

4.2. Neck Securing Device

Another critical concern about the ResQbot 1.0 design is the loading process, which begins with the casualty's head. In our previous work, presented in [19,23], we deemed the process to be safe and to satisfy the safety metrics, based on the experimental evaluations (we refer readers to [19,23] for more details on the experiments). Nevertheless, this process still raises safety concerns in terms of potential damage to the casualty's head and neck upon initial contact (see Figure 2b).

To address this safety issue, we propose a novel neck securing device (NSD) as a new safety feature of ResQbot 2.0, which required a new and innovative design. The main purpose of the NSD module is to properly secure the neck and critical parts of the head to avoid excessive bending of the cervical spine and hard impact during loading. We propose an NSD that uses inflatable material. There are at least two main works (see Figure 7a,b) that inspired our design of the inflatable NSD module. In [30,40], the authors proposed a head support system for the SAVER robot designed to grasp the injured person and stabilise the head and neck while gently pulling the casualty onboard. This mechanical system comprises a linear actuator, a tension spring, a string-rigged pulley differential mechanism, and a pair of head pads. The string-rigged pulley differential mechanism is a device that connects the head support pads in order to apply force equally to the right and left sides of the head. Each pad can assume an asymmetric final position, which gives the device the ability to stabilise the head and neck in the position in which the patient is originally encountered. The tension string minimises discomfort and allows a safe and stable hold while restricting the motion of the head. The concept for this head support mechanism is shown in Figure 7a.

Inflatable systems such as airbags and life jacket mechanisms were investigated for possible implementation in the proposed NSD system. The Hovding inflatable cycling helmet presented in [59,60] is one of the inflatable systems considered for the proposed NSD system. It has a collar that is worn around the neck that inflates when sensors indicate that a crash has occurred. Another option is the work presented in [61], which shows various inflatable objects that can be fabricated by the Printflatables platform.

This work combines the design concept inspired by [30,40] and the inflatable object technology shown in [59,61] to develop the NSD for ResQbot 2.0. Figure 7c illustrates the proposed concept, which uses inflatable material. We investigated several designs, all with the same fundamental concept that the inflatable device should cover and surround the casualty's head and neck up to the shoulders, as illustrated in Figure 7c.

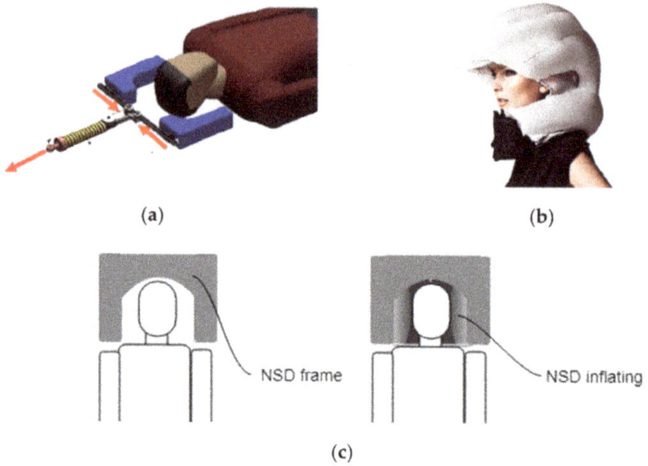

Figure 7. Several concepts of protecting the head from possible injuries. (**a**) The head support system proposed in [30,40] to immobilise the head during casualty extraction. (**b**) Hovding inflatable cycling helmet, as presented in [59,60], that protects the cyclist's head during a crash. (**c**) Proposed neck securing device (NSD), using the inflatable mechanism proposed for the ResQbot 2.0 design.

The requirement for the integration of the NSD in the stretcher bed (belt conveyor) system was that the NSD should be able to slide down along the bed to the ground and slide back up around the bed end. The NSD does not require a separate actuator for the sliding movement. The movement of the NSD relies on friction between the NSD and the conveyor belt. The NSD is placed on top of the conveyor and attached to linear guides on both sides of the NSD to constrain its movement only along the stretcher bed. Figure 8 illustrates the updated procedure to load casualties by means of the NSD system. This procedure is an extension of Step 4 in Figure 6.

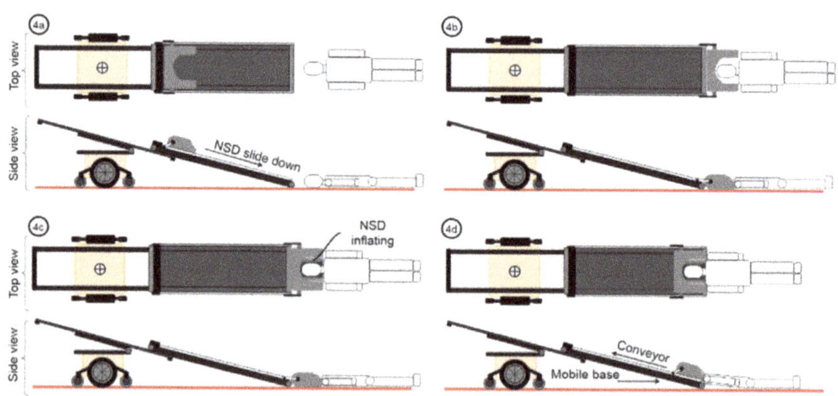

Figure 8. Illustration of the updated procedure to load casualties by means of the NSD system. This procedure is an extension of Step 4 in Figure 6. (**4a**) Once the robot achieves the desired loading configuration, the NSD module is sliding down, approaching the casualty. (**4b**) The NSD frame is placed in the desired position so that the casualty's head and neck are in the NSD frame centre. (**4c**) The inflatable system of the NSD then slowly inflates and stabilises the casualty's head and neck. (**4d**) Initiating the casualty loading process by synchronising the movement of the base and the motion of the belt conveyor.

5. Results and Discussion

5.1. ResQbot 2.0 Design Results

Figure 9 shows the final CAD design for ResQbot 2.0—a novel mobile robot stretcher bed with an inflatable neck securing device. ResQbot 2.0 consists of six main modules:

- A differential-drive mobile base that provides mobility and flexible manoeuvrability in typical urban and indoor terrains.
- A stretcher bed tilting module that enables ResQbot 2.0 to adjust the optimal loading angle for safe casualty extraction procedures. This module is driven by a linear actuator, and a bar linkage mechanism adjusts the bed's tilting angle.
- A stretcher bed sliding module that enables the robot's stretcher bed to slide up and down in order to switch between the loading configuration (for loading a casualty) and the compact configuration (for general robot navigation and casualty transportation). This module consists of rail mechanisms at both sides of the robot's stretcher bed that allow the bed to slide smoothly along its frame and a lead-screw mechanism that drives the bed's linear movement.
- A motorised stretcher bed conveyor module is essential in order to enable ResQbot 2.0 to gently load a casualty by synchronising the conveyor belt's loading movement with the movement of the mobile base (see the loco-manipulation approach presented in [19,23]).
- A pair of motorised stretcher strap modules that enable ResQbot 2.0 to secure, stabilise, and immobilise the casualty on the robot's stretcher bed in order to prevent any undesired movement that could cause additional harm to the casualty.
- A neck securing device module that consists of a rigid frame and several components that inflate and surround the casualty's head and neck up to the shoulders in order to prevent any undesired impact during the casualty loading procedure. It also stabilises and immobilises the casualty's head and neck during the extraction process.

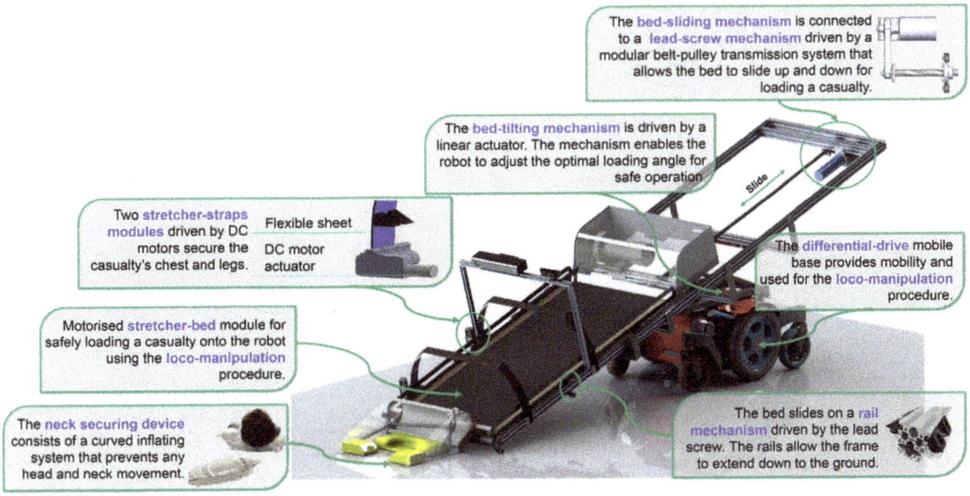

Figure 9. The ResQbot 2.0 design comprises seven main novel components: A differential-drive mobile base; a stretcher bed tilting mechanism; a stretcher bed sliding mechanism; a motorised stretcher bed conveyor module; a pair of motorised stretcher strap modules; and a neck securing device module.

5.2. The ResQbot 2.0 Assembly

Figure 10 shows the fully assembled ResQbot 2.0 in the compact (Figure 10a) and loading configurations (Figure 10b). In the following subsections, we discuss each main module and mechanism.

(a) (b)

Figure 10. The fully assembled ResQbot 2.0. (**a**) The compact configuration is used for general navigation and casualty transportation. (**b**) ResQbot 2.0 in the loading configuration, allowing the robot to adjust its configuration to the optimal loading angle to perform the casualty extraction procedure safely.

5.2.1. Differential-Drive Mobile Base

ResQbot 2.0 is designed to use a differential-drive mobile base module for mobility. This module provides fast and flexible manoeuvrability in flat terrain typically found in urban and indoor environments. The mobile base platform is a customised version of a commercially available powered wheelchair—Quickie Salsa-M—manufactured by Sunrise Medical [62]. This mobile base has a versatile design and is stable since it was designed to transport a disabled person both indoors and outdoors. The differential-drive wheels are located at the centre to enable a compact turning circle. In order to ensure its stability and safety while manoeuvring, this platform is equipped with an all-wheel independent suspension and anti-pitch technology suitable for use on rough or uneven terrain. Moreover, this platform was designed to carry loads up to 140 kg, which makes it suitable for the ResQbot 2.0 application: To carry an average-size casualty weighing approximately 80–100 kg [62]. Other important reasons for choosing this mobile platform are its mission range and its maximum operational speed. This mobile base can cover up to 32 km with a 60 Ah battery [62], and it has a maximum speed of 10 kph [62]. Figure 11a shows the differential-drive mobile base module.

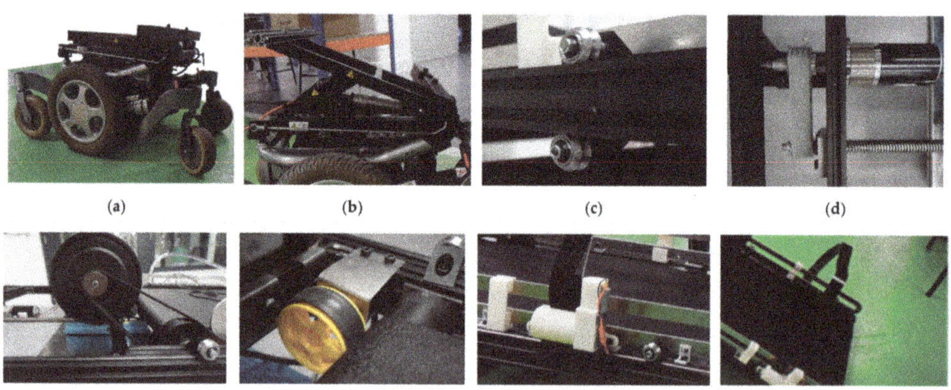

(a) (b) (c) (d)

(e) (f) (g) (h)

Figure 11. ResQbot 2.0 main modules. (**a**) Differential-drive mobile base module. (**b**) Stretcher bed tilting module. (**c**) The bed's sliding-rail mechanism. (**d**) The motor drive and the lead-screw mechanism that enable the stretcher bed to slide. (**e**) The motor drive and transmission system of the conveyor module. (**f**) The rotary encoder of the conveyor module, which allows feedback control to synchronise the conveyor and mobile base speed during the casualty loading process. (**g**) DC motor to drive the strap fastening and securing mechanism. (**h**) The stretcher strap module attached to the stretcher bed.

5.2.2. Stretcher Bed Tilting Module

The bed's tilting mechanism is the powered seat-tilt mechanism used on the Quickie Salsa-M. This tilting mechanism can be adjusted to an angle between 0 and 30 degrees [62]. The tilting mechanism is driven by a linear actuator connected to bar linkage systems. This mechanism enables ResQbot 2.0 to assume the optimal loading angle for safe casualty extraction. Figure 11b shows the bed's tilting mechanism.

5.2.3. Stretcher Bed Sliding Module

The bed's sliding module is designed to allow the stretcher bed to slide down (loading configuration) and slide up (compact transport configuration) during the casualty extraction process (see Figure 6). This module consists of two main mechanisms: The rail mechanism (see Figure 11c) and the lead-screw mechanism (see Figure 11d).

The sliding rail mechanism holds the stretcher bed on the fixed frame and allows it to slide along one axis. The mechanism bears the maximum load of the bed. Eight pairs of chrome steel metal dual V-wheels are used for the roller mechanism that holds the aluminium V-slot extrusion (i.e., the rail). This mechanism is recommended for accurate linear motion and has a high load capacity [63]. We installed four pairs of V-wheels on each side of the bed (eight pairs in total) to support a maximum load of 120 kg, including the casualty's weight.

To drive the sliding module, the bed's sliding-rail mechanism is connected to a lead-screw mechanism driven by a motor with a modular belt-pulley transmission system that allows the bed to slide up and down in order to load a casualty. During the design process, we considered several different mechanisms for this module: A ball screw, rack and pinion, lead screw, and winch were considered. Insights from the literature and expert opinions were obtained in order to evaluate the advantages and disadvantages of the different options, particularly with regards to cost-effectiveness, reliability and size, and the lead-screw mechanism was considered to be the most suitable. Even though it is not the most efficient in terms of friction, it allows for precise movement control, easy assembly, and is the most cost-effective solution, given budget constraints. Obviously, precise movement control is a critical parameter in the ResQbot 2.0 design specification since it is directly linked to safety.

5.2.4. Stretcher Bed Conveyor Belt Mechanism

Similar to its predecessor (ResQbot 1.0), ResQbot 2.0 is equipped with an active stretcher bed module that actively pulls the casualty's body up during the casualty extraction procedure. While the stretcher-bed module in ResQbot 1.0 is only capable of loading half of the casualty's body, the stretcher bed module in ResQbot 2.0 is designed to load the entire body of an average-size human casualty.

This stretcher-bed module incorporates a conveyor belt capable of transporting a maximum payload of approximately 100 kg at its maximum power. The conveyor belt is powered by a 240 V DC motor controlled through a driver module powered by a 240 V AC onboard power inverter. A pulse-width modulation (PWM) control signal is used to control the motor's speed. Figure 11e shows the electric motor and pulley-belt transmission system driving the conveyor belt.

In order to allow a closed-loop control for the conveyor belt's speed, the conveyor belt module is equipped with an incremental rotary encoder connected to the conveyor's pulley (see Figure 11f). This closed-loop control is essential in order to synchronise the conveyor belt's speed and the speed of the mobile base during the casualty loading process, which uses the loco-manipulation method, as explained in Section 4.

5.2.5. Stretcher Strap Mechanism

For safe transportation, the casualty has to be safely placed onto the stretcher bed, and the stretcher strap modules secure the casualty during transportation. The ResQbot 1.0 stretcher strap design was effective, catering to many different body sizes as well as being safe. However, the ResQbot 2.0 two stretcher strap modules are attached to the stretcher bed module to accommodate a casualty's entire body. One strap module secures and stabilises the casualty's torso, and the other secures the legs. Each strap module is driven by a 24 V DC motor (see Figure 11g). These motors are controlled to fasten and unfasten the straps during the casualty-extraction procedure. The straps' fastening force is controlled by regulating the motors' power so that it is sufficient to secure the casualty without exerting too much pressure on the casualty's body. Figure 11h shows the stretcher's strap module.

5.2.6. Neck Securing Device Module

The neck securing device (NSD) features three main components/mechanisms: (1) A frame designed to be as compact as possible and encase all the safety mechanisms implemented on the NSD; (2) inflatable components that are the main protection against any cervical spine and head injuries during the loading procedure; and (3) a sliding rail mechanism that guides the NSD when it moves up or down along the belt.

The NSD frame was designed to cover the casualty's head and neck up to the shoulders, as illustrated in Figure 7c. The frame's design allows sufficient space between the casualty's head, neck, and the frame in order to prevent any direct contact. The NSD slides along the bed to the ground and slides back up around the bed end. For that reason, the back of the device is curved to allow a smooth up and down motion (see Figure 12a).

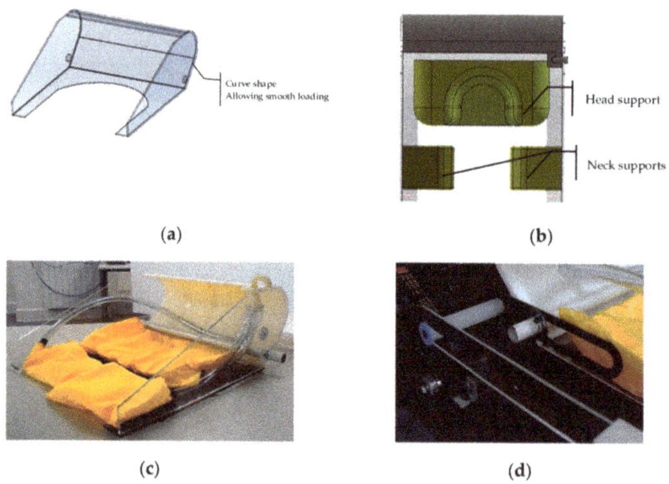

Figure 12. Design of the NSD module. The NSD module features three main components: An NSD frame, inflatable components, and a sliding guide mechanism. (**a**) The design has a curved shape to enable smooth loading up and down during the operation. (**b**) Design of the NSD inflatable components that support the casualty's head and neck during the extraction procedure. (**c**) The NSD module. (**d**) The NSD sliding rail mechanism that allows it to slide up and down along the stretcher bed module.

The NSD inflatable components consist of three separate inflatable modules: One lifts and supports the back of the head, and two support the neck, as shown in Figure 12b. The inflatable modules are inflated by means of a pressurised air chamber that is controlled by a solenoid valve. In order to fully support the neck and head, the inflatables curve around the neck to stabilise the head and prevent unwanted movement. The neck support modules support the neck from each side, and during inflation, the modules slightly lift the neck and support it from the bottom. This allows the head to bend back slightly to facilitate unblocking the respiratory tracts. Figure 12c shows the inflatable components of the NSD module.

In order to guide the NSD up and down the stretcher bed, a pair of sliding guides were developed. They consist of two long rails attached to each side of the stretcher bed, a rod that slides up and down along the rails, and two pin joints connecting each side of the NSD frame to the sliding rod. Figure 12d shows the NSD sliding rail mechanism.

5.3. *ResQbot 2.0 Casualty Extraction Simulation*

In order to evaluate the ResQbot 2.0. design, we built a model using the Gazebo physics engine simulator to simulate the casualty extraction procedure (illustrated in Figures 5 and 7). Figure 13 shows the sequential snapshots (a,b,c) of ResQbot 2.0 carrying out a simulated casualty extraction. The simulation demonstrates the proposed casualty extraction method introduced in Section 4. The snapshot images show that the ResQbot 2.0 design enables it to successfully carry out the complete casualty extraction procedure. Additionally, upon the publication, we will upload any additional experiment results and videos to the ResQbot web page, including the link to the open-source Gazebo simulation model of ResQbot 2.0 (https://www.imperial.ac.uk/robot-intelligence/robots/resqbot/ (accessed on 15 May 2021)) [64].

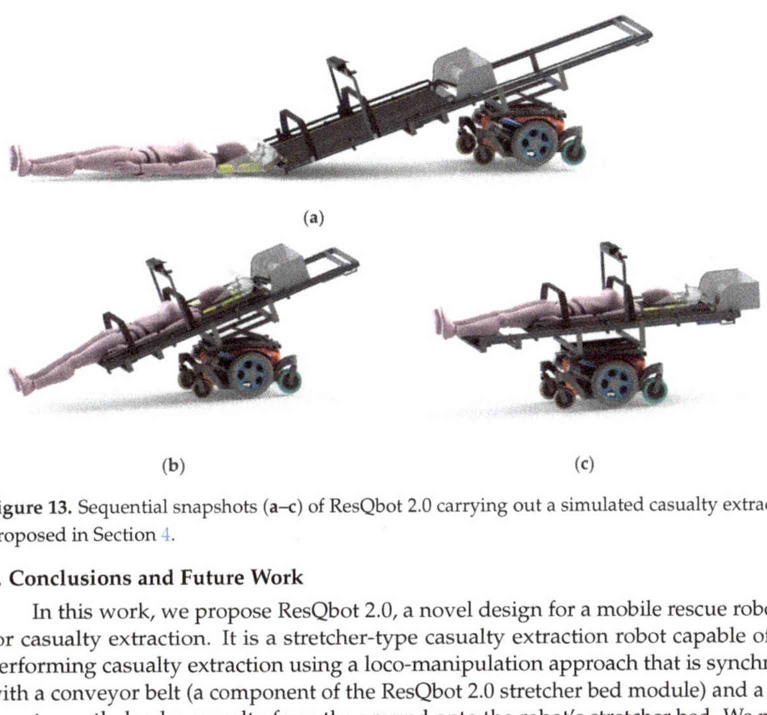

Figure 13. Sequential snapshots (**a–c**) of ResQbot 2.0 carrying out a simulated casualty extraction, as proposed in Section 4.

6. Conclusions and Future Work

In this work, we propose ResQbot 2.0, a novel design for a mobile rescue robot used for casualty extraction. It is a stretcher-type casualty extraction robot capable of safely performing casualty extraction using a loco-manipulation approach that is synchronised with a conveyor belt (a component of the ResQbot 2.0 stretcher bed module) and a mobile base to gently load a casualty from the ground onto the robot's stretcher bed. We propose a new casualty extraction procedure using the novel features in ResQbot 2.0 in order to ensure a safe casualty extraction routine. We verified the proposed design and the casualty extraction procedure by conducting simulation experiments in the Gazebo physics engine simulator. Based on the simulation results, the ResQbot 2.0 design is a feasible option to successfully carry out a safe casualty extraction.

Ongoing work has been focused on the design and development of the ResQbot 2.0 platform. Unfortunately, due to the COVID-19 pandemic, currently we are only able to verify the design and development results using Gazebo, a physics engine simulator. In future work, we are eager to conduct an extensive number of physical experiments that we have developed to evaluate the ResQbot 2.0 performance during casualty extraction tasks in terms of safety, reliability, efficiency, and its limitations under various conditions.

Human-robot teaming in several rescue scenarios could also be studied in depth in order to enable humans to successfully work together with ResQbot and increase the performance of the rescue missions. Moreover, since we have developed the open-sourced full model of the ResQbot 2.0 design for the Gazebo simulator, more advanced simulation experiments, such as a complete scenario of autonomous casualty extraction experiments, can be explored in the future works. The future generation of ResQbot evolution could support concurrent therapies, such as supplemental oxygen and other patient life supports, as well as the easy accessories attachment, such as air monitoring devices in CBRN incidents and cameras in other incidents.

Author Contributions: Conceptualisation, R.P.S., I.K., J.B., A.G., A.D., E.d.C. and P.K.; formal analysis, R.P.S., I.K., J.B., A.G., A.D. and E.d.C.; investigation, R.P.S., I.K., J.B., A.G., A.D. and E.d.C.; methodology, R.P.S. and P.K.; software, R.P.S.; supervision, P.K.; validation, R.P.S., N.R., I.K., J.B., A.G., A.D., E.d.C., S.R. and R.H.; visualisation, R.P.S. and J.B.; writing—original draft, R.P.S. and N.R.; writing—review and editing, N.R. and P.K. All authors have read and agreed to the published version of the manuscript.

Funding: Roni Permana Saputra received a fully financial support for his Ph.D. program through the Indonesia Endowment Fund for Education (LPDP).

Institutional Review Board Statement: Not applicable.

Informed Consent Statement: Not applicable.

Data Availability Statement: Not applicable.

Acknowledgments: The authors would like to thank you for the feedback received from the NHS London Ambulance Service team at the arranged meeting on Tuesday 4 June 2019 at the Dyson School of Design Engineering, Imperial College London.

Conflicts of Interest: The authors declare no conflict of interest.

Abbreviations

Abbreviations

The following abbreviations are used in this manuscript:

SAR	Search and Rescue
CBRN	Chemical, Biological, Radiological, and Nuclear
TAGS	Tactical Amphibious Ground Support System
REX	Robotic Extraction
REV	Robotic Evacuation
NSD	Neck Securing Device
UAV	Unmanned Aerial Vehicle
UGV	Unmanned Ground Vehicle

References

1. Dinh, M.M.; Bein, K.; Roncal, S.; Byrne, C.M.; Petchell, J.; Brennan, J. Redefining the Golden Hour for Severe Head Injury in an Urban Setting: The Effect of Prehospital Arrival Times on Patient Outcomes. *Injury* **2013**, *44*, 606–610. [CrossRef] [PubMed]
2. Harmsen, A.M.K.; Giannakopoulos, G.F.; Moerbeek, P.R.; Jansma, E.P.; Bonjer, H.J.; Bloemers, F.W. The Influence of Prehospital Time on Trauma Patients Outcome: A Systematic Review. *Injury* **2015**, *46*, 602–609. [CrossRef] [PubMed]
3. Lerner, E.B.; Moscati, R.M. The Golden Hour: Scientific Fact or Medical "Urban Legend"? *Acad. Emerg. Med.* **2001**, *8*, 758–760. [CrossRef] [PubMed]
4. Williams, A.; Sebastian, B.; Ben-Tzvi, P. Review and Analysis of Search, Extraction, Evacuation, and Medical Field Treatment Robots. *J. Intell. Robot. Syst.* **2019**, *96*, 401–418. [CrossRef]
5. Murphy, R.R.; Tadokoro, S.; Nardi, D.; Jacoff, A.; Fiorini, P.; Choset, H.; Erkmen, A.M. Search and Rescue Robotics. In *Springer Handbook of Robotics*; Siciliano, B., Khatib, O., Eds.; Springer: Berlin/Heidelberg, Germany, 2008; pp. 1151–1173. ISBN 978-3-540-30301-5.
6. Yoo, A.C.; Gilbert, G.R.; Broderick, T.J. Military Robotic Combat Casualty Extraction and Care. In *Surgical Robotics: Systems Applications and Visions*; Rosen, J., Hannaford, B., Satava, R.M., Eds.; Springer: Boston, MA, USA, 2011; pp. 13–32. ISBN 978-1-4419-1126-1.
7. Theobald, D. Mobile Extraction-Assist Robot. U.S. Patent 7719222B2, 18 May 2010.
8. Hu, J.; Lim, Y.-J. Robotic First Responder System and Method. U.S. Patent 20140150806A1, 5 June 2014.

9. Delmerico, J.; Mintchev, S.; Giusti, A.; Gromov, B.; Melo, K.; Horvat, T.; Cadena, C.; Hutter, M.; Ijspeert, A.; Floreano, D.; et al. The Current State and Future Outlook of Rescue Robotics. *J. Field Robot.* **2019**, *36*, 1171–1191. [CrossRef]
10. West, C.; Arvin, F.; Cheah, W.; West, A.; Watson, S.; Giuliani, M.; Lennox, B. A Debris Clearance Robot for Extreme Environments. In *Towards Autonomous Robotic Systems*; Althoefer, K., Konstantinova, J., Zhang, K., Eds.; Springer International Publishing: Cham, Switzerland, 2019; pp. 148–159.
11. Mandow, A.; Serón, J.; Pastor, F.; García-Cerezo, A. Experimental Validation of a Robotic Stretcher for Casualty Evacuation in a Man-Made Disaster Exercise. In Proceedings of the 2020 IEEE International Symposium on Safety, Security, and Rescue Robotics (SSRR), Virtual, 4–6 November 2020; pp. 241–245.
12. Yeong, S.P.; King, M.; Dol, S.S. A Review on Marine Search and Rescue Operations Using Unmanned Aerial Vehicles. *World Acad. Sci. Eng. Technol. Int. J. Mar. Environ. Sci.* **2015**, *9*, 396–399.
13. Mishra, B.; Garg, D.; Narang, P.; Mishra, V. Drone-Surveillance for Search and Rescue in Natural Disaster. *Comput. Commun.* **2020**, *156*, 1–10. [CrossRef]
14. Quan, A.; Herrmann, C.; Soliman, H. Project Vulture: A Prototype for Using Drones in Search and Rescue Operations. In Proceedings of the 15th International Conference on Distributed Computing in Sensor Systems (DCOSS), Santorini, Greece, 29–31 May 2019; pp. 619–624. [CrossRef]
15. Karaca, Y.; Cicek, M.; Tatli, O.; Sahin, A.; Pasli, S.; Beser, M.F.; Turedi, S. The Potential Use of Unmanned Aircraft Systems (Drones) in Mountain Search and Rescue Operations. *Am. J. Emerg. Med.* **2018**, *36*, 583–588. [CrossRef]
16. McRae, J.N.; Gay, C.J.; Nielsen, B.M.; Hunt, A.P. Using an Unmanned Aircraft System (Drone) to Conduct a Complex High Altitude Search and Rescue Operation: A Case Study. *Wilderness Environ. Med.* **2019**, *30*, 287–290. [CrossRef]
17. Hayley Drones for Search and Rescue. Available online: https://coptrz.com/drone-in-sar-operations/ (accessed on 13 May 2021).
18. Dukowitz, Z. Drones in Search and Rescue: 5 Stories Show Casing Ways Search and Rescue Uses Drones to Save Lives. Available online: https://uavcoach.com/search-and-rescue-drones/ (accessed on 13 May 2021).
19. Saputra, R.P.; Kormushev, P. Resqbot: A Mobile Rescue Robot with Immersive Teleperception for Casualty Extraction. In Proceedings of the Annual Conference Towards Autonomous Robotic Systems, Bristol, UK, 25–27 July 2018; Springer: Berlin/Heidelberg, Germany, 2018; pp. 209–220.
20. Saputra, R.P.; Rakicevic, N.; Kormushev, P. Sim-to-Real Learning for Casualty Detection from Ground Projected Point Cloud Data. *arXiv* **2019**, arXiv:1908.03057.
21. Saputra, R.P.; Rakicevic, N.; Chappell, D.; Wang, K.; Kormushev, P. Hierarchical Decomposed-Objective Model Predictive Control for Autonomous Casualty Extraction. *IEEE Access* **2021**, 1. [CrossRef]
22. Saputra, R.P.; Kormushev, P. Casualty Detection from 3D Point Cloud Data for Autonomous Ground Mobile Rescue Robots. In Proceedings of the 2018 IEEE International Symposium on Safety, Security, and Rescue Robotics, SSRR 2018, Philadeplhia, PA, USA, 6–8 August 2018.
23. Saputra, R.P.; Kormushev, P. ResQbot: A Mobile Rescue Robot for Casualty Extraction. In Proceedings of the Companion of the 2018 ACM/IEEE International Conference on Human-Robot Interaction, Chicago, IL, USA, 5–8 March 2018; pp. 239–240.
24. Wikimedia Commons Spanish Army IRobot PackBot 510 IED Robot. Available online: https://commons.wikimedia.org/wiki/File:IRobot_PackBot_510_E.T..JPG (accessed on 4 June 2021).
25. Gilbert, G.R.; Beebe, M.K. *United States Department of Defense Research in Robotic Unmanned Systems for Combat Casualty Care*; U.S. Department of Defense: Washington, DC, USA, 2010.
26. Gilbert, G.; Turner, T.; Marchessault, R. *Army Medical Robotics Research*; United States Army Medical Research and Materiel Command: Fort Detrick, MD, USA, 2007.
27. Quick, D. Battlefield Extraction-Assist Robot to Ferry Wounded to Safety. Available online: https://newatlas.com/battlefield-extraction-assist-robot/17059/1/9%0AROBOTICS%0ABattle (accessed on 13 May 2021).
28. Ning, M.; Ma, Z.; Chen, H.; Cao, J.; Zhu, C.; Liu, Y.; Wang, Y. Design and Analysis for a Multifunctional Rescue Robot with Four-Bar Wheel-Legged Structure. *Adv. Mech. Eng.* **2018**, *10*, 1687814017747399. [CrossRef]
29. Choi, B.; Lee, W.; Park, G.; Lee, Y.; Min, J.; Hong, S. Development and Control of a Military Rescue Robot for Casualty Extraction Task. *J. Field Robot.* **2019**, *36*, 656–676. [CrossRef]
30. Williams, A.; Sebastian, B.; Ben-Tzvi, P. A Robotic Head Stabilization Device for Medical Transport. *Robotics* **2019**, *8*, 23. [CrossRef]
31. Iwano, Y.; Osuka, K.; Amano, H. Evaluation of Rescue Support Stretcher System. In Proceedings of the 2012 IEEE RO-MAN: The 21st IEEE International Symposium on Robot and Human Interactive Communication, Paris, France, 9–13 September 2012; pp. 245–250.
32. Yamauchi, B. PackBot: A Versatile Platform for Military Robotics. In Proceedings of the Unmanned Ground Vehicle Technology VI; SPIE: Bellingham, WA,USA, 2004; Volume 5422, pp. 228–237.
33. Hstar Technology. Hstar Technology Research Programs. Available online: http://hstar.ishopworld.com/solutions/research_programs.html (accessed on 11 May 2021).
34. Lee, W.; Lee, Y.; Park, G.; Hong, S.; Kang, Y. A Whole-Body Rescue Motion Control with Task-Priority Strategy for a Rescue Robot. *Auton. Robot.* **2017**, *41*, 243–258. [CrossRef]
35. Choi, B.; Park, G.; Lee, Y. Practical Control of a Rescue Robot While Maneuvering on Uneven Terrain. *J. Mech. Sci. Technol.* **2018**, *32*, 2021–2028. [CrossRef]

36. Iwano, Y.; Osuka, K.; Amano, H. Development of Stretcher Component Robots for Rescue Activity. In Proceedings of the IEEE Conference on Robotics, Automation and Mechatronics, Singapore, 1–3 December 2004; Volume 2, pp. 915–920.
37. Iwano, Y.; Osuka, K.; Amano, H. Development of Rescue Support Stretcher System. In Proceedings of the 2010 IEEE Safety Security and Rescue Robotics, Bremem, Germany, 26–30 July 2010; pp. 1–6.
38. Iwano, Y.; Osuka, K.; Amano, H. Development of Rescue Support Stretcher System with Stair-Climbing. In Proceedings of the 2011 IEEE International Symposium on Safety, Security, and Rescue Robotics, Kyoto, Japan, 1–5 November 2011; pp. 245–250.
39. Miyazawa, K. Fire Robots Developed by the Tokyo Fire Department. *Adv. Robot.* **2002**, *16*, 553–556. [CrossRef]
40. Sebastian, B.; Williams, A.; Ben-Tzvi, P. Control of a Head Stabilization System for Use in Robotic Disaster Response. In Proceedings of the ASME 2017 International Mechanical Engineering Congress and Exposition IMECE2017, Tampa, FL, USA, 3–9 November 2017.
41. Sebastian, B. Traversability Estimation Techniques for Improved Navigation of Tracked Mobile Robots. Ph.D. Thesis, Virginia Polytechnic Institute and State University, Blacksburg, VA, USA, 2019.
42. Williams, A.J. A Robotic Head Stabilization Device for Post-Trauma Transport. Master's Thesis, Virginia Tech, Blacksburg, VA, USA, 2018.
43. Engsberg, J.R.; Standeven, J.W.; Shurtleff, T.L.; Tricamo, J.M.; Landau, W.M. Spinal Cord and Brain Injury Protection: Testing Concept for a Protective Device. *Spinal Cord* **2009**, *47*, 634–639. [CrossRef]
44. Payne, A.R.; Patel, S. Injury Mechanisms & Injury Criteria: Head Injury Mechanisms And Injury Criteria. Available online: http://www.eurailsafe.net/subsites/operas/HTML/Section3/Section3.3frm.htm (accessed on 15 March 2021).
45. Herring, C.L. *Rescue Carriers and Drags*; FETI Drill Guide 14-09; Louisiana State University: Baton Rouge, LA, USA, 2014.
46. Fire Engineering Training Minutes: Webbing and Victim Drag Techniques. Available online: https://www.fireengineering.com/training/training-minutes-webbing-and-victim-drag-techniques/ (accessed on 1 April 2021).
47. Finazzo, S. Carrying & Dragging Techniques—Lets Get Carried Away. Available online: https://www.offgridweb.com/preparation/carrying-dragging-techniques-lets-get-carried-away/ (accessed on 1 April 2021).
48. Hann, A. *A Photographic Guide to Prehospital Spinal Care*; Emergency Technologies: Raleigh, NC, USA, 2004.
49. Coen, S.D. Spinal Cord Injury: Preventing Secondary Injury. *AACN Clin. Issues Crit. Care Nurs.* **1992**, *3*, 44–54. [CrossRef]
50. Abram, S.; Bulstrode, C. Routine Spinal Immobilization in Trauma Patients: What Are the Advantages and Disadvantages? *Surgeon* **2010**, *8*, 218–222. [CrossRef]
51. Vaillancourt, C.; Stiell, I.G.; Beaudoin, T.; Maloney, J.; Anton, A.R.; Bradford, P.; Cain, E.; Travers, A.; Stempien, M.; Lees, M. The Out-of-Hospital Validation of the Canadian C-Spine Rule by Paramedics. *Ann. Emerg. Med.* **2009**, *54*, 663–671. [CrossRef]
52. Connell, R.A.; Graham, C.A.; Munro, P.T. Is Spinal Immobilisation Necessary for All Patients Sustaining Isolated Penetrating Trauma? *Injury* **2003**, *34*, 912–914. [CrossRef]
53. Holla, M.; Huisman, J.M.R.; Verdonschot, N.; Goosen, J.; Hosman, A.J.F.; Hannink, G. The Ability of External Immobilizers to Restrict Movement of the Cervical Spine: A Systematic Review. *Eur. Spine J.* **2016**, *25*, 2023–2036. [CrossRef] [PubMed]
54. Stuke, L.E.; Pons, P.T.; Guy, J.S.; Chapleau, W.P.; Butler, F.K.; McSwain, N.E. Prehospital Spine Immobilization for Penetrating Trauma—Review and Recommendations from the Prehospital Trauma Life Support Executive Committee. *J. Trauma Acute Care Surg.* **2011**, *71*, 763–770. [CrossRef] [PubMed]
55. Eastridge, B.J.; Mabry, R.L.; Seguin, P.; Cantrell, J.; Tops, T.; Uribe, P.; Mallett, O.; Zubko, T.; Oetjen-Gerdes, L.; Rasmussen, T.E.; et al. Death on the Battlefield (2001–2011): Implications for the Future of Combat Casualty Care. *J. Trauma Acute Care Surg.* **2012**, *73*, S431–S437. [CrossRef] [PubMed]
56. Kwan, I.; Bunn, F.; Roberts, I.G. Spinal Immobilisation for Trauma Patients. *Cochrane Database Syst. Rev.* **2001**. [CrossRef] [PubMed]
57. Peck, G.E.; Shipway, D.J.H.; Tsang, K.; Fertleman, M. Cervical Spine Immobilisation in the Elderly: A Literature Review. *Br. J. Neurosurg.* **2018**, *32*, 286–290. [CrossRef]
58. Head Injury Criteria Tolerance Levels. Available online: http://www.eurailsafe.net/subsites/operas/HTML/Section3/Page3.3.1.4.htm (accessed on 15 March 2021).
59. Carlström, V. This New Airbag Collar Is a Safer Alternative to the Traditional Bike Helmet—And It'll Soon Be Available across the Globe. Available online: https://www.businessinsider.com/a-new-inflatable-collar-is-safer-than-the-traditional-bike-helmet-2018-8?r=US&IR=T (accessed on 19 March 2021).
60. Nextcrave Hovding Invisible Bike Helmet. Available online: http://www.nextcrave.com/content/999-hovding-invisible-bike-helmet (accessed on 14 May 2021).
61. Sareen, H.; Umapathi, U.; Shin, P.; Kakehi, Y.; Ou, J.; Ishii, H.; Maes, P. Printflatables: Printing Human-Scale, Functional and Dynamic Inflatable Objects. In Proceedings of the 2017 CHI Conference on Human Factors in Computing Systems, Denver, CO, USA, 6–11 May 2017; Association for Computing Machinery: New York, NY, USA, 2017; pp. 3669–3680.
62. Sunrise Medical Salsa M2 Powered Wheelchair. Available online: https://www.sunrisemedical.co.uk/powered-wheelchairs/quickie/power-wheelchairs/quickie-salsa-m2 (accessed on 23 March 2021).
63. Ooznest Metal Dual V Wheel. Available online: https://ooznest.co.uk/product/metal-dual-v-wheel (accessed on 23 March 2021).
64. ResQbot. Available online: https://www.imperial.ac.uk/robot-intelligence/robots/resqbot/ (accessed on 13 May 2021).

applied sciences

MDPI

Article

Task-Based Design Approach: Development of a Planar Cable-Driven Parallel Robot for Upper Limb Rehabilitation

Ferdaws Ennaiem [1,2,*], Abdelbadiâ Chaker [2], Med Amine Laribi [1,*], Juan Sandoval [1], Sami Bennour [2], Abdelfattah Mlika [2], Lotfi Romdhane [2,3] and Saïd Zeghloul [1]

[1] Department of Mechanical Engineering and Complex Systems (GMSC), Pprime Institute French National Centre for Scientific Research (CNRS), National Higher School of Mechanics and Aeroengineering (ENSMA), University of Poitiers, UPR 3346 Poitiers, France; juan.sandoval@univ-poitiers.fr (J.S.); said.zeghloul@univ-poitiers.fr (S.Z.)

[2] Mechanical Laboratory of Sousse (LMS), National Engineering School of Sousse, University of Sousse, Sousse 4000, Tunisia; abdelbadia.chaker@eniso.u-sousse.tn (A.C.); sami.bennour.meca@eniso.u-sousse.tn (S.B.); abdelfattah.mlika@eniso.u-sousse.tn (A.M.); lromdhane@aus.edu (L.R.)

[3] Department of Mechanical Engineering, American University of Sharjah, Sharjah P.O. Box 26666, United Arab Emirates

* Correspondence: ferdaws.ennaiem@univ-poitiers.fr (F.E.); med.amine.laribi@univ-poitiers.fr (M.A.L.)

Abstract: This paper deals with the optimal design of a planar cable-driven parallel robot (CDPR), with three degrees of freedom, intended for assisting the patient's affected upper limb along a prescribed movement. A Qualisys motion capture system was used to record the prescribed task performed by a healthy subject. For each pose taken by the center of mass of the end-effector, the cable tensions, the elastic stiffness and the dexterity were optimized while satisfying a set of constraints. First, a multiobjective formulation of the optimization problem was adopted. Since selecting a single solution among the multiple ones given by the Pareto front presents an issue, a mono-objective formulation was chosen, where the objective function was defined as a weighted sum of the chosen criteria. The appropriate values of the weighted coefficients were studied with the aim of identifying their influence on the optimization process and, thus, a judicious choice was made. A prototype of the optimal design of the CDPR was developed and validated experimentally on the prescribed workspace using the position control approach for the motors. The tests showed promising reliability of the proposed design for the task.

Keywords: planar CDPR; prescribed task; optimization problem; multiobjective formulation; mono-objective formulation; position control approach; validated experimentally

Citation: Ennaiem, F.; Chaker, A.; Laribi, M.A.; Sandoval, J.; Bennour, S.; Mlika, A.; Romdhane, L.; Zeghloul, S. Task-Based Design Approach: Development of a Planar Cable-Driven Parallel Robot for Upper Limb Rehabilitation. *Appl. Sci.* **2021**, *11*, 5635. https://doi.org/10.3390/app11125635

Academic Editor: Manuel Armada

Received: 24 May 2021
Accepted: 14 June 2021
Published: 18 June 2021

1. Introduction

Functional rehabilitation aims to recover as much as possible of the patient's locomotion independence. It requires the assistance of a therapist to perform repetitive exercises for an injured member [1]. Task-oriented protocols, where the patient is assisted to perform a specific prescribed movement, such as kicking a ball or standing up and walking, show promising outcomes compared to the conventional training based on passively moving the impaired joints in the limits of their range of motion [2].

Rehabilitation sessions can last up to several weeks [3]. In addition, to guarantee a better quality of the followed protocol, one-to-one assistance is needed [4]. However, the limited number of available therapists influences the high-intensity and the repetition of the assistance. Given these issues, researchers have developed robotic devices to assist practitioners' tasks [5–9]. They also provide the opportunity to assess the patient's recovery progress and monitor protocol efficiency; for instance, by using ARMin [10], the ARM Guide [11], and the MIME [12].

The robotic devices used for clinical practice allow mainly the rehabilitation of the knee, the shoulder and the elbow [13]. Cable-driven parallel robots (CDPRs) [14] can be used to extend the coverage of existing robotic platforms since they allow rehabilitation of other joints [13]. They have also a larger translational workspace, less dynamic inertia and higher flexibility compared to serial manipulators [13]. Thanks to their simple reconfiguration and light weight, they have no setup constraints and can be used at the patients' personal spaces to reduce the need to move to rehabilitation centers.

Various studies have been carried out dealing with the design [15], control [16–18] and the structural optimization of CDPRs. In [19], Lorenzo et al. studied the optimal position of the cable exit points allowing minimization of CDPR size for fully constrained and suspended configurations. In this work, the location of the cable exit points was supposed to be fixed, which was an unrealistic assumption. In [20], Hussein et al. optimized the CDPR geometry by minimizing the maximum cable tensions. This criterion could not guarantee minimum cable tensions since only the maximum value was optimized. Abbas et al. [21] studied the optimal design of a suspended CDPR using first the workspace area, then the global condition index, which describes the robot dexterity, as two separated objective functions and, thus, only one criterion in each optimization was used. Yangmin et al. [22] studied first, the optimal design of a CDPR taking into consideration the dexterity then the elastic stiffness as criteria, then a multiobjective optimization approach was used mixing the two characteristics. The selection of each criterion weight was chosen in a way that either the system was preferred to be more dexterous or stiffer. The authors chose to give the same importance to the two criteria. Such a choice can be improved by studying the influence of the variation of each coefficient on the objective function and on the criteria.

The gaps presented above are taken into consideration in this paper. The goal is to design a planar CDPR for upper limb rehabilitation. The prescribed exercise consists of tracing the number eight shape with the hand. This form was used among others since it involves shoulder and elbow motions [23]. The rehabilitation of the wrist joint was also considered since the hand orientation was not fixed. The cable tensions, the elastic stiffness and the dexterity were selected as the problem criteria. The choice of each criterion weight was justified, and the design parameters were set in a way to obtain a realistic design. This design problem can be formulated as a constrained optimization problem. Thus, this paper aims to design an optimal planar CDPR, based on the above-mentioned criteria, able to mobilize the patient's affected upper limb along the prescribed task. The design optimization problem was formulated after considering the disadvantages of some proposed approaches introduced above. Once the optimal structure was selected, an experimental prototype was developed allowing a check of solution feasibility and reliability.

The paper is organized as follows: Section 2 details the experimental protocol followed to record the prescribed task. The planar CDPR model is presented as well as the criteria and the constraints adopted to seek the optimal design. In Section 3, the optimization results, using a multiobjective formulation then an adapted mono-objective formulation, are discussed and the experimental validation using a developed prototype is shown. The last section concludes the paper.

2. Materials and Methods

2.1. Prescribed Exercise Analysis

Rehabilitation aims to recover the functional abilities of the affected member by performing intensive and repetitive training [1]. The chosen exercise for this study was commonly performed for upper limb rehabilitation. It consisted of tracing with the hand an "8" curve. This exercise allows the rehabilitation of the three upper limb joints.

A healthy subject was asked to perform the prescribed drill in order to obtain a normal trajectory to use as a reference. This exercise involved five joints' movements, namely the three rotations of the shoulder and the flexion/extension motion of the elbow and the wrist. The participant's gestures were recorded using a Qualisys motion capture system with five infra-red cameras and five reflective markers attached to his hand as illustrated in Figure 1. Figure 2 illustrates the steps followed for the data acquisition.

(a) (b)

Figure 1. (**a**) Motion capture setup; (**b**) Marker locations.

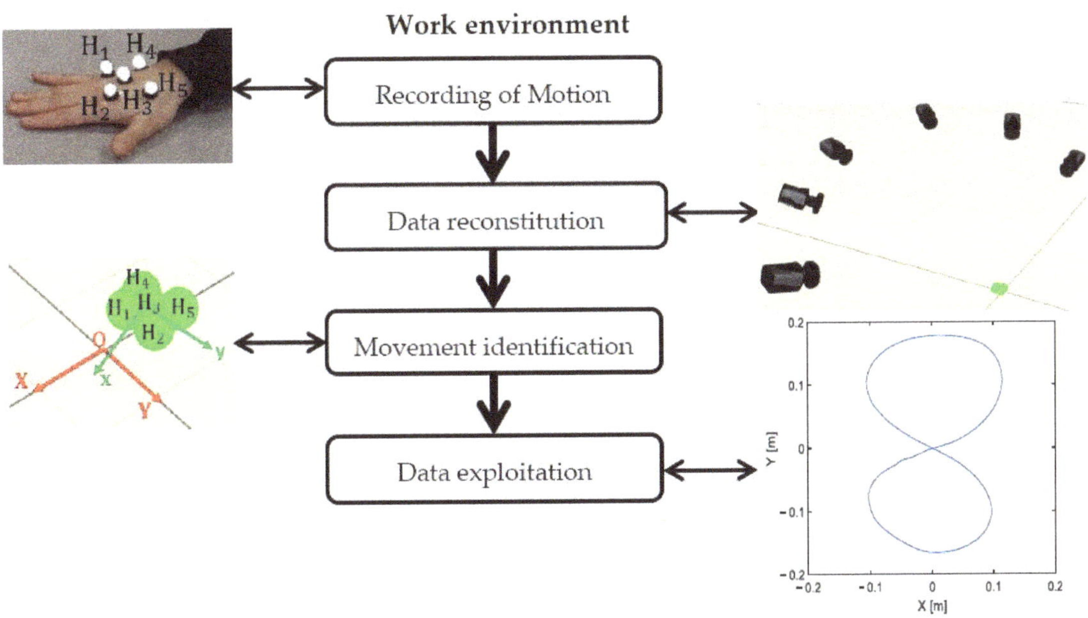

Figure 2. Data acquisition steps.

The hand trajectory was defined by tracking the successive positions taken by the marker H_3 during the prescribed exercise. Hand orientation was delimited by computing the rotation angle between a local frame (H_3, **x**, **y**) attached to the participant's hand and a global frame (O, **X**, **Y**) attached to the table as illustrated in Figure 3. Since the patient's hand was attached to the robot end-effector, the recorded data, given in Figure 4, were used to define the robot's prescribed workspace.

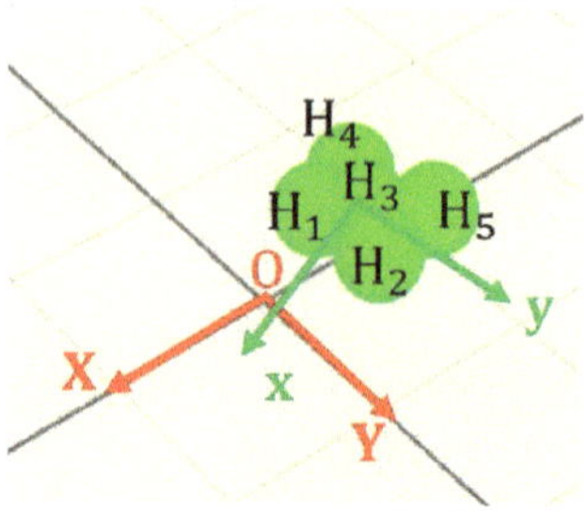

Figure 3. Local and global frames.

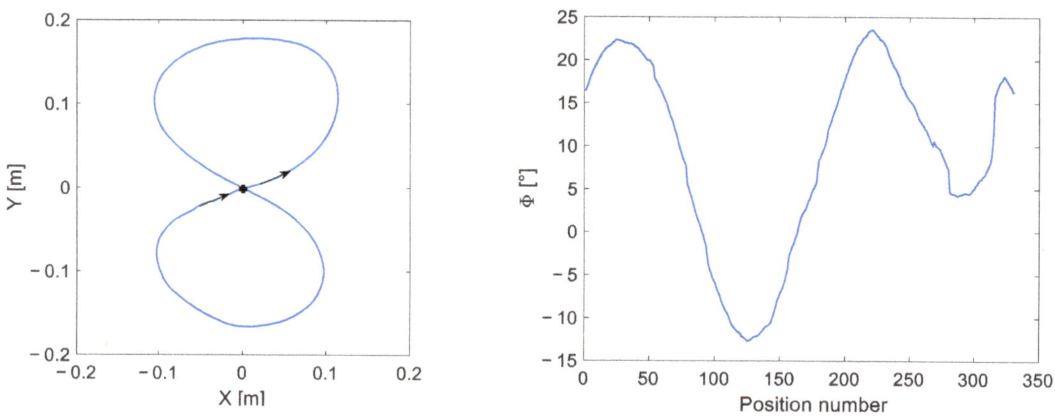

(a) (b)

Figure 4. (a) Hand trajectory (composed of 331 equidistant points); (b) hand orientation. "+" denotes the starting and the ending position.

2.2. Optimal Synthesis Problem and Its Formulation

A planar cable-driven parallel robot with three degrees of freedom (DOFs) is considered in this paper. Since at least one more cable than the DOFs was needed to fully constrain the robot, four cables were used [24]. Their mass and elasticity were neglected. They were modeled as straight lines. The design process consisted of finding the optimal position of each actuator and the end-effector size satisfying a set of criteria and constraints. The motors' positions were defined using the parameters $\mathfrak{a}_i$ and $\mathfrak{b}_i$, which represent the coordinates of the center of the pulley $\mathcal{P}_i$, fixed on the ith actuator. The mobile platform was considered as a square of side $\mathfrak{c}$ (see Figure 5). The design vector is given by Equation (1). The global and the local frames $(\mathbf{X}, \mathbf{Y})$ and $(\mathbf{x}, \mathbf{y})$, respectively, matched those represented in Figure 2, used for the motion capture analysis.

$$\mathbf{I} = [\mathfrak{a}_1, \mathfrak{b}_1, \mathfrak{a}_2, \mathfrak{b}_2, \mathfrak{a}_3, \mathfrak{b}_3, \mathfrak{a}_4, \mathfrak{b}_4, \mathfrak{c}], \tag{1}$$

$\mathfrak{n}_i$ denotes the unit vector along the ith cable. It is defined as follows, where ψ_i and θ_i are expressed as given in Equation (2). L_i is the length of the ith cable and R_p is the pulleys' radius. R_p is considered to be fixed, since the cable radius and the coiling effects are neglected.

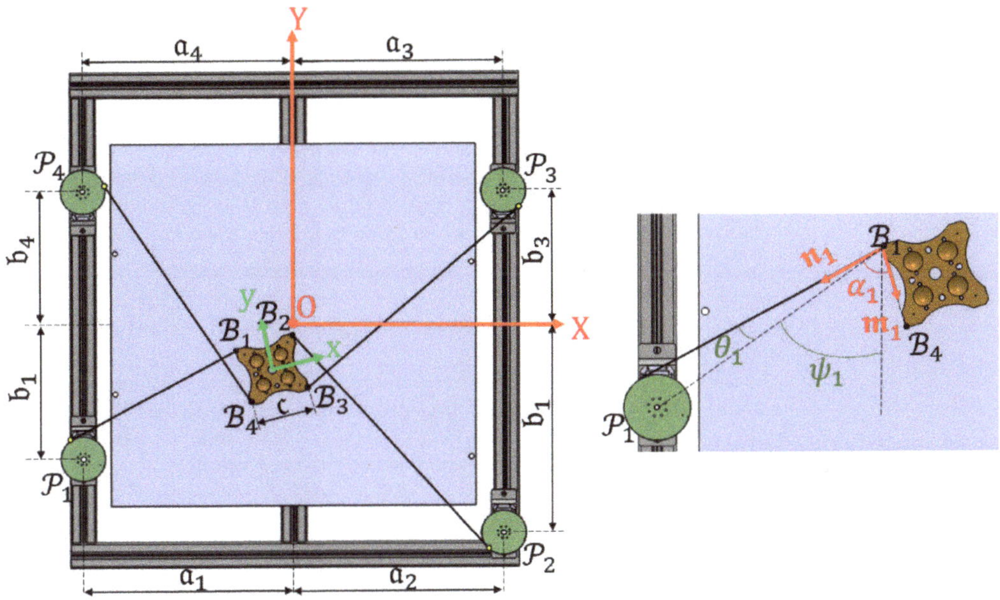

Figure 5. Geometric representation of the planar robot.

$$\mathfrak{n}_1 = \begin{bmatrix} -\sin(\psi_1+\theta_1) \\ -\cos(\psi_1+\theta_1) \end{bmatrix}, \mathfrak{n}_2 = \begin{bmatrix} \sin(\psi_2-\theta_2) \\ -\cos(\psi_2-\theta_2) \end{bmatrix}, \mathfrak{n}_3 = \begin{bmatrix} \sin(\psi_3+\theta_3) \\ \cos(\psi_3+\theta_3) \end{bmatrix}, \mathfrak{n}_4 = \begin{bmatrix} -\sin(\psi_4-\theta_4) \\ \cos(\psi_4-\theta_4) \end{bmatrix}, \tag{2}$$

$$\psi_i = \tan^{-1}\left(\frac{|\mathcal{P}_{ix} - \mathcal{B}_{i_x}|}{|\mathcal{P}_{iy} - \mathcal{B}_{i_y}|}\right), \tag{3}$$

$$\theta_i = \tan^{-1}\left(\frac{R_p}{L_i}\right), \tag{4}$$

The end-effector dynamic model was obtained using Newton-Euler formulation with the assumption of neglecting the cables mass and the dynamics of the pulleys. Its expression is given by Equation (5).

$$\begin{bmatrix} \sum \mathbf{F} \\ \sum \mathcal{M} \end{bmatrix} = \mathbf{M}\ddot{\chi} + \mathbf{C}\dot{\chi} = \mathbf{J}^{\mathrm{T}}\mathbf{T} + \mathbf{f_g} + \mathbf{F}_{\mathbf{ext/EE}}, \tag{5}$$

where $\mathbf{M} = \begin{bmatrix} m_p\mathbb{I}_{3\times 3} & 0_{3*3} \\ 0_{3\times 3} & \mathbf{R_\Phi}\,\mathbb{I}_\mathbf{p}\,\mathbf{R_\Phi}^{\mathrm{T}} \end{bmatrix}$ and $\mathbf{C} = \begin{bmatrix} 0_{3\times 1} \\ \omega \times \mathbf{R_\Phi}\mathbb{I}_\mathbf{p}\,\mathbf{R_\Phi}^{\mathrm{T}}\omega \end{bmatrix}$ are the mass and the Coriolis matrices, respectively. m_p and ω are the mass and the angular velocity of the end-effector, $\mathbf{R_\Phi}$ is the rotation matrix, $\mathbb{I}_\mathbf{p}$ denotes the inertial matrix of the mobile platform written in its center of mass, $\chi = [x\ y\ \Phi]^{\mathrm{T}}$ is the pose vector of the end-effector, $\mathbf{J}^{\mathrm{T}}$ is the transpose of the Jacobian matrix given in Equation (6), and $\mathbf{T}$, $\mathbf{f_g}$, and $\mathbf{F}_{\mathbf{ext/EE}}$ are the cable tensions vector, the gravity force and the external forces applied on the end-effector, respectively.

$$\mathbf{J}^{\mathrm{T}} = \begin{bmatrix} \mathfrak{n}_\mathrm{i} & \mathbf{Z}\cdot(\mathbf{R_\Phi}\cdot\mathbf{B_i}) \times \mathfrak{n}_\mathrm{i} \end{bmatrix}^{\mathrm{T}}, \; i = 1..4, \tag{6}$$

$\mathbf{B_i}$ is the vector containing the anchor points coordinates ($\mathcal{B}_i$) in the local frame.

Thus, the expression of the cable tensions vector **T** can be deduced using Equation (5) as follows:

$$\mathbf{T} = \mathbf{T_p} + \mathbf{T_h} = \left(\mathbf{J^T}\right)^+ (\mathbf{M}\ddot{\mathbb{X}} + \mathbf{C}\dot{\mathbb{X}} - \mathbf{f_g} - \mathbf{F_{ext/O_L}}) + \lambda\,\mathrm{Null}\left(\mathbf{J^T}\right) \tag{7}$$

where $\mathbf{T_p}$ and $\mathbf{T_h}$ are the particular and the homogenous solutions, $(\mathbf{J^T})^+$ is the Moore-Penrose pseudoinverse and λ is an arbitrary scalar. The cable tensions must be bounded between a minimum positive and a maximum value in order to avoid slack and over-tensioned cables. This condition forms the first problem constraint, which is formulated as follows:

$$0 < T_{min} \leq T_i(j) \leq T_{max}, \;\; i = 1..4, \;\; j = 1..n \tag{8}$$

where n is the number of points composing each trajectory, $T_i(j)$ is the tension of the ith cable at the jth position.

The second constraint concerns the collisions between the cables and the end-effector. to prevent this issue, the angle α_i between the mobile platform and the ith cable is computed for each position of the prescribed trajectories. α_i must remain higher than a limit angle α_{lim}. The formulation of the collision constraint is given as follows:

$$\alpha_i = \cos^{-1}\left(\mathfrak{n}_i \cdot \frac{\mathfrak{m}_i}{\|\mathfrak{m}_i\|}\right) > \alpha_{lim}, \qquad i = 1..4, \tag{9}$$

where $\mathfrak{m}_1 = -\mathfrak{m}_4 = \mathcal{B}_1\mathcal{B}_4$, $\mathfrak{m}_2 = -\mathfrak{m}_3 = \mathcal{B}_2\mathcal{B}_3$, and $\mathfrak{n}_i$ is the unit vector along the ith cable as illustrated in Figure 5.

The last constraint concerns the location of the points $\mathcal{P}_i$. They must be located on the edges of the square forming the robot fixed frame. This constraint facilitates the robot reconfiguration when any other trajectory, included in the robot workspace, is selected.

The 4×3 Jacobian matrix **J** of the considered robot includes components of different physical units since the latter has mixed DOFs. To have a meaningful value of the condition number, the Jacobian matrix must be normalized allowing assessment of the closeness of a pose to a singularity. In order to settle the dimensional inhomogeneity, several methods have been suggested based on dividing the rotational elements of the Jacobian matrix **J** by a conventional length L. Lee et al. defined L in [25] as a nominal length represented by the distance between the origins of the global and the local frames. Angeles introduced in [26] the notion of natural length, which is the value of L that minimizes the condition number. This length is approximated to the radius of the end-effector [27]. The latter method was adopted in this paper. The homogenous Jacobian matrix, $\mathbf{J}_h$, of size 4×3, is calculated as given by Equation (10), where $\mathfrak{c}$ is the side length of the mobile platform.

$$\mathbf{J_h} = \mathbf{J}\cdot diag\left(1,\, 1,\, \frac{2}{\mathfrak{c}\sqrt{2}}\right), \tag{10}$$

Three criteria are considered for the optimization problem formulation, namely minimizing the tension in each cable and maximizing both the elastic stiffness and the dexterity of the end-effector. The first, the second, and the third criteria are characterized using the parameters C_1, C_2, and C_3, respectively, as follows:

$$C_1 = \frac{1}{4}\sum_{i=1}^{4} \frac{\sum_{j=1}^{n} T_{i(j)}}{n \,. \max\limits_{j=1..n} T_{i(j)}}, \tag{11}$$

$$C_2 = 1 - \frac{1}{n}\sum_{j=1}^{n} \frac{\lambda_{min}\,(j)}{\lambda_{max}\,(j)}, \tag{12}$$

$$C_3 = 1 - \frac{1}{n}\sum_{j=1}^{n} \frac{\sigma_{min}\,(j)}{\sigma_{max}\,(j)}, \tag{13}$$

where λ and σ are the eigenvalue of the cable stiffness matrix and the singular value of the normalized Jacobian matrix $\mathbf{J_h}$. The maximum value of the ratios $\frac{\lambda_{min}}{\lambda_{max}}$ and $\frac{\sigma_{min}}{\sigma_{max}}$ is 1, thus, maximizing the elastic stiffness and the dexterity leads to minimizing the parameters C_2 and C_3.

3. Results and Discussions

3.1. Multiobjective Formulation

The multiobjective formulation is given in Equation (14). A penalty formulation was adopted to handle the problem constraints.

$$\begin{cases} \min(C_1(\mathbf{I}) + \wp_1 + \wp_2 + \wp_3), \\ \min(C_2(\mathbf{I}) + \wp_1 + \wp_2 + \wp_3), \\ \min(C_3(\mathbf{I}) + \wp_1 + \wp_2 + \wp_3), \end{cases} \tag{14}$$

$$\wp_1 = \begin{cases} 0 & \text{if } \; 0 < T_{min} \leq T_i(j) \leq T_{max} \\ \psi & \text{otherwise} \end{cases}, \tag{15}$$

$$\wp_2 = \begin{cases} 0 & \text{if } \; \alpha_i > \alpha_{lim} \\ \psi & \text{otherwise} \end{cases}, \tag{16}$$

$$\wp_3 = \begin{cases} 0 & \text{if } \; \mathcal{P}_i \in \text{fixed frame} \\ \psi & \text{otherwise} \end{cases}, \tag{17}$$

where $\wp_{i,\ i=1,2,3}$ are the penalty functions and ψ is a large scalar.

Multiple solutions coexist forming the Pareto front displayed in Figure 6. Without adding supplementary information about the desired result, all the nondominated points form a potential optimal solution.

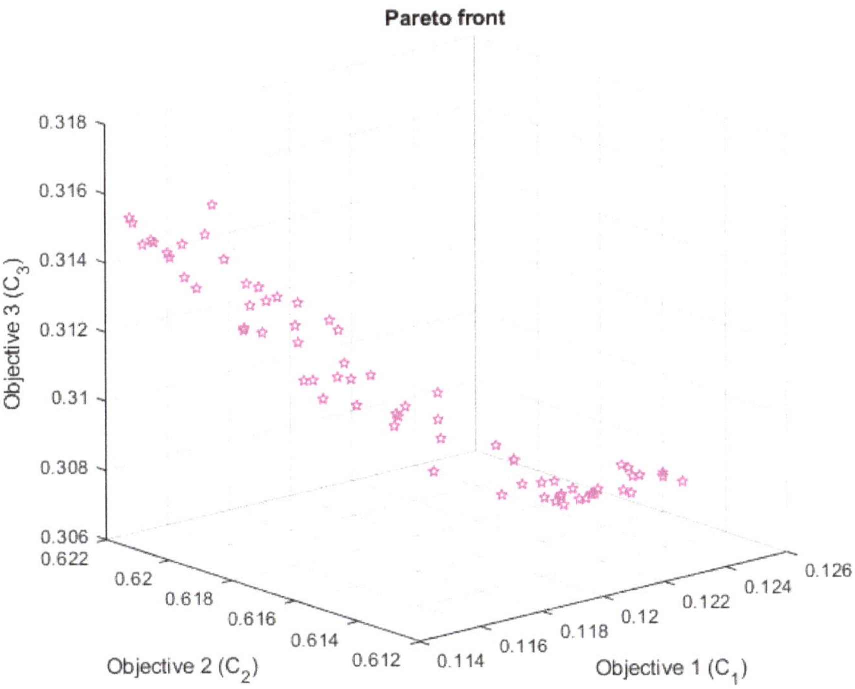

Figure 6. The Pareto front.

In order to facilitate its interpretation, two dimensions representations of the Pareto front are given. Figure 7 shows the cable tension vs the elastic stiffness criteria, the cable tension vs dexterity criteria, the dexterity vs the elastic stiffness criteria and the histogram of each criterion. The three criteria are normalized, and their values vary from 0 to 1.

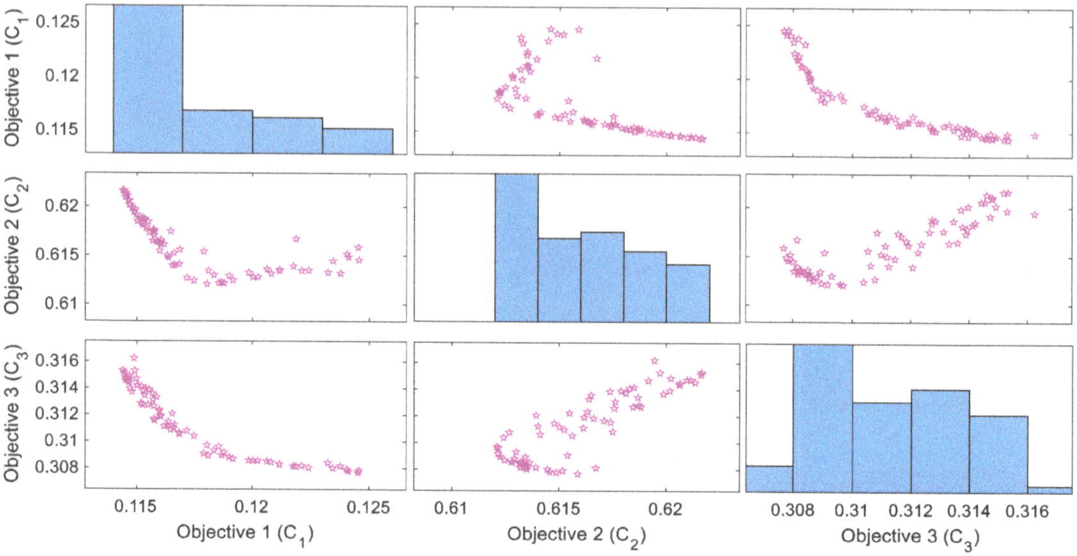

Figure 7. Bi-objective visualizations of the Pareto front.

The nondominated solutions for each two criteria present the preferable solutions if only the two corresponding objectives are considered. For the overall problem, the most desirable results are those which lay within the nondominated solutions of all the pairs illustrated in Figure 7. For the Pareto front presented above, the selection of the optimal solution was complicated since each solution minimized no more than two criteria and maximized those remaining. Additional subjective preference information was, therefore, needed to facilitate the decision-making.

3.2. Mono-Objective Formulation

Another method can be used for the resolution of a multiobjective optimization problem. This technique, called the scalarization method [28], consists of attributing a weight to each criterion that corresponds to the priority given to this objective in the optimization process. This method allows a single solution. Based on this technique, the modified optimization aims to find the optimal design vector $\mathbf{I}^*$ which minimizes the new objective function $\mathcal{F}$, defined as a weighted sum of the three subfunctions introduced above. The optimization problem is then formulated as follows:

$$\min(\mathcal{F}(\mathbf{I})),$$

Subject to

$$\begin{gathered} 0 < T_{min} \leq T_i(j) \leq T_{max}, \;\; i = 1..4, \;\; j = 1..n \\ \alpha_i > \alpha_{lim}, \qquad i = 1..4 \\ \mathcal{P}_i \; \epsilon \text{ fixed frame}, \;\; i = 1..4 \end{gathered} \tag{18}$$

The problem constraints are handled using a penalty approach. $\mathcal{F}$ is then expressed as follows:

$$\mathcal{F}(\mathbf{I}) = \beta_1\mathcal{C}_1(\mathbf{I}) + \beta_2\mathcal{C}_2(\mathbf{I}) + \beta_3\mathcal{C}_3(\mathbf{I}) + \wp_1 + \wp_2 + \wp_3, \quad (19)$$

where $\wp_{i,\ i=1,2,3}$ are the penalty functions given by the Equations (15)–(17) and β_1, β_2 and β_3 define the weight of each criterion. These coefficients must verify Equation (20).

$$\sum_{i=1}^{3} \beta_i = 1, \quad (20)$$

Three strategies have been used in the literature to compute the values of the coefficients β_i [29], namely, the equal weights method [30], the rank order centroid weights method [31] and the rank-sum weights method [31].

The equal weights method was used in our previous work [32]. In this paper, a deeper study of the appropriate values of these weighting coefficients is conducted in order to identify the impact of each coefficient on the quality of the final solution and the optimization process. A mapping of the global objective function and those corresponding to the three criteria is computed for all the possible coefficient values inside the interval (0,1) with respect to Equation (20), using the Particle Swarm Optimization algorithm (PSO) [33]. The computed variations are illustrated in Figure 8.

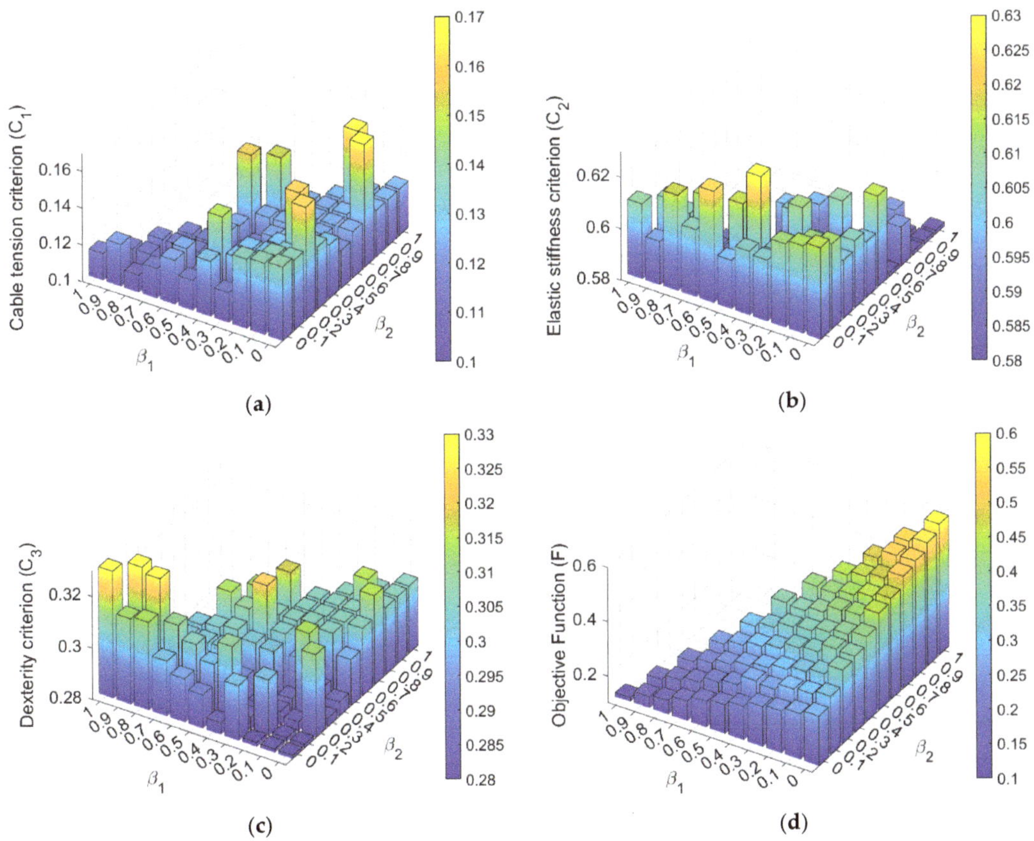

Figure 8. Variation of (**a**) the cable tension criterion, (**b**) the elastic stiffness criterion, (**c**) the dexterity criterion and (**d**) the objective function along the prescribed movement vs β_1 and β_2 ($\beta_3 = 1 - (\beta_1 + \beta_2)$).

Referring to Figure 8d, the objective function decreases with β_1. This means that the cable tension criterion $\mathcal{C}_1$ has the greatest impact on the objective function $\mathcal{F}$ compared to the other coefficients β_2 and β_3. Thus, having a structure with a minimum cable tension distribution is easier than generating good dexterity and elastic stiffness.

As illustrated in Figure 8c, the dexterity criterion $\mathcal{C}_3$ is nearly equal to 0.3 ($\mathcal{C}_3 \in [0.3, 0.33]$ for $\beta_3 = 0$ and $\mathcal{C}_3 = 0.28$ for $\beta_3 = 1$). Thus, according to Equation (13), the robot dexterity remains between 0.67 and 0.72. In other words, it takes good values even when β_1 and β_2 are large (β_3 is close to zero). This is contrary to the elastic stiffness criterion $\mathcal{C}_2$,whose minimum value is equal to 0.58 for $\beta_2 = 1$, which corresponds to an elastic stiffness equal to 0.42 (according to Equation (12)).

Based on this mapping study, since the dexterity has acceptable values for any chosen β_3, the lowest coefficient is given to the criterion $\mathcal{C}_3$. The remaining weight is divided equally between the two other criteria. Thus, $\beta_3 = 0.1$, $\beta_1 = \beta_2 = 0.45$.

Several methods exist in the literature allowing the resolution of optimization problems [34,35]. Particle Swarm Optimization (PSO) was used in this paper to select the optimal design vector. The different parameters used for the problem resolution are listed in Table 1. The lower and the upper boundaries of each design parameter are listed in Table 2.

Table 1. Optimization problem parameters.

Parameter	Value
Mobile platform weight [kg]	0.5
Pulley radius [mm]	37
Cable radius [mm]	0.3
T_{min}[N]	0.5
T_{max}[N]	15
$\alpha_{lim}[°]$	2
$\beta_1 = \beta_2$	0.45
β_3	0.1
Population size	100

Table 2. Boundaries of the design parameters.

Parameter	$\mathfrak{a}_1$	$\mathfrak{b}_1$	$\mathfrak{a}_2$	$\mathfrak{b}_2$	$\mathfrak{a}_3$	$\mathfrak{b}_3$	$\mathfrak{a}_4$	$\mathfrak{b}_4$	$\mathfrak{c}$
Lower bounds [m]	−0.375	−0.385	0	−0.385	0	0	−0.375	0	0.1
Upper bounds [m]	0	0	0.375	0	0.375	0.385	0	0.385	0.15

The obtained solution, as well as the corresponding values of the objective function and the three criteria, are given in Table 3.

Table 3. Optimization results.

Parameter	Value
Optimal design vector $\mathbf{I}^*$	$[-0.375, -0.39, 0.375, -0.26, 0.375, 0.18, -0.375, 0.39, 0.1]$
Objective function $\mathcal{F}$	0.35
Cable tension criterion $\mathcal{C}_1$	0.124
Elastic stiffness criterion $\mathcal{C}_2$	0.588
Dexterity criterion $\mathcal{C}_3$	0.3

The maximum cable tension, the elastic stiffness and the dexterity of the optimal robot structure computed along the task workspace are illustrated in Figure 9. Figure 10 displays the free collision static workspace of the optimal structure for different values of Φ.

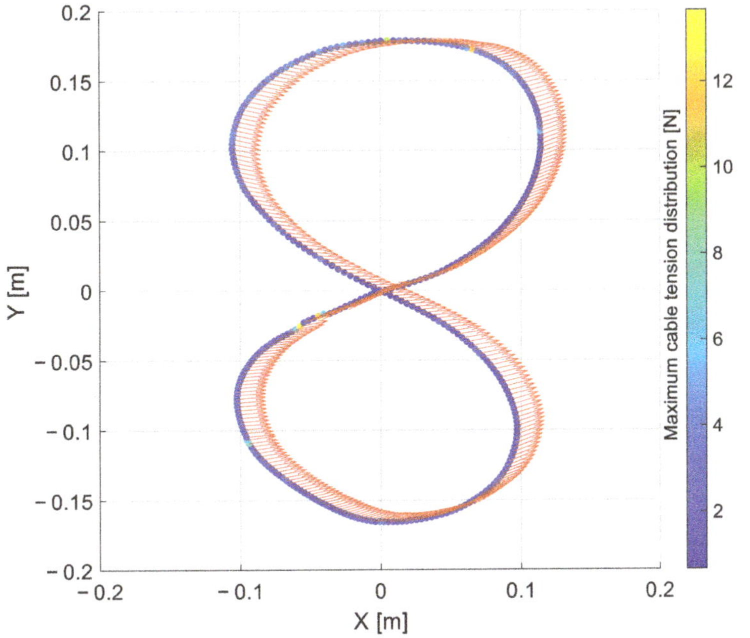

(**a**)

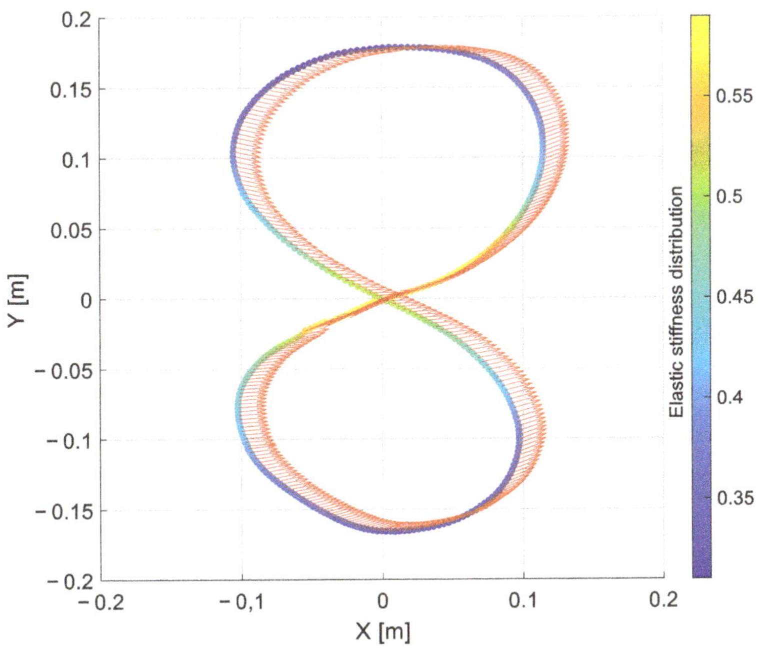

(**b**)

Figure 9. *Cont.*

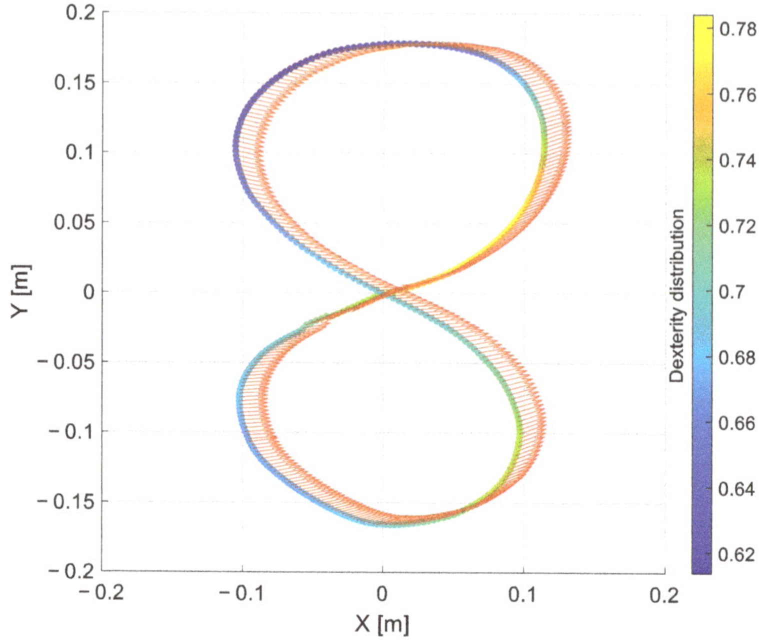

(c)

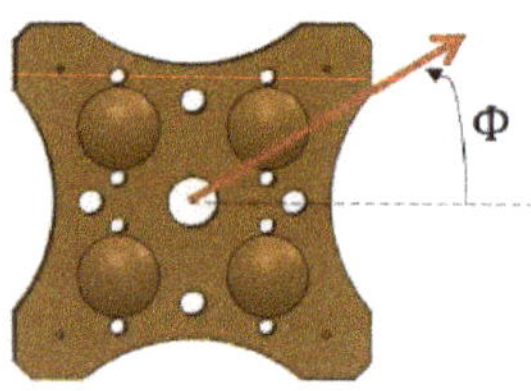

(d)

Figure 9. Optimization results (**a**) the cable tension distribution, (**b**) the elastic stiffness distribution, (**c**) the dexterity distribution along the task workspace and (**d**) the arrows signification represented in (**a**–**c**).

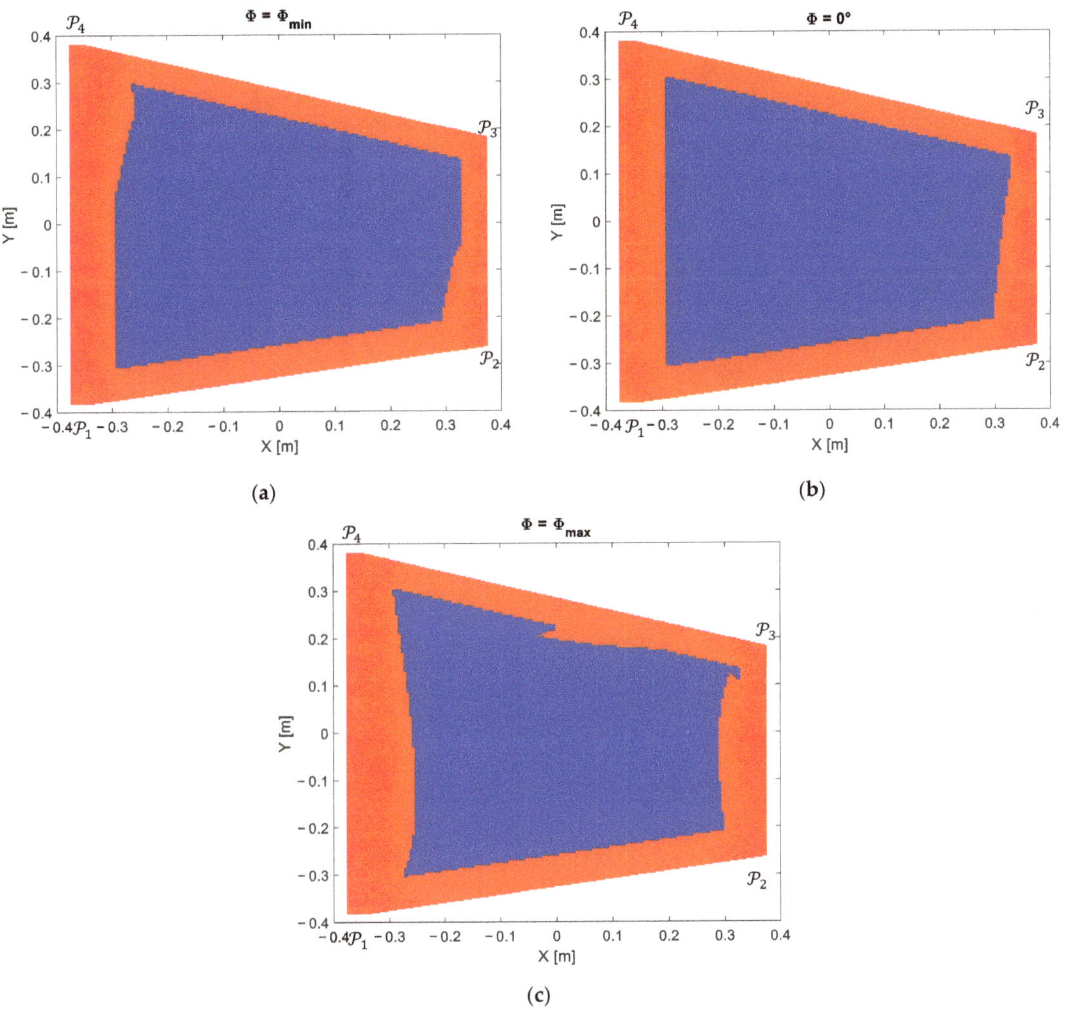

Figure 10. The free collision static workspace (in blue) for (**a**) $\Phi = \Phi_{min} = -13°$, (**b**) $\Phi = 0°$ and (**c**) $\Phi = \Phi_{max} = 24°$.

4. Experimental Validation

Following the optimization study detailed above, a CDPR prototype, presented in Figure 11a, was developed. The actuators' positions and the end-effector size were adjusted according to the optimal design vector resulting in Table 2. It was composed of a rigid frame with four actuated pulleys controlling the end-effector poses by means of four cables. The patient sits near to the robot and grabs the mobile platform with his hand. Figure 11b shows the prototype representation using QTM (Qualisys Track Manager) software recording the passive markers' locations along the end-effector movement.

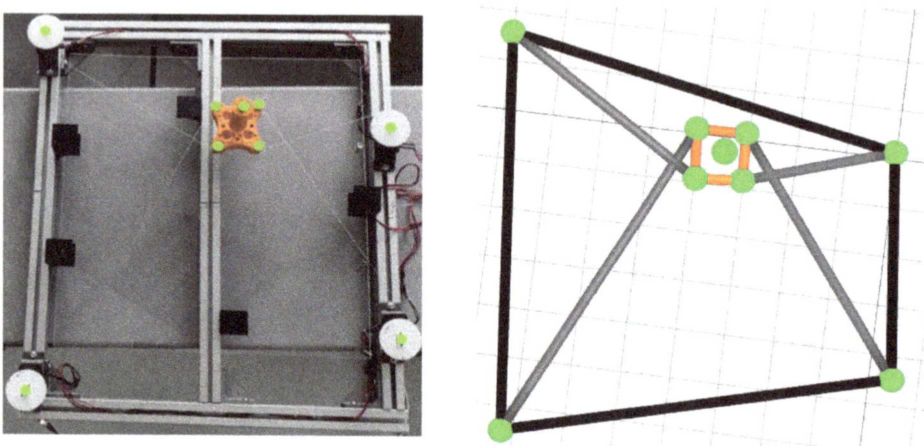

(a) (b)

Figure 11. (**a**) Robot prototype and (**b**) its representation using QTM (Qualisys Track Manager) software. Markers are represented in green.

The prototype was actuated using the Dynamixel MX-106T and equipped with a feedforward, a PID controller and a gear reducer. The pulleys and the end-effector were produced using 3D printing. The cables were made of Dacron.

The extended position control mode was used for the actuator commands. The desired angular positions of the motors were computed using the inverse kinematic model. Figure 12 presents the block diagram of the robot control.

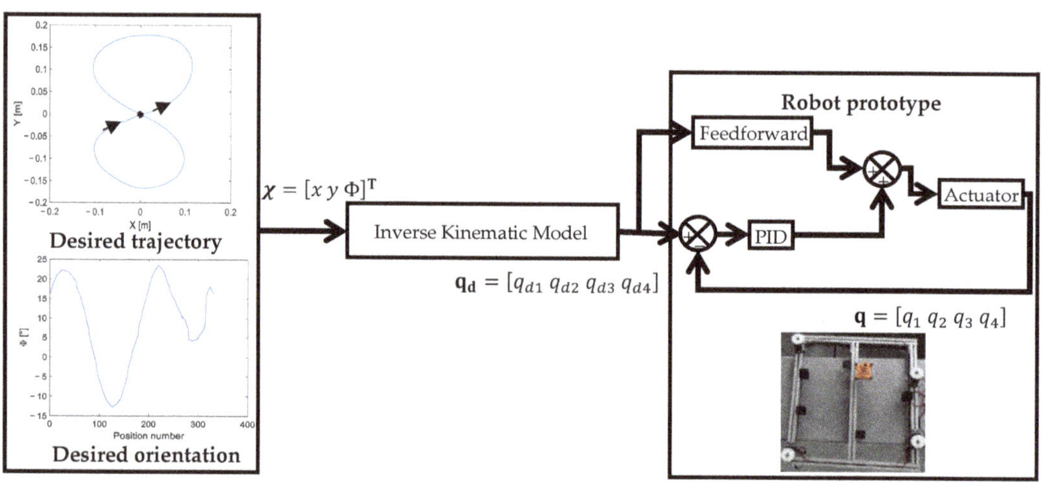

Figure 12. Control block diagram.

The angular positions of the pulleys were controlled so that the robot end-effector could replicate the prescribed exercise. A motion capture system was also used in this phase to evaluate the performance of the robot by tracking the mobile platform poses along its movement so that a comparison, illustrated in Figure 13, between the target and the performed trajectories could be made. This external and global approach allowed consideration of all error sources inside and outside the control loop, such as anchor point errors and the cable's behavior.

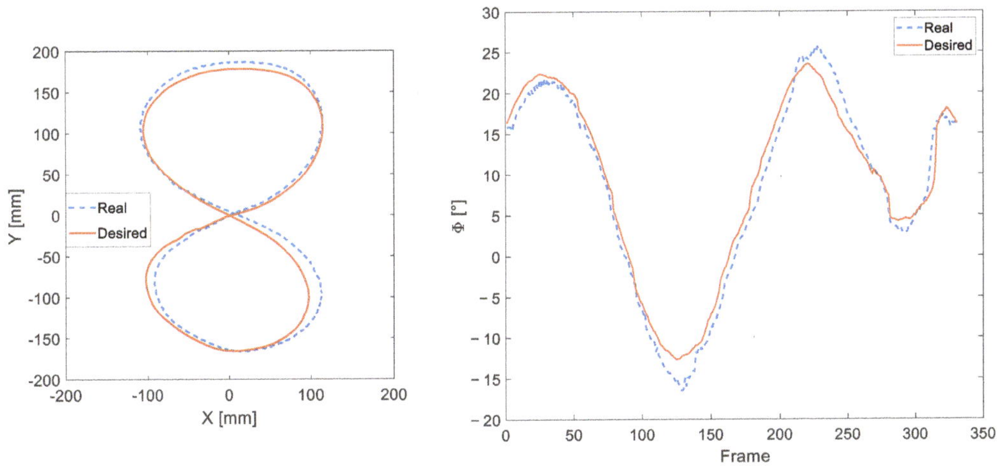

(**a**) (**b**)

Figure 13. Variations between the desired and the real (**a**) trajectories and (**b**) orientations.

In order to assess the robot prototype performance, the mean absolute, the root-mean-square and the standard deviation position and orientation errors were computed as given in Table 4.

Table 4. Measured errors between the desired and the performed end-effector poses.

Parameter	Equation	Value	
Mean absolute error	$\frac{1}{n}\sum_{i=1}^{n} \lvert \mathbf{X_t}(i) - \mathbf{X_m}(i) \rvert$	Position [mm] Orientation [°]	10.63 1.64
Root-mean-square error	$\sqrt{\frac{\sum_{i=1}^{n}(\mathbf{X_t}(i) - X_m(i))^2}{n}}$	Position [mm] Orientation [°]	13 1.95
Error standard deviation	$\sqrt{\frac{\sum_{i=1}^{n}(\epsilon(i) - \overline{\epsilon})^2}{n-1}}$	Position [mm] Orientation [°]	12.57 1.06

Where $\mathbf{X_t}(i)$ and $\mathbf{X_m}(i)$ are the true and the measured values corresponding to the ith position, $\epsilon(i) = \mathbf{X_t}(i) - \mathbf{X_m}(i)$ is the measured error at the ith position, $\overline{\epsilon}$ is the mean of ϵ and n is the number of points forming the trajectory.

5. Conclusions

This paper aims to find the optimal design of a planar cable-driven parallel robot for upper limb rehabilitation purposes. For that, an optimization problem was formulated in which the dexterity, the elastic stiffness and the tension in the cables were the chosen criteria. Bounding the cable tensions, avoiding collisions with the mobile platform and controlling the potential positions of the pulleys were the selected constraints. A multi-objective formulation of the optimization design problem showed different nondominated solutions for each pair of criteria, which complicated the selection of a single solution. A mono-objective formulation was then adopted. Weighting coefficients were studied, and appropriate values were selected. Finally, the validation of the robot optimal design was carried out using a real prototype to verify its behavior along the given trajectory. Different computed errors between the desired and the performed trajectories showed promising results regarding prototype reliability.

Author Contributions: Conceptualization, methodology, supervision, writing—review and editing, M.A.L., A.C., J.S., A.M., L.R., S.B., and S.Z.; software and writing—original draft preparation, F.E.; project administration, M.A.L., and A.C. All authors have read and agreed to the published version of the manuscript.

Funding: This work was financially supported by the PHC Utique program of the French Ministry of Foreign Affairs and Ministry of higher education, research and innovation and the Tunisian Ministry of higher education and scientific research in the CMCU, project number 19G1121.

Institutional Review Board Statement: Not applicable.

Informed Consent Statement: Informed consent was obtained from all subjects involved in the study.

Data Availability Statement: Data is contained within the article.

Conflicts of Interest: The authors declare no conflict of interest.

References

1. Babaiasl, M.; Mahdioun, S.H.; Jaryani, P.; Yazdani, M. A review of technological and clinical aspects of robot-aided rehabilitation of upper-extremity after stroke. *Disabil. Rehabil. Assist. Technol.* **2016**, *11*, 263–280. [CrossRef] [PubMed]
2. Khan, S.B.; Rahman, M.H.; Haque, M.O.; Rahman, E. Effectiveness of Task Oriented Physiotherapy Along with Conventional Physiotherapy for Patients with Stroke. *IJNPT* **2019**, *5*, 37–41. [CrossRef]
3. Le Guiet, J.L.; Le Claire, G. Pendant combien de temps doit-on pratiquer la rééducation du membre supérieur chez l'hémiplégique? *Ann. Réadaptation Méd. Phys.* **1998**, *41*, 107–113. [CrossRef]
4. Prange, G.B.; Jannink, M.J.A.; Groothuis-Oudshoorn, C.G.M.; Hermens, H.J.; Ijzerman, M.J. Systematic review of the effect of robot-aided therapy on recovery of the hemiparetic arm after stroke. *JRRD* **2006**, *43*, 171–184. [CrossRef]
5. Gassert, R.; Dietz, V. Rehabilitation robots for the treatment of sensorimotor deficits: A neurophysiological perspective. *JNER* **2018**, *15*, 1–15. [CrossRef] [PubMed]
6. Garcia, D.A.; Arredondo, R.; Morris, M.; Tosunoglu, S. A review of rehabilitation strategies for stroke recovery. In Proceedings of the ASME Early Career Technical Conference, Atlanta, Georgia, USA, 2–3 November 2012; Volume 11, pp. 24–31.
7. Oña, E.D.; Cano-de La Cuerda, R.; Sánchez-Herrera, P.; Balaguer, C.; Jardón, A. A review of robotics in neurorehabilitation: Towards an automated process for upper limb. *J. Healthc. Eng.* **2018**, *2018*. [CrossRef]
8. Maciejasz, P.; Eschweiler, J.; Gerlach-Hahn, K.; Jansen-Troy, A.; Leonhardt, S. A survey on robotic devices for upper limb rehabilitation. *JNER* **2014**, *11*, 1–29. [CrossRef]
9. Qassim, H.M.; Wan Hasan, W.Z. A Review on Upper Limb Rehabilitation Robots. *Appl. Sci* **2020**, *10*, 6976. [CrossRef]
10. Nef, T.; Riener, R. ARMin-design of a novel arm rehabilitation robot. In *9th ICORR*; IEEE: Chicago, IL, USA, 2005; pp. 57–60. [CrossRef]
11. Reinkensmeyer, D.J.; Kahn, L.E.; Averbuch, M.; McKenna-Cole, A.; Schmit, B.D.; Rymer, W.Z. Understanding and treating arm movement impairment after chronic brain injury: Progress with the ARM guide. *JRRD* **2000**, *37*, 653–662.
12. Lum, P.S.; Burgar, C.G.; Van der Loos, M.; Shor, P.C.; Majmundar, M.; Yap, R. MIME robotic device for upper-limb neurorehabilitation in subacute stroke subjects: A follow-up study. *J. Rehabil. Res. Dev.* **2006**, *43*, 631. [CrossRef]
13. Xiong, H.; Diao, X. A review of cable-driven rehabilitation devices. *Disabil. Rehabil. Assist. Technol.* **2020**, *15*, 885–897. [CrossRef]
14. Qian, S.; Zi, B.; Shang, W.W.; Xu, Q.S. A Review on Cable-driven Parallel Robots. *Chin. J. Mech. Eng.* **2018**, *31*, 66. [CrossRef]
15. Scalera, L.; Gasparetto, A.; Zanotto, D. Design and experimental validation of a 3-dof underactuated pendulum-like robot. *IEEE ASME Trans. Mechatron.* **2019**, *25*, 217–228. [CrossRef]
16. Ji, H.; Shang, W.; Cong, S. Adaptive Control of a Spatial 3-Degree-of-Freedom Cable-Driven Parallel Robot with Kinematic and Dynamic Uncertainties. In Proceedings of the 2020 5th International Conference on Advanced Robotics and Mechatronics (ICARM 2020), Shenzhen, China, 14 September 2020; pp. 142–147. [CrossRef]
17. Mottola, G.; Gosselin, C.; Carricato, M. Dynamically-feasible elliptical trajectories for fully constrained 3-DOF Cable-suspended parallel robots. In *Cable-Driven Parallel Robots*; Springer: Cham, Switzerland, 2018; pp. 219–230. [CrossRef]
18. Xiang, S.; Gao, H.; Liu, Z.; Gosselin, C. Dynamic transition trajectory planning of three-DOF cable-suspended parallel robots via linear time-varying MPC. *Mech. Mach. Theory* **2020**, *146*, 103715. [CrossRef]
19. Gagliardini, L.; Caro, S.; Gouttefarde, M.; Wenger, P.; Girin, A. Optimal design of cable-driven parallel robots for large industrial structures. In Proceedings of the ICRA 2014, Hong Kong, China, 31 May–7 June 2014; pp. 5744–5749. [CrossRef]
20. Hussein, H.; Santos, J.C.; Gouttefarde, M. Geometric Optimization of a Large Scale CDPR Operating on a Building Façade. In Proceedings of the IROS 2018, Madrid, Spain, 1–5 October 2018; pp. 5117–5124. [CrossRef]
21. Fattah, A.; Agrawal, S.K. On the Design of Cable-Suspended Planar Parallel Robots. *J. Mech. Des.* **2005**, *127*, 1021–1028. [CrossRef]

22. Li, Y.; Xu, Q. GA-Based Multi-Objective Optimal Design of a Planar 3-DOF Cable-Driven Parallel Manipulator. In Proceedings of the IEEE Robio 2006, Kunming, China, 17–20 December 2006; pp. 1360–1365. [CrossRef]
23. Chaparro-Rico, B.D.M.; Cafolla, D.; Castillo-Castaneda, E.; Ceccarelli, M. Design of arm exercises for rehabilitation assistance. *J. Eng. Res.* **2020**, *8*, 204–218. [CrossRef]
24. Kawamura, S.; Ito, K. A new type of master robot for teleoperation using a radial wire drive system. In Proceedings of the IROS 1993, Yokohama, Japan, 26–30 July 1993; pp. 55–60. [CrossRef]
25. Lee, J.H.; Eom, K.S.; Suh, I.I. Design of a new 6-DOF parallel haptic device. In Proceedings of the ICRA 2001, Seoul, Korea, 21–26 May 2001; Volume 1, pp. 886–891. [CrossRef]
26. Ma, O.; Angeles, J. Optimum architecture design of platform manipulators. In Proceedings of the 5th ICARR 'Robots in Unstructured Environments, Atlanta, GA, USA, 2–6 May 1993; Volume 2, pp. 1130–1135. [CrossRef]
27. Khan, S.; Andersson, K.; Wikander, J. Jacobian Matrix Normalization—A Comparison of Different Approaches in the Context of Multi-Objective Optimization of 6-DOF Haptic Devices. *J. Intell. Robot. Syst.* **2015**, *79*, 87–100. [CrossRef]
28. Gunantara, N. A review of multi-objective optimization: Methods and its applications. *Cogent Eng.* **2018**, *5*, 1502242. [CrossRef]
29. Jia, J.; Fischer, G.W.; Dyer, J.S. Attribute weighting methods and decision quality in the presence of response error: A simulation study. *J. Behav. Decis. Mak.* **1998**, *11*, 85–105. [CrossRef]
30. Dawes, R.M.; Corrigan, B. Linear models in decision making. *Psychol. Bull.* **1974**, *81*, 95. [CrossRef]
31. Einhorn, H.J.; McCoach, W. A simple multiattribute utility procedure for evaluation. *Behav. Sci.* **1977**, *22*, 270–282. [CrossRef]
32. Ferdaws, E.; El Golli, H.; Chaker, A.; Laribi, M.A.; Sandoval, J.; Bennour, S.; Mlika, A.; Romdhane, L.; Zeghloul, S. Design Optimization and Dynamic Control of a 3-d.o.f. Planar Cable-Driven Parallel Robot for Upper Limb Rehabilitation. In Proceedings of the New Trends in Medical and Service Robotics MESROB, Virtual Conference, 7–9 June 2021.
33. Eberhart, R.; Kennedy, J. A new optimizer using particle swarm theory. In Proceedings of the Sixth International Symposium on Micro Machine and Human Science MHS'95, Nagoya, Japan, 4–6 October 1995; pp. 39–43. [CrossRef]
34. Sands, T. Optimization provenance of whiplash compensation for flexible space robotics. *Aerospace* **2019**, *6*, 93. [CrossRef]
35. Shorman, S.M.; Pitchay, S.A. Significance of parameters in genetic algorithm, the strengths, its limitations and challenges in image recovery. *ARPN J. Eng. Appl. Sci.* **2006**, *10*, 585–593.

applied sciences

MDPI

Article

A 4-DOF Upper Limb Exoskeleton for Physical Assistance: Design, Modeling, Control and Performance Evaluation

Muhammad Ahsan Gull [1,*], Mikkel Thoegersen [2], Stefan Hein Bengtson [3], Mostafa Mohammadi [2], Lotte N. S. Andreasen Struijk [2], Thomas B. Moeslund [3], Thomas Bak [4] and Shaoping Bai [1,*]

1 Department of Materials and Production, Aalborg University, 9220 Aalborg, Denmark
2 Rehabilitation Engineering and Robotics, Center for Rehabilitation Robotics, Department of Health Science and Technology, Aalborg University, 9220 Aalborg, Denmark; mt@hst.aau.dk (M.T.); mostafa@hst.aau.dk (M.M.); naja@hst.aau.dk (L.N.S.A.S.)
3 Department of Architecture and Media Technology, Aalborg University, 9000 Aalborg, Denmark; shbe@create.aau.dk (S.H.B.); tbm@create.aau.dk (T.B.M.)
4 Department of Electronics Systems, Aalborg University, 9220 Aalborg, Denmark; tba@es.aau.dk
* Correspondence: mag@mp.aau.dk (M.A.G.); shb@mp.aau.dk (S.B.); Tel.: +45-81931727 (M.A.G.)

Abstract: Wheelchair mounted upper limb exoskeletons offer an alternative way to support disabled individuals in their activities of daily living (ADL). Key challenges in exoskeleton technology include innovative mechanical design and implementation of a control method that can assure a safe and comfortable interaction between the human upper limb and exoskeleton. In this article, we present a mechanical design of a four degrees of freedom (DOF) wheelchair mounted upper limb exoskeleton. The design takes advantage of non-backdrivable mechanism that can hold the output position without energy consumption and provide assistance to the completely paralyzed users. Moreover, a PD-based trajectory tracking control is implemented to enhance the performance of human exoskeleton system for two different tasks. Preliminary results are provided to show the effectiveness and reliability of using the proposed design for physically disabled people.

Keywords: wheelchair upper limb exoskeleton robot; ADL assistance; PD control; dynamic modeling of an upper limb exoskeleton; trajectory tracking; wearable exoskeleton

Citation: Gull, M.A.; Thougersen, M.; Bengtson, S.H.; Mohammadi, M.; Andreasen Struijk, L.N.S.; Moeslund, T.B.; Bak, T.; Bai, S. A 4-DOF Upper Limb Exoskeleton for Physical Assistance: Design, Modeling, Control and Performance Evaluation. *Appl. Sci.* **2021**, *11*, 5865. https://doi.org/10.3390/app11135865

Received: 19 May 2021
Accepted: 21 June 2021
Published: 24 June 2021

1. Introduction

Cervical spinal cord injury (SCI) may result in incomplete or complete tetraplegia and lead to paralysis of all four extremities. Upper limb onset is one of the most profound impairments that significantly degrades the life of individuals with tetraplegia by compromising independence and social interactions. Moreover, it imposes a substantial financial burden on society in the long run. While advanced medical and surgical techniques, such as stem cell therapy, nerve transfer surgery, etc., have been used to restore the upper limb functionality, in some severe cases, it is hard to achieve desired results. Emerging technologies, such as assistive robots, can provide an alternative way to facilitate individuals with physical impairments in activities of daily living (ADL) [1,2] or therapeutic exercises [3,4].

During the past few decades, upper limb exoskeletons used for power amplification and rehabilitation have attracted intensive attention from the health care and engineering sectors [5]. However, given the utility and growing demand of exoskeletons for physical assistance, the technology still faces challenges in mechanical design, controls, and human–robot interaction. Of them, the mechanical design of a shoulder exoskeleton, including kinematic and kinetic analysis, is a major issue in developing an ergonomic system [6]. Christensen et al. [7] proposed a new three degrees of freedom (DOF) spherical mechanism to comply with the human glenohumeral joint movements. The proposed mechanism takes advantage of the double parallelogram (DPL) mechanism, which connects two revolute joints to achieve a spherical workspace and maintains a remote center of motion (RCM).

The results from the biomechanical analysis of the DPL mechanism presented in [7] have shown its significance for the exoskeleton applications [1]. Similarly, Castro et al. [8] presented a novel 3-DOF curved scissor mechanism that connects two revolute joints. The proposed mechanism complies with the human shoulder movements by maintaining the instantaneous center of rotation. Since the above mechanisms can support complex shoulder movements and provide a singularity-free workspace, the passive internal rotation has made it difficult to use for individual with tetraplegia. Alternatively, several other designs, including fully active or hybrid mechanisms to comply with the shoulder anatomical movements, were proposed [9–12]. These exoskeletons support the full range of shoulder girdle movement by preserving the remote center of rotation, but their effects on supporting the physically impaired people in common ADL have not yet been evaluated [13]. Moreover, flexible and parallel mechanisms have also been investigated to reduce inertial problem, but their size and complexity remain issues to be further addressed. Apart from the shoulder exoskeletons, exoskeletons that can support human forearm [4,14–17] and wrist movements [18–20] were developed. Among the existing mechanisms, a direct drive method and a C-ring mechanism are commonly used to support human forearm extension/flexion movements and wrist rotation, as reported in [2,5,20,21].

The feasibility of using an upper limb exoskeleton cannot only be proved by its design. Selection of a control method for improved physical human–robot interaction (pHRI) is essential for successful implementation and user acceptance. Regarding the trajectory tracking problem, proportional-derivative (PD) and proportional-integral-derivative (PID) controllers have been widely investigated for the different types of exoskeletons. Ease of implementation without having prior knowledge of robot dynamics and an ability to independently tune the control parameters have made the PD/PID control method among the most widely used control schemes [22]. However, in the PID controller, an integrator usually reduces the bandwidth of a closed loop system and removes the steady state error caused by extensive disturbance and uncertainties. Alternatively, a high value of the integrator gain may compromise the transient performance and destroys the system's stability. Therefore, many robotic manipulators, including exoskeletons, use purely PD control or PD control with relatively small integral gain [1,22–24]. It is known that a PD controller can guarantee a semi global asymptotic stability after appropriately tuning the gains [23,24].

Several studies have been conducted to modify the linear PID controller that can guarantee an asymptotic stability. For example, PD control with sliding mode compensation [25], PD-based fuzzy sliding mode control [1], PD control with neural compensation [22] and so on. It is well understood that the PD controller can guarantee the stability for the robotic manipulators, but the asymptotic stability cannot be achieved if the robot dynamic contains gravitational torque. The exoskeleton presented in this study is designed to safely support the user in their ADL, especially the C-ring mechanism designed for shoulder and wrist rotation and the worm gear used to drive the elbow joint exoskeleton to hold the output position without energy consumption because of its non-backdrivability [26]. Moreover, hard constraints in the joint mechanisms may not allow the users to move beyond the safety limits.

In this paper, we present a PD control in the joint space to control the four degrees of freedom (DOF) upper limb exoskeleton robot [2] and investigate its effect as an assistive device to support individuals with physical impairments of the upper limbs in a set of ADLs. The contribution of the article can be summarized as follows.

1. The proposed design can support the human upper limb musculoskeletal structure in basic ADL by providing a kinematically safe and singularity-free workspace. The deign along with the PD control is able to provide a satisfactory tracking performance. It is hypothesized that the trajectory tracking for C-ring mechanism and worm gear mechanism is less prone to the variation in payload, weight of human arm, and exoskeleton due to its ability to hold the output position without energy consumption.

2. The integration of the upper limb exoskeleton with the CarbonHand glove (BioServo Technologies AB, Kista, Sweden) offers a new paradigm that not only supports the user in manipulation but facilitates them also in hand opening and closing. The experimental evaluation has shown that the proposed design with the PD control scheme is appropriate in performing several ADLs, such as eating/drinking.

The paper is organized as follows. The mechanical design of a wheelchair exoskeleton is presented in Section 2 together with the kinematic modeling required to fulfil the task requirements in operational space. The dynamic model of the upper limb exoskeleton along with the PD control scheme is presented in Section 3. Moreover, the PD controller implementation along with the experimental results on the wheelchair exoskeletons is illustrated in Section 4. Subsequently, a discussion on the exoskeleton performance and its potential future directions are presented in Section 5. The work is finally concluded in Section 6.

2. Upper Limb Exoskeleton Robot

2.1. Mechanical Design

This section presents a design of an adaptive 4-DOF upper limb wheelchair mounted exoskeleton that can actively support the wearer in performing their activities of daily living, such as eating and drinking. The exoskeleton was designed after carefully analyzing the human upper limb biomechanics. To reduce the complexity of the human biomechanics, several studies have modeled the human arm as 7 degrees of freedom kinematics system by enforcing the simplifications to the upper limb joints and segments [27]. However, we have noticed that the 4-DOF exoskeleton is sufficient for the most common ADLs and keeps the workspace of the human upper extremity intact.

The exoskeleton in Figure 1b is designed as an open-chain structure to replicate the anatomy of human right upper limb and provides a controllable assistive torque to each joint. To describe the design and complete functioning of a robotic exoskeleton, we have separated the design into three sub-modules, i.e., shoulder joint mechanism, elbow module and a wrist module.

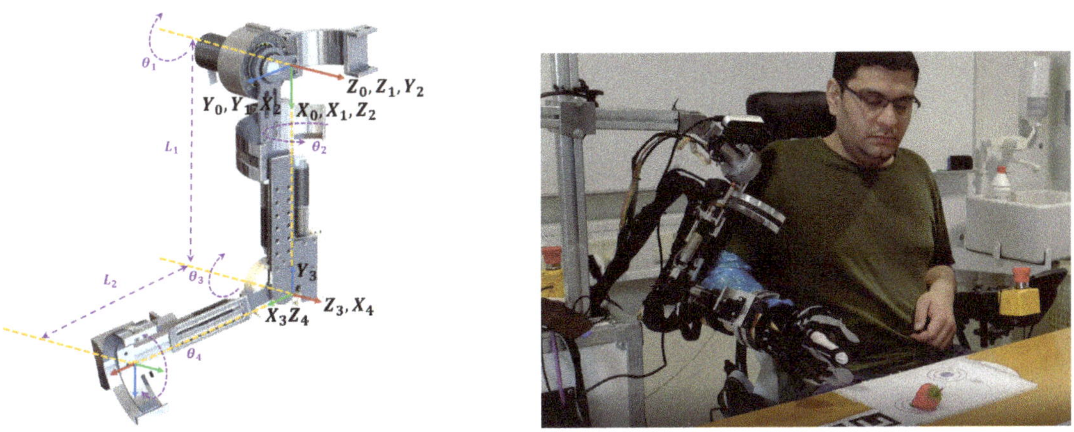

(a) (b)

Figure 1. Overview of a 4-DOF wheelchair exoskeleton. (**a**) Mechanical model of 4-DOF upper limb exoskeleton. (**b**) A prototype of a wheel chair exoskeleton with carbon hand developed from SEM glove (Supplementary Materials).

The human shoulder (Glenohumeral) joint is modeled as a 3-DOF spherical joint that describes the orientation of the human upper arm. These three successive rotations are abduction/adduction, extension/flexion, and internal/external rotation. Hence, an open chain serial mechanism with three revolute joints whose axes of rotation intersect at a common point is kinematically equivalent to a spherical joint. Based on this observation, we have designed a shoulder mechanism that can actively support the 2-DOF glenohumeral joint movements such as shoulder extension/flexion movement and shoulder internal/external rotation, as shown in Figure 1a. The shoulder abduction/adduction movement is passively adjustable. Locking the upper arm abduction movement will prevent the user from moving beyond the wheelchair workspace, causing uncomfortable interaction with an external environment. The complete design of the shoulder mechanism, shown in Figure 1a, is able to preserve the dynamic center of rotation throughout its workspace. The exoskeleton's extension/flexion is achieved by a direct drive brushless DC motor (EC-i40) and a CSD-17-80-2A-R harmonic drive to amplify the motor torque. A dovetail C-ring mechanism is used to actively support the human upper arm internal/external rotation. Furthermore, a 4 pole EC Maxon motor and a speed reducer drive the C-ring mechanism through a spur gear set.

The elbow joint module consists of a normal revolute joint. A Maxon EC-4 pole motor with a speed reducer located near the elbow joint controls the forearm extension/flexion through a worm gear set. The length of the exoskeleton's upper link is adjustable to adapt the user with different anthropomorphic parameters. Moreover, an upper arm support prevents the offset between the exoskeleton and human anatomical joints, i.e., shoulder and elbow joint, causing an uncomfortable interaction between the two systems. Finally, the wrist module consists of a C-ring mechanism that is designed to support the human wrist rotation (radial/ulnar deviation). A 4 pole EC Maxon motor and a speed reducer located along the forearm likewise actuate the C-ring of the wrist joint.

2.2. Kinematics

The kinematic model of the exoskeleton robot is developed by using Denavit–Hartenberg (DH) parameters defined in Table 1, where L_1 and L_2 represent the lengths of the upper arm and forearm links, respectively. Based on the DH parameters, the transformation matrix is given by

$$\mathrm{T}_{i-1,i} = \begin{bmatrix} c\theta_i & -s\theta_i c\alpha & s\theta_i s\alpha_i & a_i c\theta_i \\ s\theta_i & c\theta_i c\alpha_i & -c\theta_i s\alpha_i & a_i s\theta_i \\ 0 & s\alpha_i & c\alpha_i & d_i \\ 0 & 0 & 0 & 1 \end{bmatrix} \tag{1}$$

where s and c represent the sine and the cosine functions, respectively.

The forward kinematics is obtained by computing the overall matrix of transformation from the base frame to the wrist

Table 1. Denavit–Hartenberg (DH) parameters.

Joints	α_i	a_i	d_i	θ_i
1	$\pi/2$	0	0	$\pi/2-\theta_1$
2	$\pi/2$	0	L_1	$\pi+\theta_2$
3	$-\pi/2$	0	0	θ_3
4	0	0	L_2	θ_4

$$\mathrm{T}_{0,4} = \begin{bmatrix} m_{11} & m_{12} & m_{13} & n_{14} \\ m_{21} & m_{22} & m_{23} & n_{24} \\ m_{31} & m_{32} & m_{33} & n_{34} \\ 0 & 0 & 0 & 1 \end{bmatrix} \tag{2}$$

where all entries are given in the Appendix A.

The inverse kinematics is derived from the transformation matrix (2). The joint angles can be obtained as:

$$\begin{aligned} \theta_2 &= \pi + \arctan 2(n_{34}, n_{14}) \\ \theta_3 &= \pm \arcsin\left(\frac{n_{34}}{L_2 \sin\theta_2}\right) \\ \theta_1 &= \frac{\pi}{2} + \frac{1}{2}\arcsin \frac{n_{14}-n_{24}}{L_1+L_2\cos\theta_3+L_2\cos\theta_2\sin\theta_3} \\ \theta_4 &= \arccos\left(\frac{m_{32}\cos\theta_2+m_{31}\cos\theta_3\sin\theta_2}{\cos^2\theta_2-\cos\theta_3\sin\theta_2}\right) \end{aligned} \quad (3)$$

2.3. Workspace and Singularity Analysis

The two most important properties that influence the geometrical design of a robotic exoskeleton are workspace and singularity analysis [28]. The kinematic model is used to analyze the workspace of the human upper limb and exoskeleton robot. Given the position of any point in the workspace, it is important to determine whether it belongs to the actual workspace or not, and helps to verify if at least one solution for the joint angles exists [2]. Therefore, a direct search method is employed to essentially evaluate the existence of an inverse kinematics solution for the human and robotic exoskeleton, shown in Figure 2b,c. The kinematic properties selected for this study are given in Appendix C.

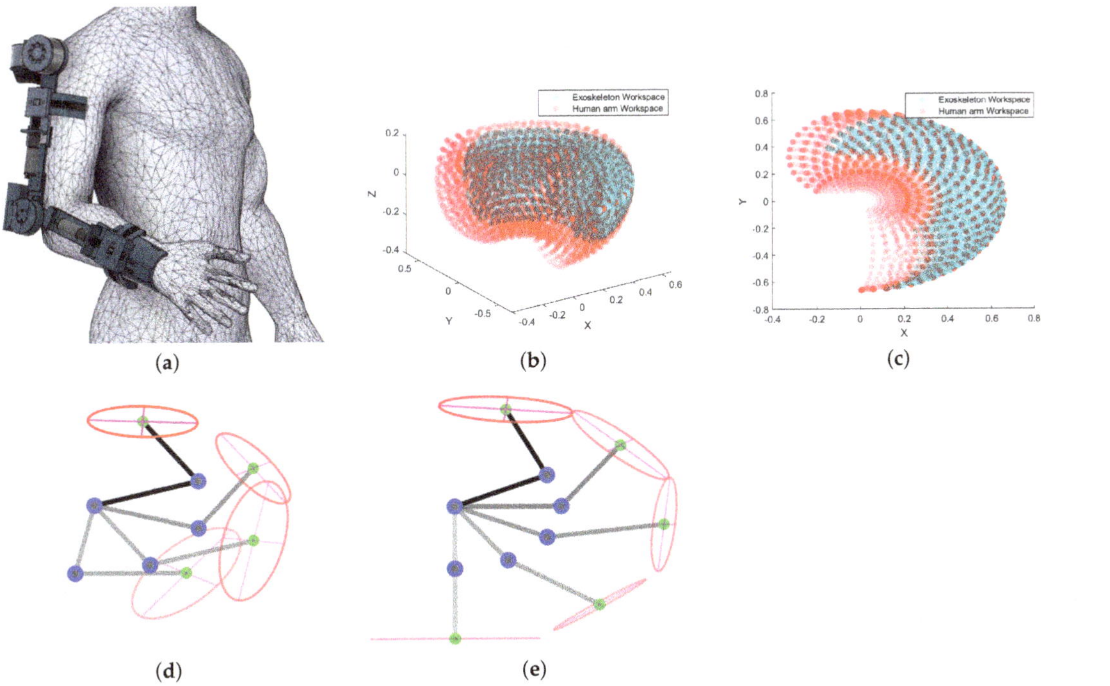

Figure 2. Workspace analysis of the 4-DOF upper limb exoskeleton (cyan) within the human arm workspace (red) (measured in meters): (**a**) human–exoskeleton system, (**b**) isotropic view of upper limb exoskeleton and human arm workspace, (**c**) sagittal plane view of upper limb exoskeleton and human arm workspace, (**d**) different configurations of human–exoskeleton system in high manipulability region, (**e**) configurations of human–exoskeleton system in low manipulability region.

Apart from analyzing the reachable workspace, implementation of safe and stable operation is also required due to kinematic singularities within the workspace. Hence, it is necessary to identify all singular configurations while planning trajectories for the robotic exoskeleton. The manipulability ellipsoid and determinant of the Jacobian matrix are the two important indices that characterize the degree of singularity [29]. Our study determines the kinematic performance of the exoskeleton system by analyzing the manipulability index, which gives us information about the low and high manipulability regions, shown in Figure 2d,e.

In the manipulability analysis, we look at the position of the wrist only. Thus, we take Jacobian in the form of

$$\mathbf{J} = \frac{\partial \boldsymbol{n}}{\partial \boldsymbol{\theta}_i} = \begin{bmatrix} -n_{24} + L_1 c\theta_1 & -n_{34} c\theta_1 & -L_2(s\theta_1 s\theta_3 + c\theta_1 c\theta_2 c\theta_3) \\ n_{14} + L_1 s\theta_1 & -n_{34} s\theta_1 & L_2(c\theta_1 s\theta_3 - s\theta_1 c\theta_2 c\theta_3) \\ 0 & -L_2 c\theta_2 s\theta_3 & -L_2 s\theta_2 c\theta_3 \end{bmatrix} \tag{4}$$

where $\boldsymbol{n} = [n_{14}\ n_{24}\ n_{34}]^T$, $\boldsymbol{\theta}_i = [\theta_1\ \theta_2\ \theta_3]^T$. The manipulability index can be determined after computing the Jacobian as follows:

$$\mu(\mathbf{J}) = \sqrt{|\mathbf{J}\mathbf{J}^T|} \tag{5}$$

where μ is the manipulability index. Figure 2d,e display different configurations of the human upper limb and exoskeleton system and their corresponding manipulability ellipses in high and low region of manipulability. Moreover, the manipulability analysis gives us information about the uniform distribution of the forces and torques applied by the exoskeleton system to the human upper limb [30]. Another important aspect of analyzing the manipulability ellipse is to identify the singular configuration of the exoskeleton system in the workspace. If the determinant of the Jacobian matrix is zero, the robot encounters singularity or exhibit zero manipulability. Hence, this analysis can be used for the robot path planning where it will try to avoid the low region of manipulability.

3. Exoskeleton Control System

The control architecture for the upper limb exoskeleton and carbon hand is shown in Figure 3. The control system is implemented in the robotic operating system (ROS), which includes task planning for activities of daily living, a complete path planning for the robotic exoskeleton, computing the inverse kinematics, trajectory generation, and controller design. Furthermore, input control signals are used to perform an ADL while wearing the robotic exoskeleton and carbon hand.

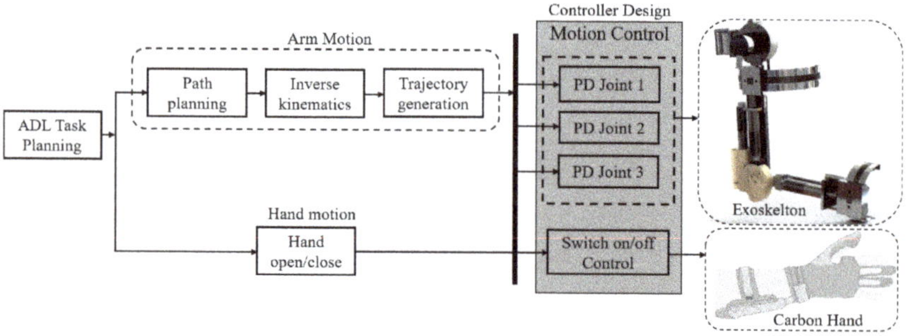

Figure 3. An overall system control architecture. The tasks for the activities of daily living (ADL) is predefined. The PD control method is implemented for the individual joint control. A carbon hand developed by SEM glove is adopted to control the hand opening/closing movement (switch on/off control).

The control system consists of four motors, controllers (Maxon EPOS4 Compact 50/8 CAN) and encoders. Moreover, a graphical interface was developed using a combination of PyQt4, Python and ROS that can be used to choose the various types of control modes, tune control parameters, sending high level control commands and real-time logging of data. A CAN bus communication is adopted as the communication method between ROS and Maxon EPOS4.

PD Control Scheme for Upper Limb Exoskeleton Robot

The dynamic model of an exoskeleton can be derived using the Lagrange formulation and can be expressed by the following equation

$$M(q)\ddot{q} + C(q,\dot{q})\dot{q} + \tau_g = \tau \tag{6}$$

where $q \in \Re$ is a position vector, $M(q)$ is inertia matrix, $C(q,\dot{q})$ represents the Coriolis forces and τ_g is the torque due to gravity. Although we have used the model-free PD/PID control scheme, the dynamic model of the system is still provided in (4) to simulate the dynamic response of the robotic exoskeleton. All entries of the dynamic Equation (4) can be found in Appendix B.

In this article, a PD-based trajectory tracking control problem is investigated, where the joint angle trajectories q are bound to track the desired trajectories q_d (Algorithm 1).

The PD control law can be expressed as:

$$\begin{aligned} u &= K_p\tilde{q} + K_d\dot{\tilde{q}} \\ \dot{\tilde{q}} &= \dot{q}_d - \dot{q} \end{aligned} \tag{7}$$

where $\tilde{q} = q_d - q$. K_p and K_d are the proportional and differential gains, respectively.

We stabilize the open loop robotic system (6) by using the stability property of the PD control scheme (9) and form a stable closed loop system as follows:

$$\begin{aligned} M(q)\ddot{q} + C(q,\dot{q})\dot{q} + \tau_g &= K_p\tilde{q} + K_d\dot{\tilde{q}} \\ \dot{\tilde{q}} &= \dot{q}_d - \dot{q} \end{aligned} \tag{8}$$

The equation can be written in the matrix form as:

$$\begin{bmatrix} \dot{\tilde{q}} \\ \ddot{\tilde{q}} \end{bmatrix} = \begin{bmatrix} \dot{q}_d - \dot{q} \\ \ddot{q}_d + \frac{1}{M}(C\dot{q} + g - K_p\tilde{q} - K_d\dot{\tilde{q}}) \end{bmatrix} \qquad \because \ddot{q} = \ddot{q}_d - \ddot{\tilde{q}} \tag{9}$$

Algorithm 1 PD-based trajectory tracking for each joint

Given:
- Sampling time: T_s
- User define parameters: k_p, k_d
- Desired trajectory: $q_d(k)$

Initialization:
- $\tilde{q}(0) = 0$
- $k \leftarrow 0$

Repeat:
- $\tilde{q}(k) = q_d(k) - q(k)$, $\dot{\tilde{q}}(k) = \frac{\tilde{q}(k)-\tilde{q}(k-1)}{T_s}$
- $Output = k_p\tilde{q}(k) + k_d\dot{\tilde{q}}(k)$
- $\tilde{q}(k-1) \leftarrow \tilde{q}(k)$
- $k \leftarrow k+1$

4. Control Implementation in the Upper Limb Exoskeleton and Experimental Evaluation

The challenge of the human–exoskeleton system lies in its complicated interaction in which the robotic motion is coupled with the human upper limb musculoskeletal system. Thus, we have selected joint angle trajectories to evaluate the system's performance, which helps us to analyze the influence of the kinematic/kinematic properties of the human–exoskeleton system for different manipulation activities. In this section, a model-free PD-based trajectory tracking is implemented to demonstrate the performance of the wheelchair exoskeleton. The architecture of the control scheme is presented in Figure 3.

In our study, we selected two tasks to evaluate the effectiveness of using a wheelchair exoskeleton, shown in Figures 4 and 5. Several positions in the task space were preliminarily defined via human demonstration, and the trajectories were generated in the joint space corresponding to each task. Sixteen trials were recorded from the two subjects for each task, where they were instructed to sit in the wheelchair by wearing the exoskeleton and forced to follow the desired joint angle trajectories, shown in Figure 6. The joint angle trajectories were recorded, and the whole system was evaluated upon the tracking performance of all joints represented by the root mean square (RMSE) value from the 16 trials shown in Figure 7. The detailed statistics representing the human–exoskeleton system's performance are listed in Table 2. For the normal drinking task, it was noted that the human–exoskeleton system was able to satisfactorily track the reference trajectories and shown average RMSE values of 0.0247 rad, 0.0210 rad, and 0.0207 rad for the three joints, respectively. Moreover, the variation in the RMSE values among the 16 trials was also in the acceptable range, i.e., 0.0184 rad, 0.0027 rad, and 0.0071 rad for all three joints, which shows that the human–exoskeleton was able to perform the task during different trials satisfactorily.

(a) (b) (c) (d) (e)

Figure 4. Demonstration of normal drinking task: (**a**) initial position $t = 0$ s, (**b**) exoskeleton moves to the grasping position $t = 11$ s, (**c**) the drinking position $t = 31$ s, (**d**) drop the bottle to the table $t = 51$ s, (**e**) get back to the initial position $t = 72$ s.

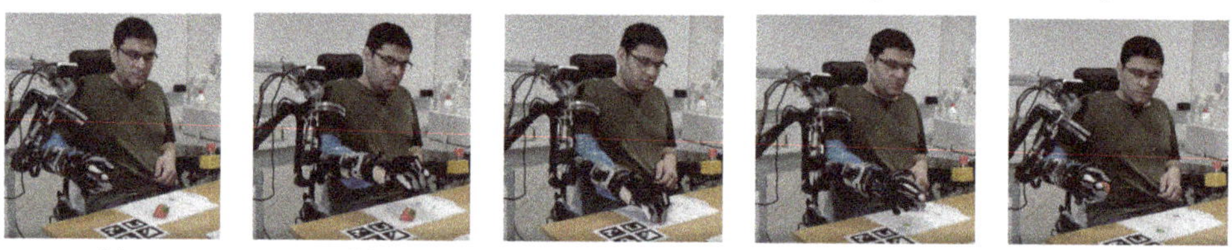

(a) (b) (c) (d) (e)

Figure 5. Demonstration of an object picking-up task: (**a**) initial position $t = 0$ s, (**b**) moving over the target $t = 8$ s, (**c**) exoskeleton moves to a grasping position and hold the object $t = 16$ s, (**d**) pick the object up $t = 24$ s, (**e**) get back to the initial position $t = 42$ s.

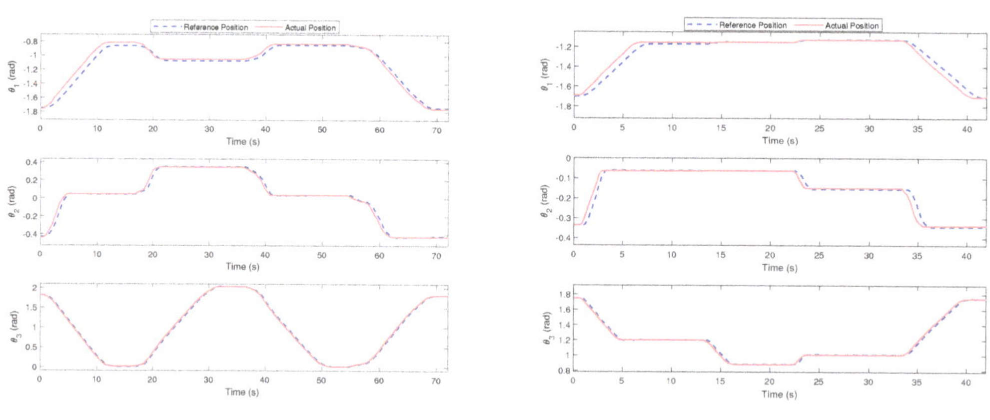

(a) (b)

Figure 6. Exoskeleton's trajectory tracking control performance assessment for two different ADL. (**a**) PD-based trajectory tracking controls for drinking task. (**b**) PD-based trajectory tracking control for an object picking-up task.

Table 2. Statistical analysis of PD control method using RMSE value for the performance assessment of the wheelchair exoskeleton.

Joints	Average RMSE	Max RMSE	Min RMSE	Variance of RMSE	Average RMSE	Max RMSE	Min RMSE	Variance RMSE
	Drinking Task				Object Picking Task			
Joint 1	0.0247	0.0382	0.0198	0.0184	0.0360	0.0402	0.0323	0.0079
Joint 2	0.0210	0.0223	0.0196	0.0027	0.0146	0.0155	0.0131	0.0024
Joint 3	0.0207	0.0238	0.0167	0.0071	0.0184	0.0213	0.0150	0.0062

The robotic system and the control algorithms can be designed to fulfil the requirement for a particular task, but sometimes it is hard to achieve generality. Thus, to maximize the functional reliability of the presented system, we selected a second task to evaluate the system's performance, shown in Figure 5. It is noted that the tracking performance of the shoulder joint was reduced, while the tracking accuracy of Joint 2 and Joint 3 was increased compared to the normal drinking task, shown in Figure 7. In general, the variation in the RMSE values among the 16 trials was in an acceptable range, i.e., 0.0079 rad, 0.0024 rad, and 0.0062 rad for all three joints.

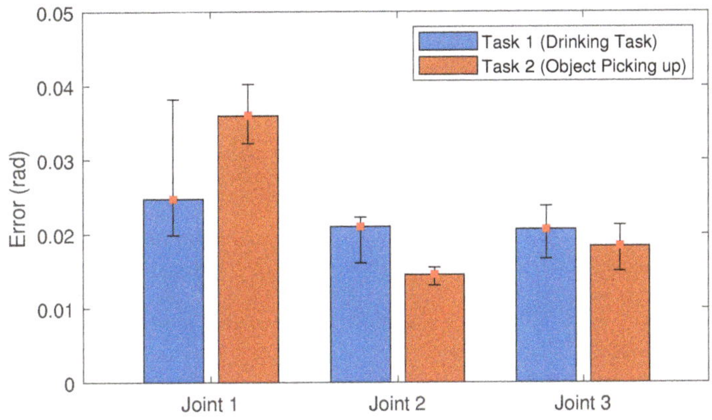

Figure 7. Bar diagram of the RMSE with variance from 16 trials for each task.

5. Discussion

Table 2 summarizes the results of two experiments and presents a statistical analysis of the joint trajectories to demonstrate the effectiveness of using an exoskeleton system for basic ADLs. The data illustrated in Figure 7 show the variations in mean RMSE values among two tasks in joint space. Several parameters, such as mechanism design, mode of actuation, selection of control method to accommodate variations in payload, and human anthropomorphic parameters, may influence the functional reliability of the human–exoskeleton system.

Implementation of a basic PD control method without gravity compensation/human arm weight compensation and a backdrivability of shoulder mechanism had made it difficult to achieve a more precise control compared to the other joint mechanisms. Therefore, average RMSE values for the shoulder joint were comparatively higher than the other two joints. Teng et al. [1] implemented a PD control with gravity compensation and analyzed the effect of uncertain dynamics and external disturbances on the human–exoskeleton system. Data presented in [1,31,32] have shown that the performance of an exoskeleton driving human shoulder joint was comparatively lower than the elbow joint exoskeleton because the gravity torque induced by variable payload and human arm weight may affect the relative precision. Alternatively, the C-ring and worm gear mechanisms responsible for supporting a human shoulder joint rotation and elbow joint movements take advantage of a large reduction ratio. The design also facilitates holding the output position without energy consumption because of its non-backdrivability. Moreover, the two joints are relatively less affected by the variation in the payload and upper limb anatomy as can be seen from error bar diagram presented in Figure 7. Deyby et al. [12] presented a similar mechanism and analyzed the position and orientation synchronization between the human upper limb and exoskeleton.

The study demonstrates that the exoskeleton presented is applicable for motion assistance of physically impaired people in their ADLs. In our future work, we will extend this study to clinically evaluate the system to examine comfort, patient acceptance, and functional use of the system with severe to moderate upper limb impairment. We will look into more advanced control methods to compensate for ill effects caused by uncertain dynamics and external disturbances and study their implications for assistive applications. Manipulability/singularity free workspace is another important factor that should be considered during the path planning of an exoskeleton robot. Future work will focus on developing a method to optimize the exoskeleton's trajectory in the task space and attempt to maximize likelihood of manipulation in the high manipulability region; thereby, it guarantees uniform distribution of forces and the torques and improves the physical human–robot interaction.

6. Conclusions

In this article, we present the mechanical design, control, and performance evaluation of the wheelchair exoskeleton for physical assistance. The design takes advantage of non-backdrivable mechanisms and holds the output position of the exoskeleton without energy consumption. Furthermore, an overall structure of the exoskeleton system offers compatible kinematics and provides a safer ROM that generates a variety of unconstrained motions for active assistance.

The experiments performed evaluated the system's response to shoulder extension/flexion, shoulder internal/external rotation, and elbow extension/flexion for two different ADLs. The statistical analysis of the joint angle trajectories shows that the proposed system and the implementation of PD-based control method are appropriate for performing several essential tasks. Upon the data presented, it is expected that the system will be able to support the tetraplegia users in different ADLs, such as drinking/eating, which helps them in maintaining an independent lifestyle.

Supplementary Materials: The following are available at https://www.mdpi.com/2076-3417/11/13/5865/s1.

Author Contributions: Conceptualization, M.A.G., S.B., M.T. and L.N.S.A.S.; methodology, M.A.G.; software, M.T., S.H.B. and M.M.; validation, M.A.G.; formal analysis, M.A.G.; investigation, M.A.G.; writing—original draft preparation, M.A.G.; writing—review and editing, M.A.G., S.B., M.T., L.N.S.A.S. and T.B.M.; supervision, S.B. and T.B.M., T.B. All authors have read and agreed to the published version of the manuscript.

Funding: This work has been supported by AAU EXOTIC project.

Institutional Review Board Statement: Not applicable.

Informed Consent Statement: The experiments presented in this study were the part of system testing thus did not require consent statement.

Data Availability Statement: Data is contained within the article.

Conflicts of Interest: The authors declare no conflict of interest.

Appendix A

$$m_{11} = c\theta_4(s\theta_1 s\theta_3 + c\theta_1 c\theta_2 c\theta_3) - c\theta_1 s\theta_2 s\theta_4$$

$$m_{12} = -s\theta_4(s\theta_1 s\theta_3 + c\theta_1 c\theta_2 c\theta_3) - c\theta_1 c\theta_4 s\theta_2$$

$$m_{13} = c\theta_3 s\theta_1 - c\theta_1 c\theta_2 s\theta_3$$

$$m_{21} = -c\theta_4(c\theta_1 s\theta_3 - c\theta_2 c\theta_3 s\theta_1) - s\theta_1 s\theta_2 s\theta_4$$

$$m_{22} = s\theta_4(c\theta_1 s\theta_3 - c\theta_2 c\theta_3 s\theta_1) - c\theta_4 s\theta_1 s\theta_2$$

$$m_{23} = -c\theta_1 c\theta_3 - c\theta_2 s\theta_1 s\theta_3$$

$$m_{31} = c\theta_2 s\theta_4 + c\theta_3 c\theta_4 s\theta_2$$

$$m_{32} = c\theta_2 c\theta_4 - c\theta_3 s\theta_2 s\theta_4$$

$$m_{33} = -s\theta_2 s\theta_3$$

$$n_{14} = L_2(c\theta_3 s\theta_1 - c\theta_1 c\theta_2 s\theta_3) + L_1 s\theta_1$$

$$n_{24} = -L_2(c\theta_1 c\theta_3 + c\theta_2 s\theta_1 s\theta_3) - L_1 c\theta_1$$

$$n_{34} = -L_2 s\theta_2 s\theta_3$$

Appendix B

$$M_{11} = I_1 + L_1^2 m_3 + L_{c1}^2 m_1 + L_{c1}^2 m_2 + L_{c2}^2 m_3 c\theta_2{}^2 + L_{c2}^2 m_3 c\theta_3{}^2 + 2L_1 L_{c2} m_3 c\theta_3 - L_{c2}^2 m_3 c\theta_2{}^2 c\theta_3{}^2$$

$$M_{12} = L_{c2} m_3 s\theta_2 s\theta_3 (L_1 + L_{c2} c\theta_3)$$

$$M_{13} = -L_{c2} m_3 c\theta_2 (L_{c2} + L_1 c\theta_3)$$

$$M_{21} = L_{c2} m_3 s\theta_2 s\theta_3 (L_1 + L_{c2} c\theta_3)$$

$$M_{22} = -m_3 L_{c2}^2 c\theta_3{}^2 + m_3 L_{c2}^2 + I_2$$

$$M_{23} = 0$$

$$M_{31} = -L_{c2} m_3 c\theta_2 (L_{c2} + L_1 c\theta_3)$$

$$M_{32} = 0$$

$$M_{33} = m_3 L_{c2}^2 + I_3$$

$$C_1 = L_{c2} m_3 (L_1 \dot\theta_2^2 c\theta_2 s\theta_3 + L_1 \dot\theta_3^2 c\theta_2 s\theta_3 - 2L_1 \dot\theta_1 \dot\theta_3 s\theta_3 - L_{c2} \dot\theta_1 \dot\theta_2 s2\theta_2 - L_{c2} \dot\theta_1 \dot\theta_3 s2\theta_3$$
$$+ L_{c2} \dot\theta_2^2 c\theta_2 c\theta_3 s\theta_3 + 2L_1 \dot\theta_2 \dot\theta_3 c\theta_3 s\theta_2 + 2L_{c2} \dot\theta_2 \dot\theta_3 c\theta_3{}^2 s\theta_2 + 2L_{c2} \dot\theta_1 \dot\theta_2 c\theta_2 c\theta_3{}^2 s\theta_2 + 2L_{c2} \dot\theta_1 \dot\theta_3 c\theta_2{}^2 c\theta_3 s\theta_3)$$
$$C_2 = (L_{c2}^2 m_3 (-2c\theta_2 s\theta_2 \dot\theta_1^2 c\theta_3{}^2 + s2\theta_2 \dot\theta_1^2 + 4\dot\theta_3 s\theta_2 \dot\theta_1 c\theta_3{}^2 - 4\dot\theta_3 sin\theta_2 \dot\theta_1 + 2\dot\theta_2 \dot\theta_3 sin2\theta_3))/2$$
$$C_3 = (L_{c2} m_3 (2L_1 \dot\theta_1^2 s\theta_3 + L_{c2} \dot\theta_1^2 sin2\theta_3 - L_{c2} \dot\theta_2^2 s2\theta_3 + 4L_{c2} \dot\theta_1 \dot\theta_2 s\theta_2 - 2L_{c2} \dot\theta_1^2 c\theta_2{}^2 c\theta_3 s\theta_3 - 4L_{c2} \dot\theta_1 \dot\theta_2 c\theta_3{}^2 s\theta_2))/2$$

$$G_1 = L_1 m_3 s\theta_1 + L_{c1} m_1 s\theta_1 + L_{c1} m_2 s\theta_1 + L_{c2} m_3 c\theta_3 s\theta_1 - L_{c2} m_3 c\theta_1 c\theta_2 s\theta_3$$

$$G_2 = L_{c2} m_3 s\theta_1 s\theta_2 s\theta_3$$

$$G_3 = L_{c2} m_3 (c\theta_1 s\theta_3 - c\theta_2 c\theta_3 s\theta_1)$$

L_{c1} is the distance of the center of mass of the exoskeleton's upper arm from the shoulder joint, and L_{c2} is the distance of the center of mass of the exoskeleton's forearm from the elbow joint. The parametric values for the above dynamic system (upper limb exoskeleton) are assumed to be: m_1 = 2.5 kg, L_1 = 0.33 m, I_1 = 0.20 kg m^2, m_2 = 1.5 kg, L_2 = 0.246 m, I_2 = 0.15 kg m^2.

Appendix C

Table A1. Mechanical properties of the exoskeleton, and the average estimated anthropomorphic parameters for human subjects.

Link	Exoskeleton		Human Subject	
	Length (m)	Weight (kg)	Length (m)	Weight (kg)
Upper arm	0.33	2.5	0.33	1.386
Forearm	0.246	1.5	0.37	0.886

References

1. Teng, L.; Gull, M.A.; Bai, S. PD-Based Fuzzy Sliding Mode Control of a Wheelchair Exoskeleton Robot. *IEEE/ASME Trans. Mechatronics* **2020**, 25, 2546–2555. [CrossRef]
2. Thøgersen, M.; Gull, M.A.; Kobbelgaard, F.V.; Mohammadi, M.; Bengtson, S.H.; Struijk, L.N.A. EXOTIC-A Discreet User-Based 5 DoF Upper-Limb Exoskeleton for Individuals with Tetraplegia. In Proceedings of the 3rd International Conference on Mechatronics, Robotics and Automation (ICMRA), Shanghai, China, 16–18 October 2020; pp. 79–83.
3. Chaparro-Rico, B.D.M.; Cafolla, D.; Castillo-Castaneda, E.; Ceccarelli, M. Design of arm exercises for rehabilitation assistance. *J. Eng. Res.* **2020**, *8, 204–218.* [CrossRef]
4. Zuccon, G.; Bottin, M.; Ceccarelli, M.; Rosati, G. Design and Performance of an Elbow Assisting Mechanism. *Machines* **2020**, *8*, 107674. [CrossRef]
5. Gull, M.A.; Bai, S.; Bak, T. A review on design of upper limb exoskeletons. *Robotics* **2020**, *9*, 16. [CrossRef]
6. Delgado, P.; Alekhya, S.; Majidirad, A.; Hakansson, N.A.; Desai, J.; Yihun, Y. Shoulder Kinematics Assessment towards Exoskeleton Development. *Appl. Sci.* **2020**, *10*, 6336. [CrossRef]
7. Christensen, S.; Bai, S. Kinematic analysis and design of a novel shoulder exoskeleton using a double parallelogram linkage. *J. Mech. Robot.* **2018**, *10*, 041008. [CrossRef]
8. Castro, M.N.; Rasmussen, J.; Andersen, M.S.; Bai, S. A compact 3-DOF shoulder mechanism constructed with scissors linkages for exoskeleton applications. *Mech. Mach. Theory* **2019**, *132*, 264–278. [CrossRef]
9. Kim, B.; Deshpande, A.D. An upper-body rehabilitation exoskeleton Harmony with an anatomical shoulder mechanism: Design, modeling, control, and performance evaluation. *Int. J. Robot. Res.* **2017**, *36*, 414–435. [CrossRef]
10. Hsieh, H.; Chen, D.; Chien, L.; Lan, C. Design of a Parallel Actuated Exoskeleton for Adaptive and Safe Robotic Shoulder Rehabilitation. *IEEE/ASME Trans. Mechatronics* **2017**, *22*, 2034–2045. [CrossRef]
11. Hunt, J.; Lee, H.; Artemiadis, P. A novel shoulder exoskeleton robot using parallel actuation and a passive slip interface. *J. Mech. Robot.* **2017**, *9*, 011002. [CrossRef]
12. Huamanchahua, D.; Vargas-Martinez, A.; Ramirez-Mendoza, R. Kinematic of the Position and Orientation Synchronization of the Posture of an DoF Upper-Limb Exoskeleton with a Virtual Object in an Immersive Virtual Reality Environment. *Electronics* **2021**, *10*, 1069. [CrossRef]
13. Galofaro, E.; D'antonio, E.; Patané, F.; Casadio, M.; Masia, L. Three-Dimensional Assessment of Upper Limb Proprioception via a Wearable Exoskeleton. *Appl. Sci.* **2021**, *11*, 2615. [CrossRef]
14. Zahedi, A.; Wang, Y.; Martinez-Hernandez, U.; Zhang, D. A wearable elbow exoskeleton for tremor suppression equipped with rotational semi-active actuator. *Mech. Syst. Signal Process.* **2021**, *157*, 107674. [CrossRef]
15. Li, Z.; Bai, S. Design and Modelling of a Compact Variable Stiffness Mechanism for Wearable Elbow Exoskeletons, In Proceedings of the 7th International Conference on Control, Mechatronics and Automation (ICCMA), Delft, The Netherlands, 6–8 November 2019; pp. 342–346.
16. Li, Z.; Chen, W.; Zhang, J.; Bai, S. Design and control of a 4-DOF cable-driven arm rehabilitation robot (CARR-4). In Proceedings of the IEEE International Conference on Cybernetics and Intelligent Systems (CIS) and IEEE Conference on Robotics, Automation and Mechatronics (RAM), Ningbo, China, 19–21 November 2017; pp. 581–586.
17. Chaparro-Rico, B.; Cafolla, D.; Ceccarelli, M.; Castillo-Castaneda, E. Design and simulation of an assisting mechanism for arm exercises. In *Advances in Italian Mechanism Science*; Springer: Vicenza, Italy, 2017; pp. 115–123.

18. Butler, N.R.; Goodwin, S.A.; Perry, J.C. Design parameters and torque profile modification of a spring-assisted hand-opening exoskeleton module. In Proceedings of the International Conference on Rehabilitation Robotics (ICORR), London, UK, 17–20 July 2017; pp. 591–596.
19. Lambelet, C.; Lyu, M.; Woolley, D.; Gassert, R.; Wenderoth, N. The eWrist—A wearable wrist exoskeleton with sEMG-based force control for stroke rehabilitation. In Proceedings of the International Conference on Rehabilitation Robotics (ICORR), London, UK, 17–20 July 2017; pp. 726–733.
20. Cui, X.; Chen, W.; Jin, X.; Agrawal, S.K. Design of a 7-DOF cable-driven arm exoskeleton (CAREX-7) and a controller for dexterous motion training or assistance. *IEEE/ASME Trans. Mechatronics* **2016**, *22*, 161–172. [CrossRef]
21. Gunasekara, M.; Gopura, R.; Jayawardena, S. 6-REXOS: Upper limb exoskeleton robot with improved pHRI. *Int. J. Adv. Robot. Syst.* **2015**, *12*, 47. [CrossRef]
22. Yu, W.; Rosen, J. Neural PID Control of Robot Manipulators With Application to an Upper Limb Exoskeleton. *IEEE Trans. Cybern.* **2013**, *43*, 673–684.
23. Yu, W.; Rosen, J. A novel linear PID controller for an upper limb exoskeleton. In Proceedings of the 49th IEEE Conference on Decision and Control (CDC), Atlanta, GA, USA, 15–17 December 2010; pp. 3548–3553.
24. Pan, Y.; Li, X.; Yu, H. Efficient PID Tracking Control of Robotic Manipulators Driven by Compliant Actuators. *IEEE Trans. Control. Syst. Technol.* **2019**, *27*, 915–922. [CrossRef]
25. Yu, W. *PID Control with Intelligent Compensation for Exoskeleton Robots*; Academic Press: Cambridge, MA, USA, 2018; pp. 13–33.
26. Paschold, C.; Sedlmair, M.; Lohner, T.; Stahl, K. Efficiency and heat balance calculation of worm gears. *Forschung Ingenieurwesen* **2020**, *84*, 115–125. [CrossRef]
27. Jung, Y.; Bae, J. Kinematic analysis of a 5-DOF upper-limb exoskeleton with a tilted and vertically translating shoulder joint. *IEEE/ASME Trans. Mechatronics* **2014**, *20*, 1428–1439. [CrossRef]
28. Carmichael, M.G.; Liu, D.; Waldron, K.J. A framework for singularity-robust manipulator control during physical human-robot interaction. *Int. J. Robot. Res.* **2017**, *36*, 861–876. [CrossRef]
29. Petrič, T.; Peternel, L.; Morimoto, J.; Babič, J. Assistive arm-exoskeleton control based on human muscular manipulability. *Front. Neurorobotics* **2019**, *13*, 30. [CrossRef] [PubMed]
30. Jin, X.; Fang, Y.; Zhang, D.; Gong, J. Design of dexterous hands based on parallel finger structures. *Mech. Mach. Theory* **2020**, *152*, 103952. [CrossRef]
31. Da Silva, L.D.; Pereira, T.F.; Leithardt, V.R.; Seman, L.O.; Zeferino, C.A. Hybrid Impedance-Admittance Control for Upper Limb Exoskeleton using Electromyography. *Appl. Sci.* **2020**, *10*, 7146. [CrossRef]
32. Liu, J.; Ren, Y.; Xu, D.; Kang, S.H.; Zhang, L.Q. EMG-based real-time linear-nonlinear cascade regression decoding of shoulder, elbow, and wrist movements in able-bodied persons and stroke survivors. *IEEE Trans. Biomed. Eng.* **2019**, *67*, 1272–1281. [CrossRef] [PubMed]

applied sciences

MDPI

Article

Virtual and Physical Prototyping of Reconfigurable Parallel Mechanisms with Single Actuation

Alexey Fomin [1,*], Daniil Petelin [1], Anton Antonov [1], Victor Glazunov [1] and Marco Ceccarelli [2,*]

1 Mechanisms Theory and Machines Structure Laboratory, Mechanical Engineering Research Institute of the Russian Academy of Sciences (IMASH RAN), 101000 Moscow, Russia; petelin_daniil@inbox.ru (D.P.); antonov.av@imash.ru (A.A.); vaglznv@mail.ru (V.G.)

2 LARM2: Laboratory of Robot Mechatronics, University of Rome "Tor Vergata", 00133 Rome, Italy

* Correspondence: alexey-nvkz@mail.ru (A.F.); marco.ceccarelli@uniroma2.it (M.C.)

Abstract: The paper presents novel models of reconfigurable parallel mechanisms (RPMs) with a single active degree-of-freedom (1-DOF). The mechanisms contain three to six identical kinematic chains, which provide three (for the tripod) to zero (for the hexapod) uncontrollable DOFs. Screw theory is applied to carry out mobility analysis and proves the existence of controllable and uncontrollable DOFs of these mechanisms. Each kinematic chain in the synthesized mechanisms consists of planar and spatial parts. Such a design provides them with reconfiguration capabilities even when the driving link is fixed. This allows reproduction of diverse output trajectories without using additional actuators. In this paper, the model of a mechanism with six kinematic chains (hexapod) has been virtually and physically prototyped. The designing and assembling algorithms are developed using the detailed computer-aided design (CAD) model, which was further used to carry out kinetostatic analysis considering complex geometry of mechanism elements and friction among all contacting surfaces of joints. The developed virtual prototype and its calculation data have been further applied to fabricate mechanism elements and assemble an actuated full-scale physical prototype for future testing.

Keywords: reconfigurable parallel mechanism (RPM); degree-of-freedom (DOF); circular guide; computer-aided design (CAD) modeling; virtual and physical prototyping; screw theory; kinetostatic analysis; digital twins; 3D printing technology

Citation: Fomin, A.; Petelin, D.; Antonov, A.; Glazunov, V.; Ceccarelli, M. Virtual and Physical Prototyping of Reconfigurable Parallel Mechanisms with Single Actuation. *Appl. Sci.* **2021**, *11*, 7158. https://doi.org/10.3390/app11157158

Academic Editor: Oscar Reinoso García

Received: 27 June 2021
Accepted: 28 July 2021
Published: 3 August 2021

1. Introduction

Currently, designing full-scale models (physical prototyping) is often associated with the development of their digital twins (virtual prototyping), which, in fact, allows for beginning the product development process. Virtual prototyping is the process of designing a mechanical system in the form of a numerical model, while physical prototyping is the process of creating a real full-scale model. A virtual prototype allows for carrying out many numerical experiments using a computer functionality, which significantly reduces the time, physical and financial costs for creating a real model.

The processes of virtual and physical prototyping complement each other and allow for creating more advanced structures, providing maximum opportunities for consideration of the design requirements [1]. Today, advanced technologies allow for creating a real prototype based on a digital model of a mechanical system in quite a short time. Rapid prototyping technologies are widely used in this direction, in particular, 3D printing technologies with application of various materials [2]. In this field, the use of 3D printing technologies is quite popular both for soft [3–5] and solid [6–8] systems.

Let us consider some examples of using 3D printing technologies for parallel mechanisms and robots. Authors of [9] present a 6-DOF parallel mechanism designed as a cm sized six-component force sensor that measures the force information in space, including three-dimensional forces and three-dimensional torques. The mechanism has eight legs, each with grooves at both ends serving as spherical joints. A 5R 2-DOF parallel robotic arm for handling paper pot seedlings in a vegetable transplanter is proposed in [10]. The elements of the arm, including a gear box for its actuation, are 3D printed. Authors of [11] demonstrate a 3-DOF spatial compliant parallel mechanism for high-precision manipulation that was 3D printed from alloy Ti-6Al-4V. Authors of [12] present a spherical 3-RRR manipulator with coaxial input shafts. The manipulator design allows for unlimited rotational capability around the vertical axis. Another 3D printed spherical 3-RRR manipulator is shown in [13]. It applies as a shoulder of a prosthetic device. Authors of [14] consider the optimal design of one more spherical 3-RRR manipulator. The structure of one leg was changed to reduce the singularities within the workspace. Authors of [15] present a reconfigurable parallel mechanism with RRPS chains, where only prismatic joints are actuated and a base-coupled revolute joint in each chain can lock or turn during the mechanism motion. Authors of [16] show a 3-DOF parallel manipulator designed as a tripod. This manipulator can be used as a leg for walking systems.

The presented research is directed at virtual and physical prototyping of the mechanism, which belongs to a class of reconfigurable parallel mechanisms (RPMs). Mechanisms of this type can change their configurations to reproduce various end-effector trajectories or vary workspace dimensions. These mechanisms can also change the number of DOFs when their structure remains unchanged [17,18]. Some RPMs provide change-type of end-effector movements when passing through singular positions [19]. Such mechanisms also have the key advantages of parallel systems: high rigidity, ability to manipulate heavy loads and high positioning accuracy [20–22]. Possible practical applications of RPMs include rehabilitation medicine [23], gripping operations [24], additive technologies [25,26], machine elements processing [27–29] and force sensors [30].

One should note that the RPMs in the studies mentioned above include multiple actuators. It makes their structure and control more complex. In this study, we present a model of a novel RPM with the least possible number of actuators, i.e., with one actuator. In contrast to the known RPMs, the presented one does not have additional kinematic chains or modular elements, which are used only for reconfiguration.

In this regard, the proposed study aims to develop the virtual and physical prototyping algorithms for the RPMs actuated from a single drive. The physical prototyping is based on the 3D printing technologies.

The rest of the paper is organized as follows. Section 2 discusses different mechanism designs of the created RPMs having three to six legs and mobility from four to one. Section 3 follows next and provides structural analysis performed via screw theory. Section 4 shows a virtual prototyping algorithm for the hexapod with a single active DOF. This section also includes CAD calculations of the moment on the driving link (kinetostatic analysis). Based on the previous parts, a physical prototyping algorithm that includes 3D printing technologies finishes Section 4. Section 5 finalizes the study and provides conclusions.

2. Mechanisms Design

This paper proposes a new kinematic chain design that provides reconfigurability. Figure 1a shows this kinematic chain that consists of links 4–13 and is coupled between the fixed (circular guide 1), driving (shaft 2 and wheel 3 forming a single element) and output (platform 14) links. This chain includes the following elements: shaft 4, gear 5 and driving pulley 6, which form a single link (rigid connection); shaft 7, driven pulley 8 and crank 9, which also organize a single link (rigid connection); slide block 10; swinging arm 11 and carriage 12, which also form a single link (rigid connection); and leg 13, which ends with the spherical joints at both sides. Figure 1b demonstrates reconfigurable capabilities of the chain. Here, the same carriage can provide different displacements δ_1, δ_2 and δ_3, which correspond to the crank angles β_1, β_2 and β_3 with the fixed driving wheel.

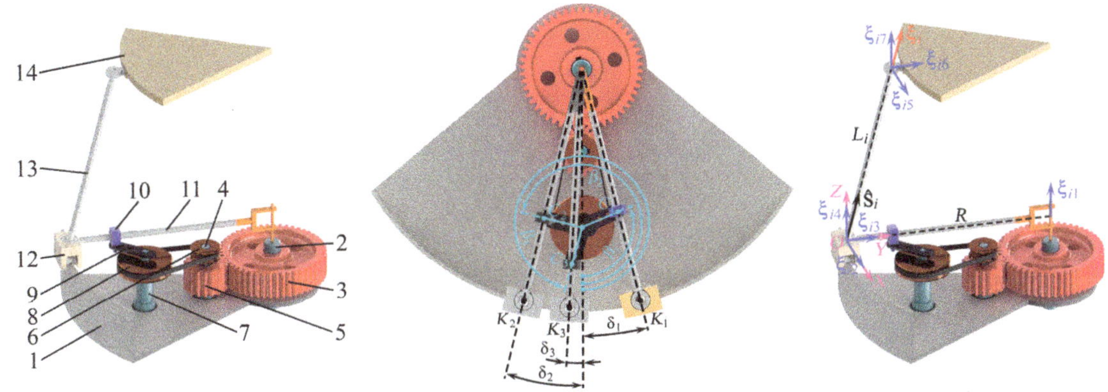

(**a**) (**b**) (**c**)

Figure 1. CAD schemes of the single reconfigurable kinematic chain: (**a**) main components: circular guide 1 (fixed link); shaft 2 and wheel 3 (driving link); shaft 4, gear 5 and driving pulley 6; shaft 7, driven pulley 8 and crank 9; slide block 10; swinging arm 11 and carriage 12; leg 13; and platform 14 (output link); (**b**) reconfigurable positions of the chain with carriage displacements δ_1, δ_2 and δ_3, corresponding to the crank angles β_1, β_2 and β_3 with the fixed driving wheel; (**c**) twists for a single kinematic chain and a reciprocal wrench for the chain with the locked drive.

We can synthesize different mechanisms by connecting the output link to the base with several kinematic chains, shown in Figure 1a. Figure 2a–d present possible mechanisms as CAD models with three to six kinematic chains. In each chain, the input motion passes from driving links 2–3 to a link group of shaft 4, gear 5 and driving pulley 6. Next, the motion passes through the belt to a link group of shaft 7, driven pulley 8 and crank 9, which actuates swinging arm 11 through slide block 10. Swinging arm 11 displaces carriage 12 along circular guide 1. After that, the motion transmits to leg 13 and then to platform 14. Thus, the dependently actuated kinematic chains change position and orientation of platform 14 in space. In the following section, we will perform a mobility analysis of the obtained mechanisms and determine their number of DOFs.

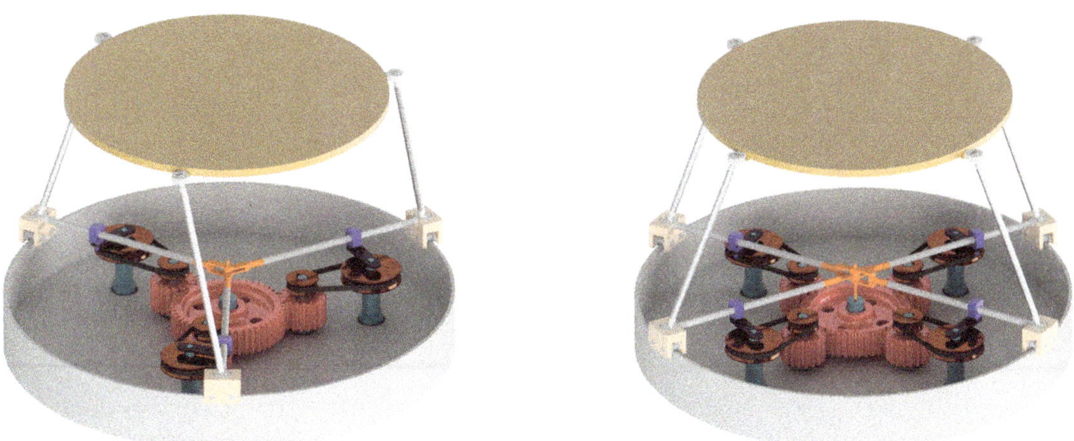

(**a**) (**b**)

Figure 2. *Cont.*

(**c**) (**d**)

Figure 2. CAD models of the synthesized mechanisms with a reconfigurable design: (**a**) tripod; (**b**) quadropod; (**c**) pentapod; (**d**) hexapod.

3. Mobility Analysis

In this section, we will apply instantaneous screw theory to analyze mechanisms mobility. This theory is common for the analysis of parallel mechanisms and highly developed in recent years. Authors of [22,31,32] thoroughly discuss the features of this approach, and we recap the basics in Appendix A. We recommend the reader novel to the subject of instantaneous screws to look through these works or at least the appendix to appreciate the material of this section.

3.1. Analysis for a Single Kinematic Chain

First, we consider one kinematic chain and the unit twists that correspond to it. We will study only the spatial part: the planar part of each chain is a 1-DOF coulisse mechanism, and the rotational angle of the driving wheel uniquely determines the carriage position.

To begin with, we define coordinate system $OXYZ$, of which center O is in the center of the carriage spherical joint, as shown in Figure 1c. Let axis X be tangent to the circular guide, axis Y be directed along the swinging arm to its rotation axis and axis Z be orthogonal to the circular guide plane. Let $\hat{\mathbf{s}}_i = \begin{bmatrix} s_i^x & s_i^y & s_i^z \end{bmatrix}^{\mathrm{T}}$ be a unit vector directed from one spherical joint to the other, where $i = 1 \ldots n_i$ is an index of a kinematic chain, and $n_i = 3 \ldots 6$ is a number of chains depending on the mechanism structure shown in Figure 2a–d. Then, the platform twist system for one kinematic chain can be written as follows in this coordinate system:

$$
\begin{aligned}
\xi_{i1} &= \begin{bmatrix} 0 & 0 & 1 & R & 0 & 0 \end{bmatrix}^{\mathrm{T}}, \\
\xi_{i2} &= \begin{bmatrix} 1 & 0 & 0 & 0 & 0 & 0 \end{bmatrix}^{\mathrm{T}}, \\
\xi_{i3} &= \begin{bmatrix} 0 & 1 & 0 & 0 & 0 & 0 \end{bmatrix}^{\mathrm{T}}, \\
\xi_{i4} &= \begin{bmatrix} 0 & 0 & 1 & 0 & 0 & 0 \end{bmatrix}^{\mathrm{T}}, \\
\xi_{i5} &= \begin{bmatrix} 1 & 0 & 0 & 0 & L_i s_i^z & -L_i s_i^y \end{bmatrix}^{\mathrm{T}}, \\
\xi_{i6} &= \begin{bmatrix} 0 & 1 & 0 & -L_i s_i^z & 0 & L_i s_i^x \end{bmatrix}^{\mathrm{T}}, \\
\xi_{i7} &= \begin{bmatrix} 0 & 0 & 1 & L_i s_i^y & -L_i s_i^x & 0 \end{bmatrix}^{\mathrm{T}},
\end{aligned}
\tag{1}
$$

where ξ_{i1} corresponds to the carriage motion along the circular guide; ξ_{i2}, ξ_{i3} and ξ_{i4} relate to the spherical motion in the carriage joint; ξ_{i5}, ξ_{i6} and ξ_{i7} relate to the spherical motion in the platform joint; R is a radius of the circular guide; and L_i is a distance between the spherical joints (leg length).

We can compose matrix $\mathbf{T}_i$, whose columns are twists (1), and examine its rank to determine if the kinematic chain constrains the platform motion or not. To obtain the rank, we can perform a Gaussian elimination procedure [33] (p. 97) and transform initial matrix $\mathbf{T}_i$ to the following upper triangle $\mathbf{U}_i$:

$$\mathbf{U}_i = \begin{bmatrix} 1 & 1 & 0 & 0 & 1 & 0 & 0 \\ 0 & -R & 0 & 0 & -R & -L_i s_i^z & L_i s_i^y \\ 0 & 0 & 1 & 0 & 0 & 1 & 0 \\ 0 & 0 & 0 & 1 & 0 & 0 & 1 \\ 0 & 0 & 0 & 0 & L_i s_i^z & 0 & -L_i s_i^x \\ 0 & 0 & 0 & 0 & 0 & L_i s_i^x & -L_i s_i^x s_i^y / s_i^z \end{bmatrix}, \text{ if } s_i^z \neq 0,$$
$$\mathbf{U}_i = \begin{bmatrix} 1 & 1 & 0 & 0 & 1 & 0 & 0 \\ 0 & -R & 0 & 0 & -R & 0 & L_i s_i^y \\ 0 & 0 & 1 & 0 & 0 & 1 & 0 \\ 0 & 0 & 0 & 1 & 0 & 0 & 1 \\ 0 & 0 & 0 & 0 & -L_i s_i^y & L_i s_i^x & 0 \\ 0 & 0 & 0 & 0 & 0 & 0 & -L_i s_i^x \end{bmatrix}, \text{ if } s_i^z = 0. \tag{2}$$

For a general chain configuration, there are no zero rows in matrix $\mathbf{U}_i$; hence, the rank equals six. This means the chain does not impose any constraints on the platform motion (see Equation (A3) in Appendix A: only a zero wrench will satisfy it). However, in some configurations, the matrix can be rank deficient. For example, when $s_i^x = 0$ (the leg projects directly on the swinging arm), the rank equals five, and the chain imposes one constraint on the platform. Such configurations are known as leg singularities [34] and should be avoided during motion planning.

Let us consider the situation when the driving wheel is fixed. As we will see in the following subsection, this case is important for analyzing the mobility of the output link concerning several kinematic chains. In this case, the carriage does not move, and we can perform a similar analysis as above, leaving twist ξ_{i1} out from matrix $\mathbf{T}_i$. Corresponding upper triangle matrix $\mathbf{U}_i$ will have a form as below:

$$\mathbf{U}_i = \begin{bmatrix} 1 & 0 & 0 & 1 & 0 & 0 \\ 0 & 1 & 0 & 0 & 1 & 0 \\ 0 & 0 & 1 & 0 & 0 & 1 \\ 0 & 0 & 0 & L_i s_i^z & 0 & -L_i s_i^x \\ 0 & 0 & 0 & 0 & -L_i s_i^z & L_i s_i^y \\ 0 & 0 & 0 & 0 & 0 & 0 \end{bmatrix}. \tag{3}$$

The matrix above has a rank equal to five, and there exists one constraint on the platform by the chain. This constraint has wrench ζ_i reciprocal to twists ξ_{i2}–ξ_{i7} that can be found using Equation (A3) from Appendix A. Using this equation, we find:

$$\zeta_i = \begin{bmatrix} \hat{\mathbf{s}}_i^{\mathrm{T}} & \mathbf{0}^{\mathrm{T}} \end{bmatrix}^{\mathrm{T}}. \tag{4}$$

The obtained wrench corresponds to a force directed along the leg and preventing the platform translation along this direction.

3.2. Analysis for Several Kinematic Chains

Now, let us consider the output link connected to the base by several kinematic chains. In the discussed mechanisms, all the chains are coupled through the driving wheel. To operate properly, the mechanism number of DOFs in a general (nonsingular) configuration should be equal to zero when the drive is fixed [22] (p. 40). If we fix the drive, each chain will impose wrench (4) on the output link. We can compose matrix $\mathbf{W}$, whose columns represent these wrenches, and the number of columns will depend on the number of kinematic chains.

In the mechanisms with three, four and five kinematic chains (Figure 2a–c), there will be three, four and five wrenches ζ_i, respectively, acting on the platform. Hence, the rank of matrix **W** will always be less than six. This means we can find reciprocal twists, which will correspond to the uncontrolled movements (or uncontrolled DOFs) of the mechanism output link. Thus, the mentioned structures shown in Figure 2a–c cannot operate correctly with a single drive.

Fixing the drive in the mechanism with six kinematic chains shown in Figure 2d will lead to six wrenches ζ_i, and matrix **W** will have six columns. If the matrix has a full rank, the platform configuration is completely determined. The mechanism operates properly and has one (controlled) DOF. In some configurations, matrix **W** can be rank deficient, and the platform will obtain additional and uncontrolled DOFs. Such configurations represent another type of singularities common for parallel mechanisms [34] that should also be avoided for proper mechanism performance.

Table 1 summarizes the results of the mobility analysis, describing the reconfigurable solutions of the proposed kinematic chain. Thus, the only workable mechanism structure of the suggested ones is the one with six kinematic chains and single DOF shown in Figure 2d. We will study exactly this mechanism in the following sections.

Table 1. Summary of the mobility analysis results.

Number of Kinematic Chains	Number of DOFs of the Output Link		
	Overall	Controllable	Uncontrollable
3	4	1	3
4	3	1	2
5	2	1	1
6	1	1	–

Though the considered mechanism has only one DOF, a simple reconfiguration procedure, implying disconnecting the belts and reorienting the cranks (Figure 1b), allows for reproducing various output link trajectories, as we will see later in the paper. This is the key feature of the proposed design and the major benefit among other known reconfigurable mechanisms mentioned earlier in the introduction.

4. Virtual and Physical Prototyping

4.1. Development of the Hexapod CAD Model (Virtual Prototyping)

Based on the kinematic scheme of the 1-DOF hexapod shown in Figure 2d, we created its virtual prototype using CAD system Autodesk Inventor. Figure 3a,b show the exploded view of the hexapod assembly, considering the alignment and arrangement of all structural elements. This assembly allows for further manufacturing of these elements by 3D printing and using the standard machine elements.

The procedure for assembling the CAD model of the hexapod is as follows. It begins with assembling circular guide 1, divided into six sectors, comprising several parts, and central ring 1a with a rolling bearing pressed into it. Then, we mount a base consisting of six columns under circular guide 1. The columns connected by metal profiles absorb the load and give additional stability to the entire system. Next, a group of links that includes shaft 2 and driving wheel 3, forming a single link, is rigidly fixed in the center of circular guide 1. After that, we connect the electric motor shaft with shaft 2 under circular guide 1.

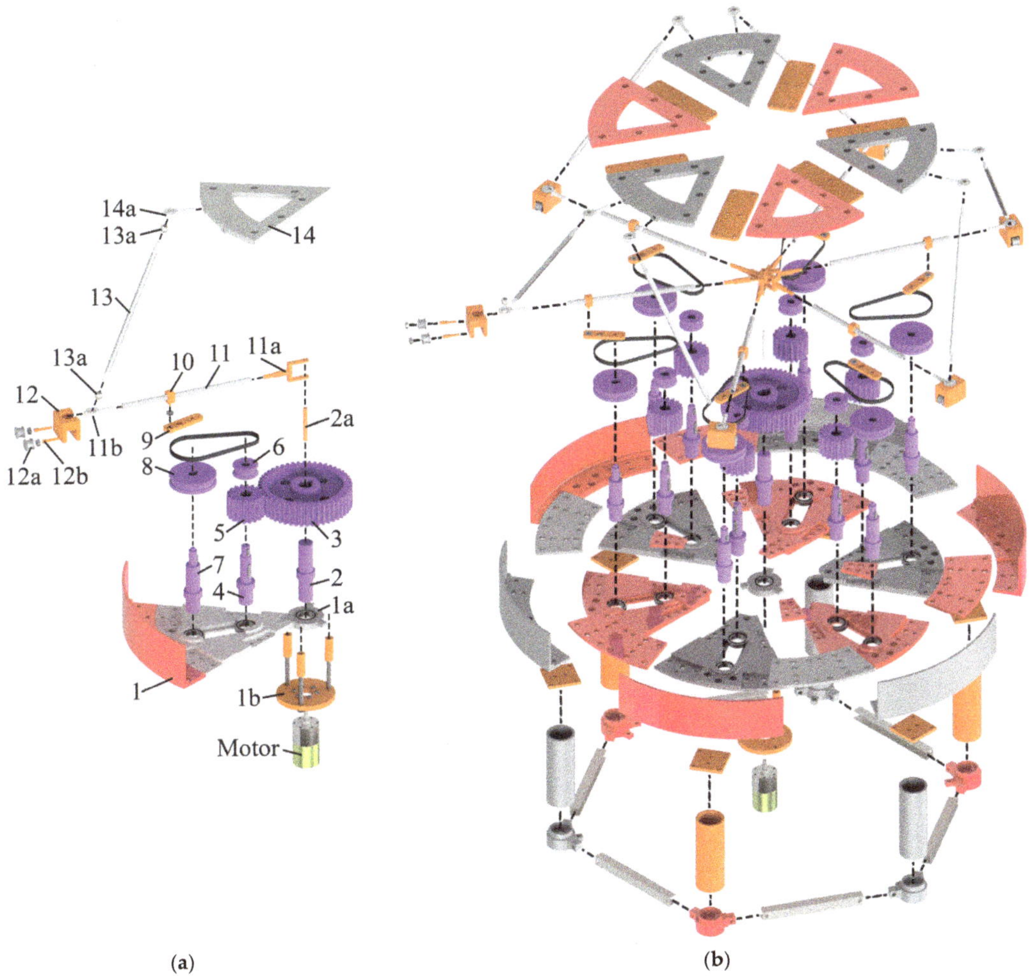

Figure 3. Exploded view of the structural elements assembly (CAD model) of the proposed hexapod: (**a**) single kinematic chain with a driving part; (**b**) complete mechanism model.

Further, we mount a group of links including shaft 4, gear 5 and driving pulley 6 in circular guide 1 through a rolling bearing. This group also forms a single link. After that, similarly, shaft 7 connects rigidly with driven pulley 8 and crank 9 and then becomes fixed in an adjacent bearing in circular guide 1. Next, we stretch the belt between driving 6 and driven 8 pulleys and install a spacer between the bearings of their shafts. Crank 9 and slide block 10 are connected by a rolling bearing.

The next group of links is a rigid connection of swinging arm 11 and carriage 12. Swinging arm 11 ends with fork 11a at one side for coupling with shaft 2a in the center of circular guide 1 and with lunule 11b of the spherical joint at another side. Fork 11a at each swinging arm has an individual geometry for locating them in one plane. Carriage 12 moves on the edge of circular guide 1 by a pair of rollers 12a, which are supported by shafts 12b. Lunule 11b is pressed into carriage 12, forming a single link of swimming arm 11 and carriage 12.

Similarly, we press balls 13a of the spherical joints into leg 13 on both sides. Platform 14, like circular guide 1, is divided into six sectors, connected by coupling plates. Further, three of these sectors connect rigidly with a pair of lunules 14a of the platform spherical joints.

Figure 4a,b show the assembled CAD model of the hexapod with all the structural elements discussed above. This model allows 3D printing its elements and using standard mechanical parts. Table 2 presents the geometrical dimensions of the hexapod elements.

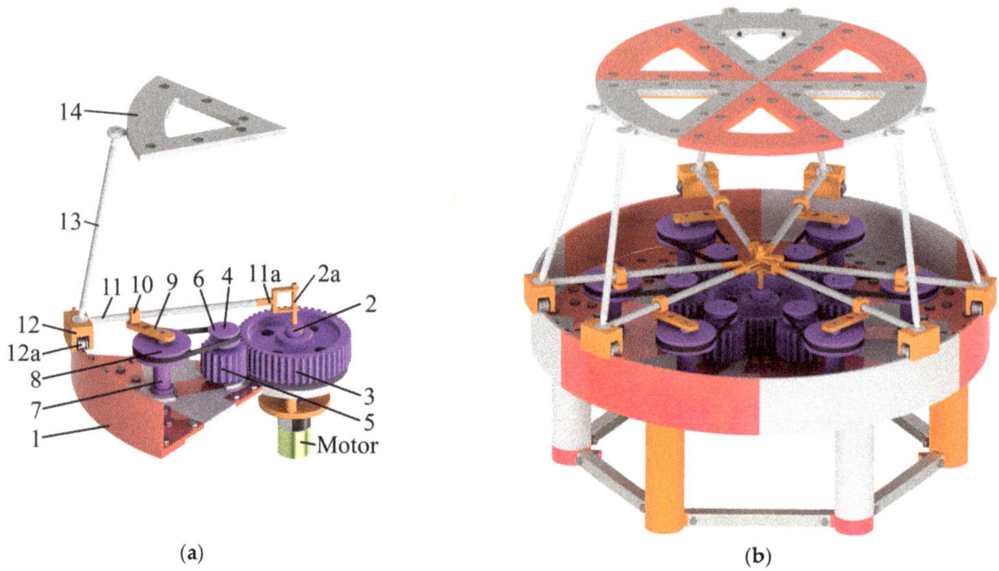

Figure 4. Assembling virtual prototype (CAD model) of the hexapod: (**a**) single kinematic chain with a driving part; (**b**) complete mechanism model.

4.2. Kinetostatic Analysis of a Hexapod Using CAD Modeling

To select an appropriate electric motor for the hexapod, we need to perform its kinetostatic analysis and define a motor torque on the driving shaft 2. Let us accept the driving wheel rotates at a speed of 5 rpm. Due to the mechanism ability to reproduce differing trajectories of the output link, we consider several cases corresponding to our previous study [35]:

- the output link performs a pure rotation around the vertical axis: $\beta_i = 255.8°$, $i = 1 \ldots 6$;
- the output link performs a simultaneous rotation and displacement relative to the vertical axis: $\beta_1 = \beta_3 = \beta_5 = 104.2°$ and $\beta_2 = \beta_4 = \beta_6 = 255.8°$;
- the output link follows a spatial trajectory changing all six coordinates: $\beta_1 = 287.6°$, $\beta_2 = 103.3°$, $\beta_3 = 278.4°$, $\beta_4 = 263.8°$, $\beta_5 = 260.6°$, $\beta_6 = 242.4°$.

We have found these values of angles β_i experimentally in CAD simulation. However, we could also apply a kind of inverse kinematics algorithm to find the angles. In this case, there exists a nontrivial numerical algorithm which can provide different solutions depending on an initial guess. We have touched on this problem in work [36]. The Supplementary Materials section presents movie of the CAD model operation with respect to the cases mentioned above.

Table 2. Geometrical dimensions of the hexapod links.

Parameter Nomenclature	Dimension
Overall dimensions of the hexapod, including the maximum displacements of the platform	530 mm × 530 mm × 457 mm (LWH dimensions)
Column height of circular guide 1	143.00 mm
Diameter of circular guide 1	500.00 mm
Height of circular guide 1	58.30 mm
Pitch diameter of driving wheel 3	125.00 mm
Number of teeth of driving wheel 3	50
Pitch diameter of gear 5	45.00 mm
Number of teeth of gear 5	18
Center distance of gear transmission 3–5	85.00 mm
Module of gears 3–5	2.50 mm
Diameter of driving pulley 6	30.00 mm
Diameter of driven pulley 8	60.00 mm
Center distance of belt transmission 6–8	76.00 mm
Length of crank 9	38.85 mm
Length of swinging arm 11 (distance from the center of circular guide 1 to the center of spherical joint 12–13)	246.05 mm
Length of leg 13	220.05 mm
Platform radius 14 excluding spherical joints 13–14	180.00 mm
Platform radius 14 including spherical joints 13–14 (distance from the center of platform 14 to the center of spherical joint 13–14)	193.00 mm
Minimum/maximum angle between paired spherical joints 13–14	20°/100°
Bearing 1,000,804 (13 pcs)	20 mm × 32 mm × 7 mm
Bearing 623ZZ (18 pcs)	3 mm × 10 mm × 4 mm
Diameter of the sphere in the spherical joints	12.00 mm

We have performed the kinetostatic calculations using an Autodesk Inventor Dynamic Simulation module, which provides dynamic modeling with various parameters of the assembling CAD model. Table 3 presents the materials and inertial parameters of the mechanism structural elements. Inertia moments of the links in the planar part of the mechanism are determined relative to their rotational axes, and inertia tensors of the links that perform a spatial movement (legs 13 and platform 14) are calculated in their mass centers relative to the axes shown in Figure A1 (Appendix B).

The calculation included the friction between all mechanism elements. We set the following friction coefficients k_f for the calculations: $k_f = 0.002$ for the rolling bearings in the planar part; $k_f = 0.02$ for sliding joints 10–11 (aluminum over ABS plastic); $k_f = 0.1$ for spherical joints 12-13 and 13-14 (steel over steel); $k_f = 0.02$ for the couplings of forks 11a with shaft 2a at the center of circular guide 1 (steel over ABS plastic). We also took efficiency ratios 0.98 and 0.95 for the gear and belt couplings, respectively.

Table 3. Inertial properties of the hexapod links.

Link/Group of Links		Material	Moment of Inertia, kg·mm^2	Mass, kg
Shaft 2 and driving wheel 3		Plastic ABS	494.19	0.257
Shaft 4, gear 5 and driving pulley 6		Plastic ABS	20.68	0.091
Shaft 7, driven pulley 8 and crank 9		Plastic ABS	32.04	0.085
Slide block 10		Plastic ABS	0.061	0.003
Swinging arm 11 (incl. elements 11a,b), and carriage 12 (incl. elements 12a,b)	#1	Aluminum, plastic ABS, steel	2545.40	0.052
	#2		2545.66	0.052
	#3		2545.76	0.053
	#4		2545.87	0.053
	#5		2545.80	0.053
	#6		2546.09	0.053
Leg 13 (incl. elements 13a)		Aluminum, plastic ABS, steel	$\begin{bmatrix} 159.87 & 0 & 0 \\ 0 & 159.87 & 0 \\ 0 & 0 & 0.33 \end{bmatrix}$	0.022
Platform 14 (incl. elements 14a)		Aluminum, plastic ABS, steel	$\begin{bmatrix} 10061.20 & 0 & 0 \\ 0 & 10061.20 & 0 \\ 0 & 0 & 20079.61 \end{bmatrix}$	1.153

Given the parameters above, we prepared the mechanism assembly (Figure 4) for dynamic modeling. Simulation results shown in Figure 5 present the calculated value of the motor torque (T_M) for the different cases of cranks initial positions. Notice how the torque increases to compensate for the weight contribution of the links when the platform displaces along the vertical axis.

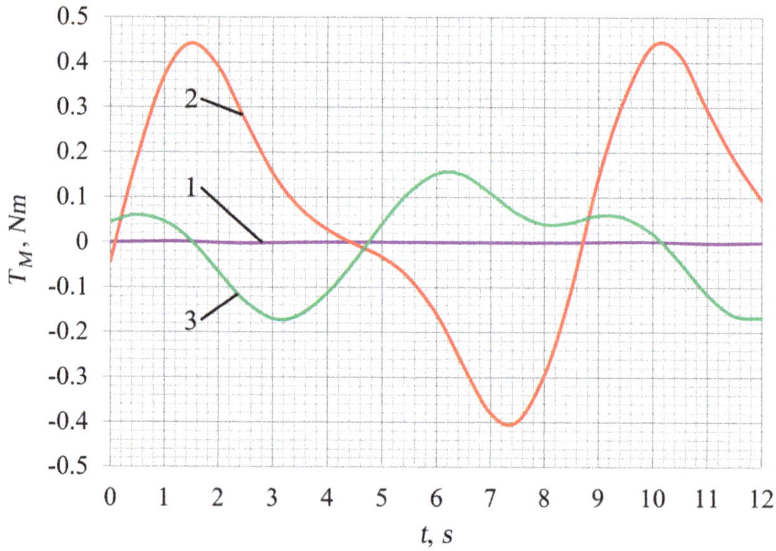

Figure 5. Time-dependent diagrams of the hexapod motor torque (T_M) for the different cranks initial positions: 1—{β_i = 255.8°}, the output link only rotates around the vertical axis; 2—{$\beta_1 = \beta_3 = \beta_5$ = 104.2° and $\beta_2 = \beta_4 = \beta_6$ = 255.8°}, the output link has simultaneous rotation and displacement relative to the vertical axis; 3—{β_1 = 287.6°, β_2 = 103.3°, β_3 = 278.4°, β_4 = 263.8°, β_5 = 260.6°, β_6 = 242.4°}, the output link performs general spatial motion changing all six coordinates.

One should note that using CAD software to perform calculations allows for considering the links geometry and material as accurately as possible. Thereby, the CAD functionality can significantly simplify the solution process and lead to more accurate results in comparison with the analytical methods.

4.3. Development of a Hexapod Full-Scale Model (Physical Prototyping)

Based on the assembling model of the mechanism shown in Figure 4, we produced its physical prototype. Figure 6 shows the prototype picture, and the Supplementary Materials section presents a movie of the prototype functioning. Thus, the virtual prototype serves as a digital twin of the produced physical prototype. Most of the prototype links and couplings were 3D printed on Flsun QQ-S. Other components were easy-to-access standard elements, including legs 13 and swinging arms 11 made of round aluminum profiles, steel spherical joints and bearings. The mechanism drive was selected based on the motor torque calculation given in Section 4.2.

Figure 6. The actuated full-scale physical prototype of the proposed hexapod.

The major advantage of such a design is in using easy-to-access materials and components. Figure 7 shows some mechanism components: the belt drive; connection of the swinging arm, carriage and leg; the platform spherical joints; swinging arms connection in the center of the circular guide. The fork-like design of the swinging arms allows positioning all the arms in one plane. Geometrical parameters of the physical prototype correspond to its digital twin (virtual prototype) presented in Figure 4.

We have two opportunities to control the hexapod. The first and the simplest one is to rotate the motor at some constant speed. The output link will perform a cyclic motion, which depends on the initial orientations of the cranks. This method was presented in Section 4.2 and [35]. The second approach implies solving an inverse kinematics problem where we set the platform trajectory and calculate the required control law for the drive. However, since the mechanism has only one DOF, we can set the motion only for one platform coordinate, while the others are calculated from the inverse kinematics together with the drive motion. We addressed this problem in [36].

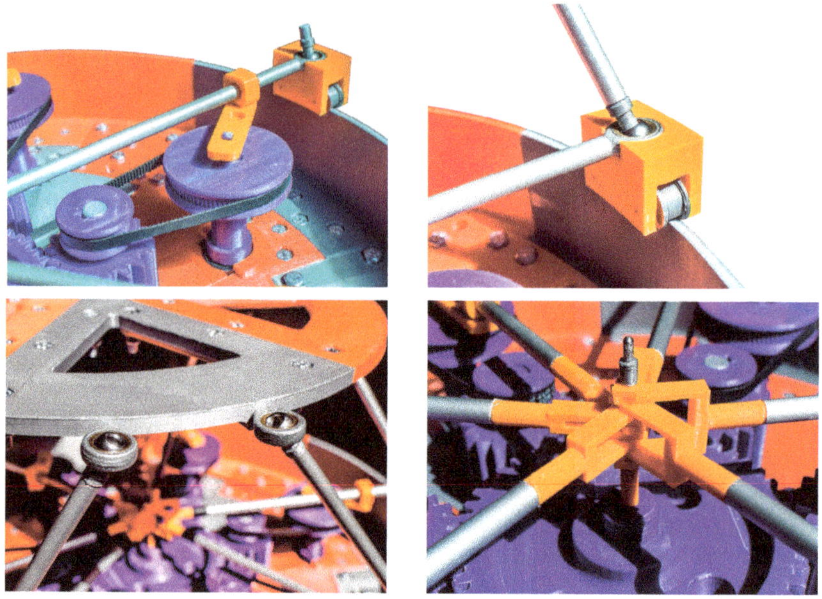

Figure 7. The main units of motion transmission of the proposed hexapod.

5. Conclusions

This paper has presented novel variations of RPMs with similar structure and a different number of kinematic chains. The mobility analysis of the mechanisms carried out via screw theory has provided a calculation of controllable and uncontrollable DOFs. It was found that 4-DOF tripod has three uncontrollable DOFs, 3-DOF quadropod has two uncontrollable DOFs, 2-DOF pentapod has one uncontrollable DOF and 1-DOF hexapod is fully controllable.

The reconfigurability of the mechanisms allows for changing the output link trajectories without using additional kinematic chains or drives. The proposed mechanisms have such design advantages as a drive fixed on the base, absence of the possibility of collisions between the adjacent carriages due to correctly chosen cranks lengths and diverse reconfigurable capabilities due to having a flexible coupling in each kinematic chain.

Based on the obtained mechanisms structures, the hexapod one (the mechanism with six kinematic chains) has been developed as an assembling CAD model, which allowed us to perform a numerical kinetostatic analysis and then fabricate an actuated physical prototype. The developed CAD model was adapted to using 3D printing technologies in prototype production.

The conducted research confirms that virtual and physical prototyping processes relate to each other, and their cooperative execution is very effective in product development. In addition, rapid prototyping technologies, which are one of the most optimal ways to develop physical prototypes, can be often realized only with virtual prototyping.

Having the physical prototype built, we can aim our further research at its experimental study, including accuracy and repeatability tests, rigidity and load capacity analyses, tests for maximum operating speed and others. Additional research can be associated with an analysis of the presented hexapod for specific practical applications, such as rehabilitation procedures, where the hexapod platform could provide various types of cyclic motions. In this regard, the analysis would aim to calculate the initial cranks configuration and the motion laws of the driving link based on the predetermined platform displacements.

Supplementary Materials: The following are available online at: https://www.mdpi.com/2076-3417/11/15/7158/s1, Video S1: Movie of the CAD model operation, Video S2: Movie of the physical prototype operation.

Author Contributions: Conceptualization, A.F. and A.A.; methodology, A.F. and A.A.; software, A.F. and D.P.; validation, A.F. and D.P.; formal analysis, A.F. and A.A.; investigation, A.F., D.P. and A.A.; resources, A.F., D.P., A.A. and V.G.; writing—original draft preparation, A.F., A.A. and M.C.; writing—review and editing, A.F. and A.A.; visualization, A.F. and D.P.; supervision, A.F., A.A. and V.G.; project administration, A.F.; funding acquisition, M.C. All authors have read and agreed to the published version of the manuscript.

Funding: This research was supported by Russian Science Foundation (RSF) under grant № 21-79-10409, https://rscf.ru/project/21-79-10409/.

Institutional Review Board Statement: Not applicable.

Informed Consent Statement: Not applicable.

Data Availability Statement: The data presented in this study are available on request from the corresponding author.

Conflicts of Interest: The authors declare that they have no conflict of interest.

Appendix A. Basics of Instantaneous Screw Theory

The major idea of screw theory relies on Chasles' theorem [32] (p. 26), by which any finite motion of a rigid body can be represented as a combination of rotation about some axis and translation parallel to it. This axis is called a screw axis. If the motion is infinitesimal, we can consider it instantaneous, and the screw axis becomes an instantaneous screw axis. This instantaneous screw motion can be characterized by three basic elements: the axis itself, the angular speed about this axis and the linear speed along it. We can combine these elements into a 6-D vector called (instantaneous) *screw* ξ [32] (p. 40):

$$\xi = \omega \begin{bmatrix} \hat{\mathbf{s}} \\ \mathbf{r} \times \hat{\mathbf{s}} \end{bmatrix} + \upsilon \begin{bmatrix} \mathbf{0} \\ \hat{\mathbf{s}} \end{bmatrix}, \tag{A1}$$

where $\hat{\mathbf{s}}$ is a unit vector directed along the screw axis; $\mathbf{r}$ is a vector pointed from the origin of the coordinate system to any point on the screw axis; ω is an angular speed relative to the screw axis; and υ is a linear speed about the screw axis.

An alternative common notation for a screw implies using scalar parameters: magnitude s and pitch $h = \upsilon/\omega$. In this case, a screw can be defined as follows [22] (p. 19):

$$\begin{aligned} \xi &= s \begin{bmatrix} \hat{\mathbf{s}} \\ \mathbf{r} \times \hat{\mathbf{s}} + h\hat{\mathbf{s}} \end{bmatrix}, \text{ if } h \neq \infty, \\ \xi &= s \begin{bmatrix} \mathbf{0} \\ \hat{\mathbf{s}} \end{bmatrix}, \text{ if } h = \infty. \end{aligned} \tag{A2}$$

Similarly, all the forces and moments acting on a rigid body can be reduced to a force directed along some axis and a moment about the axis. We can define screw ζ representing this resulting force and moment using relations similar to (A1) and (A2). To differ between these screws, terms *twist* and *wrench* are usually applied for ξ and ζ, respectively. Twists and wrenches of zero pitch ($h = 0$) correspond to pure rotations and forces. Twists and wrenches of infinite pitch ($h = \infty$) correspond to pure translations and moments.

Screws can be summed with each other and multiplied by a scalar—these properties allow screws to form vector spaces. Thus, the space of twists can define the possible velocities of the rigid body, while the space of wrenches represents all the forces and moments acting on it. If each twist ξ from the twist space and each wrench ζ from the wrench space satisfy the following condition [22] (p. 24):

$$\xi \circ \zeta = (\mathbf{P}\xi)^{\mathrm{T}}\zeta = 0, \tag{A3}$$

where $\mathbf{P}$ is a permutation matrix of the form:

$$\mathbf{P} = \begin{bmatrix} \mathbf{0}_{3\times3} & \mathbf{I}_{3\times3} \\ \mathbf{I}_{3\times3} & \mathbf{0}_{3\times3} \end{bmatrix}, \tag{A4}$$

then the wrench space represents all forces and torques that constrain the body motion. Operation "$\circ$" is known as a *reciprocal product*, and screws ξ and ζ that satisfy condition (A3) are *reciprocal*.

If two bodies are connected by a series of joints, we can associate unit twist ξ_j ($s = 1$ in expression (A2)) $j = 1 \ldots m$ with each joint, where m is a number of joints. Then, twist system T comprising twists $\xi_1, \xi_2, \ldots, \xi_m$ will define the space of possible velocities of one body relative to the other. The reciprocal wrench system W (each wrench of which is reciprocal to each twist of system T) corresponds to constraints between the bodies.

For a parallel mechanism with n kinematic chains, each having m_i joints, $i = 1 \ldots n$, we can define twists systems T_i of unit twists ξ_{ij}, $j = 1 \ldots m_i$, associated with each chain. Wrench system W_i reciprocal to T_i corresponds to constraints that the i-th chain imposes on the motion of the output link. Considering all the kinematic chains, the output link motion will be constrained such that its resulting wrench system W is a union of all wrench systems W_i. Finally, twist system T reciprocal to W will define all the possible velocities the output link can attain. One should never forget that all the screws are instantaneous, and the results of the analysis performed above depend on a particular mechanism configuration. In some configurations, twist or wrench systems can include linearly dependent screws. Such situations are known as *singularities*, and the mechanism can lose or gain degrees of freedom in singular configurations, or even change its motion type when passing through them. Authors of [34] present a thorough analysis of this topic.

Appendix B. Hexapod Elements Geometry

Figure A1 shows hexapod links with identified mass centers and axes used to calculate the inertia parameters. The links shown in Figure A1a–d,g rotate around the fixed axes, and their inertia moments are calculated relative to these axes. The inertia tensors for the links having spatial motion shown in Figure A1e,f are calculated with respect to the depicted coordinate systems placed at the mass centers. The links in Figure A1a,b have displaced mass centers because of the shaft keys.

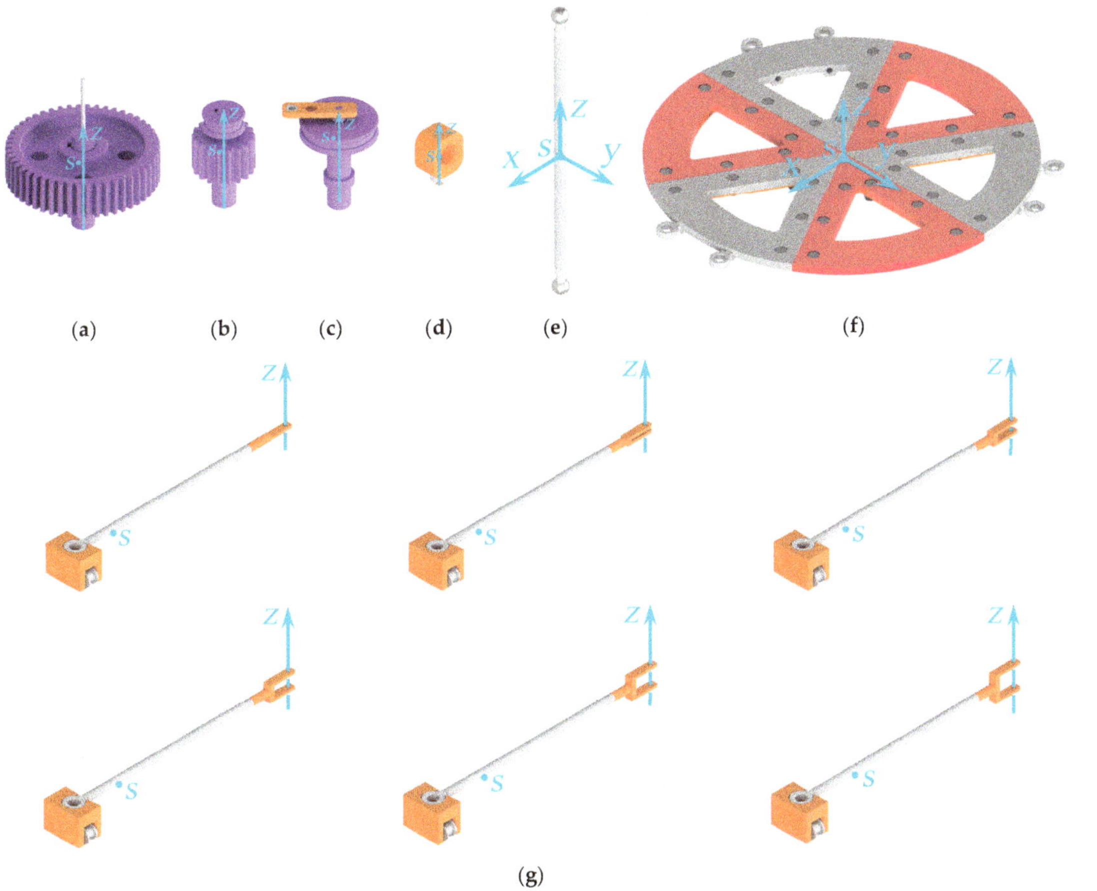

Figure A1. Definition of the inertia terms based on the hexapod elements geometry.

References

1. Gibson, I.; Gao, Z.; Campbell, I. A comparative study of virtual prototyping and physical prototyping. *Int. J. Manuf. Technol. Manag.* **2004**, *6*, 503–522. [CrossRef]
2. Chen, L.; Daly, M.C.; Sabelhaus, A.P.; Janse van Vuuren, L.A.; Garnier, H.J.; Verdugo, M.I.; Tang, E.; Spangenberg, C.U.; Ghahani, F.; Agogino, A.M.; et al. Modular elastic lattice platform for rapid prototyping of tensegrity robots. In Proceedings of the ASME 2017 International Design Engineering Technical Conferences and Computers and Information in Engineering Conference, Volume 5B: 41st Mechanisms and Robotics Conference, Cleveland, OH, USA, 6–9 August 2017; V05BT08A026. [CrossRef]
3. Wallin, T.J.; Pikul, J.; Shepherd, R.F. 3D printing of soft robotic systems. *Nat. Rev. Mater.* **2018**, *3*, 84–100. [CrossRef]
4. Gul, J.Z.; Sajid, M.; Rehman, M.M.; Siddiqui, G.U.; Shah, I.; Kim, K.-H.; Lee, J.-W.; Choi, K.H. 3D printing for soft robotics—A review. *Sci. Technol. Adv. Mater.* **2018**, *19*, 243–262. [CrossRef] [PubMed]
5. Skylar-Scott, M.A.; Mueller, J.; Visser, C.W.; Lewis, J.A. Voxelated soft matter via multimaterial multinozzle 3D printing. *Nature* **2019**, *575*, 330–335. [CrossRef]
6. Lapeyre, M.; Rouanet, P.; Grizou, J.; Nguyen, S.; Depraetre, F.; Le Falher, A.; Oudeyer, P.-Y. Poppy Project: Open-Source fabrication of 3D printed humanoid robot for science, education and art. In Proceedings of the Digital Intelligence, Nantes, France, 17–19 September 2014.
7. Chavdarov, I.; Nikolov, V.; Naydenov, B.; Boiadjiev, G. Design and control of an educational redundant 3D printed robot. In Proceedings of the 2019 International Conference on Software, Telecommunications and Computer Networks (SoftCOM), Split, Croatia, 19–21 September 2019. [CrossRef]

8. Ficht, G.; Farazi, H.; Brandenburger, A.; Rodriguez, D.; Pavlichenko, D.; Allgeuer, P.; Hosseini, M.; Behnke, S. NimbRo-OP2X: Adult-sized open-source 3D printed humanoid robot. In Proceedings of the 2018 IEEE-RAS 18th International Conference on Humanoid Robots (Humanoids), Beijing, China, 6–9 November 2018; pp. 1–9. [CrossRef]
9. Yao, J.; Zhang, H.; Xiang, X.; Bai, H.; Zhao, Y. A 3-D printed redundant six-component force sensor with eight parallel limbs. *Sens. Actuator A Phys.* **2016**, *247*, 90–97. [CrossRef]
10. Rahul, K.; Raheman, H.; Paradkar, V. Design and development of a 5R 2DOF parallel robot arm for handling paper pot seedlings in a vegetable transplanter. *Comput. Electron. Agric.* **2019**, *166*, 105014. [CrossRef]
11. Pham, M.T.; Teo, T.J.; Yeo, S.H.; Wang, P.; Nai, M.L.S. A 3-D printed Ti-6Al-4V 3-DOF compliant parallel mechanism for high precision manipulation. *IEEE/ASME Trans. Mechatron.* **2017**, *22*, 2359–2368. [CrossRef]
12. Tursynbek, I.; Niyetkaliye, A.; Shintemirov, A. Computation of unique kinematic solutions of a spherical parallel manipulator with coaxial input shafts. In Proceedings of the 2019 IEEE 15th International Conference on Automation Science and Engineering (CASE), Vancouver, BC, Canada, 22–26 August 2019; pp. 1524–1531. [CrossRef]
13. Leal-Naranjo, J.-A.; Ceccarelli, M.; Torres-San-Miguel, C.-R.; Aguilar-Perez, L.-A.; Urriolagoitia-Sosa, G.; Urriolagoitia-Calderón, G. Multi-objective optimization of a parallel manipulator for the design of a prosthetic arm using genetic algorithms. *Lat. Am. J. Solids Struct.* **2018**, *15*, e26. [CrossRef]
14. Saafi, H.; Laribi, M.A.; Arsicault, M.; Zeghloul, S. Optimal design of a new spherical parallel manipulator. In Proceedings of the 2014 23rd International Conference on Robotics in Alpe-Adria-Danube Region (RAAD), Smolenice, Slovakia, 3–5 September 2014; pp. 1–6. [CrossRef]
15. Grosch, P.; Di Gregorio, R.; Lopez, J.; Thomas, F. Motion planning for a novel reconfigurable parallel manipulator with lockable revolute joints. In Proceedings of the 2010 IEEE International Conference on Robotics and Automation (ICRA), Anchorage, AK, USA, 3–7 May 2010; pp. 4697–4702. [CrossRef]
16. Russo, M.; Ceccarelli, M. Kinematic design of a tripod parallel mechanism for robotic legs. In *Mechanisms, Transmissions and Applications, Proceedings of the Fourth MeTrApp Conference 2017, Poitiers, France, 22–24 May 2017*; Dede, M., İtik, M., Lovasz, E.C., Kiper, G., Eds.; Springer: Cham, Switzerland, 2018; pp. 121–130. [CrossRef]
17. Fanghella, P.; Galletti, C.; Giannotti, E. Parallel robots that change their group of motion. In *Advances in Robot Kinematics, Proceedings of the 10th International Symposium on Advances in Robot Kinematics, Ljubljana, Slovenia, 26–29 June 2006*; Lenarčič, J., Roth, B., Eds.; Springer: Dordrecht, The Netherlands, 2006; pp. 49–56. [CrossRef]
18. Ye, W.; Chai, X.; Zhang, K. Kinematic modeling and optimization of a new reconfigurable parallel mechanism. *Mech. Mach. Theory* **2020**, *149*, 103850. [CrossRef]
19. Kong, X.; Gosselin, C.M.; Richard. P. Type synthesis of parallel mechanisms with multiple operation modes. *J. Mech. Des.* **2007**, *129*, 595–601. [CrossRef]
20. Merlet, J.-P. *Parallel Robots*, 2nd ed.; Springer: Dordrecht, The Netherlands, 2006. [CrossRef]
21. Ceccarelli, M. *Fundamentals of Mechanics of Robotic Manipulation*; Springer: Dordrecht, The Netherlands, 2004. [CrossRef]
22. Kong, X.; Gosselin, C.M. *Type Synthesis of Parallel Mechanisms*; Springer: Berlin/Heidelberg, Germany, 2007. [CrossRef]
23. Zhang, M.; Cao, J.; Zhu, G.; Miao, Q.; Zeng, X.; Xie, S.Q. Reconfigurable workspace and torque capacity of a compliant ankle rehabilitation robot (CARR). *Robot. Auton. Syst.* **2017**, *98*, 213–221. [CrossRef]
24. Herrero, S.; Mannheim, T.; Prause, I.; Pinto, C.; Corves, B.; Altuzarra, O. Enhancing the useful workspace of a reconfigurable parallel manipulator by grasp point optimization. *Robot. Comput. Integr. Manuf.* **2015**, *31* (Suppl. C), 51–60. [CrossRef]
25. Tian, C.; Fang, Y.; Guo, S.; Qu, H. A class of reconfigurable parallel mechanisms with five-bar metamorphic linkage. *Proc. Inst. Mech. Eng. C J. Mech.* **2016**, *231*, 2089–2099. [CrossRef]
26. Ye, W.; Fang, Y.; Guo, S. Design and analysis of a reconfigurable parallel mechanism for multidirectional additive manufacturing. *Mech. Mach. Theory* **2017**, *112*, 307–326. [CrossRef]
27. Li, Q.; Xu, L.; Chen, Q.; Ye, W. New family of RPR-equivalent parallel mechanisms: Design and application. *Chin. J. Mech. Eng.* **2017**, *30*, 217–221. [CrossRef]
28. Carbonari, L.; Callegari, M.; Palmieri, G.; Palpacelli, M.-C. A new class of reconfigurable parallel kinematic machines. *Mech. Mach. Theory* **2014**, *79*, 173–183. [CrossRef]
29. Bi, Z.; Wang, L. Optimal design of reconfigurable parallel machining systems. *Robot. Comput. Integr. Manuf.* **2009**, *25*, 951–961. [CrossRef]
30. Huang, G.; Zhang, D.; Guo, S.; Qu, H. Design and optimization of a novel three-dimensional force sensor with parallel structure. *Sensors* **2018**, *18*, 2416. [CrossRef] [PubMed]
31. Huang, Z.; Li, Q.; Ding, H. *Theory of Parallel Mechanisms*; Springer: Dordrecht, The Netherlands, 2013. [CrossRef]
32. Sun, T.; Yang, S.; Lian, B. *Finite and Instantaneous Screw Theory in Robotic Mechanism*; Springer: Singapore, 2020. [CrossRef]
33. Strang, G. *Introduction to Linear Algebra*, 5th ed.; Wellesley-Cambridge Press: Wellesley, MA, USA, 2016.
34. Conconi, M.; Carricato, M. A new assessment of singularities of parallel kinematic chains. *IEEE Trans. Rob.* **2009**, *25*, 757–770. [CrossRef]

35. Fomin, A.; Antonov, A.; Glazunov, V. Forward kinematic analysis of a rotary hexapod. In *ROMANSY 23-Robot Design, Dynamics and Control, Proceedings of the 23rd CISM IFToMM Symposium, Sapporo, Japan, 20–24 September 2020*; Venture, G., Solis, J., Takeda, Y., Konno, A., Eds.; Springer: Cham, Switzerland, 2021; pp. 486–494. [CrossRef]
36. Fomin, A.; Antonov, A.; Glazunov, V.; Carbone, G. Dimensional (parametric) synthesis of the hexapod-type parallel mechanism with reconfigurable design. *Machines* **2021**, *9*, 117. [CrossRef]

applied sciences

MDPI

Article

Structural-Parametric Synthesis of the RoboMech Class Parallel Mechanism with Two Sliders

Zhumadil Baigunchekov [1,2], Med Amine Laribi [3,*], Giuseppe Carbone [4], Azamat Mustafa [2,*], Bekzat Amanov [1] and Yernar Zholdassov [1]

1 Department of Mechanics, Al–Farabi Kazakh National University, Almaty 050040, Kazakhstan; bzh47@mail.ru (Z.B.); bekzat.amanov007@gmail.com (B.A.); jera.kz@mail.ru (Y.Z.)

2 Department of Mechanical Engineering and Modeling, Satbayev University, 22a Satpaev Str., Almaty 050013, Kazakhstan

3 Department GMSC, Institut PPRIME, Université de Poitiers, CNRS, ENSMA, UPR 3346, 86962 Poitiers, France

4 Department of Mechanical, Energy and Management Engineering, University of Calabria, 87036 Rende, Italy; giuseppe.carbone@unical.it

* Correspondence: med.amine.laribi@univ-poitiers.fr (M.A.L.); mustafa_azamat@mail.ru (A.M.)

Abstract: This paper addresses the structural-parametric synthesis and kinematic analysis of the RoboMech class of parallel mechanisms (PM) having two sliders. The proposed methods allow the synthesis of a PM with its structure and geometric parameters of the links to obtain the given laws of motions of the input and output links (sliders). The paper outlines a possible application of the proposed approach to design a PM for a cold stamping technological line. The proposed PM is formed by connecting two sliders (input and output objects) using one passive and one negative closing kinematic chain (CKC). The passive CKC does not impose a geometric constraint on the movements of the sliders and the geometric parameters of its links are varied to satisfy the geometric constraint of the negative CKC. The negative CKC imposes one geometric constraint on the movements of the sliders and its geometric parameters are determined on the basis of the Chebyshev and least-square approximations. Problems of positions and analogues of velocities and accelerations of the considered PM are solved to demonstrate the feasibility and effectiveness of the proposed formulations and case of study.

Keywords: parallel mechanism; RoboMech; structural-parametric synthesis; Chebyshev and least-square approximations; kinematic analysis

Citation: Baigunchekov, Z.; Laribi, M.A.; Carbone, G.; Mustafa, A.; Amanov, B.; Zholdassov, Y. Structural-Parametric Synthesis of the RoboMech Class Parallel Mechanism with Two Sliders. *Appl. Sci.* **2021**, *11*, 9831. https://doi.org/10.3390/app11219831

Academic Editor: Manuel Armada

Received: 15 September 2021
Accepted: 14 October 2021
Published: 21 October 2021

1. Introduction

The design of manipulation robots both with serial and parallel manipulators is carried out mainly by solving inverse kinematics and developing the control systems and technical means according to the obtained laws of motions of the actuators [1–4]. In this case, the actuators of manipulation robots may work in controlled regimes of intensive accelerations and braking that worsen their dynamics and mechanical efficiency [5,6].

In order to improve the dynamic characteristics and simplify the control systems of the designed manipulators, it is advisable to set the laws of motions of the actuators along with the given laws of the end-effectors' motions. The ability to set the laws of motions of the input links improves the dynamic parameters and simplifies the control system and therefore also increases the reliability and reduces the cost of the designed manipulator. Such parallel manipulators, having the property of manipulation robots, such as reproducing the given laws of motions of end-effectors, and the property of mechanisms, such as setting the laws of motions of actuators, are called the RoboMech class parallel mechanisms or paralell manipulators (PMs) [7].

In the simultaneous setting the laws of motions of the input and output links, the RoboMech class PMs operate under certain structural schemes and geometric parameters of their links.

In this case, the control elements for the movement of the PMs, i.e., the functional relationship between the laws of motions of the input and output links, is laid in the determining structure schemes and geometric parameters of the links, i.e., in the mechanical part of the RoboMech class PMs. Such an optimal combination of mechanics and motion control of manipulation robots corresponds to the modern concept of mechatronics as a methodology for developing of simple, reliable and cheap technological automation.

The base for the structural synthesis of planar mechanisms is proposed by Assur [8], according to whom the mechanism is formed by connecting to the input link (actuator) and the base of structural groups with zero degrees of freedom (DOF). These structural groups are then called the Assur groups, which can be of different classes and orders. The existing methods of structural synthesis of mechanisms and manipulators are devoted to the determination of their structural schemes according to the given numbers of DOFs, links, kinematic pairs and their types [9–15]. A review of research on the synthesis of types of parallel robotic mechanisms was undertaken in [16]. All of these methods do not take into account the functional purposes of mechanisms or manipulators.

In kinematic synthesis (dimensional or parametric synthesis) of mechanisms, with their known structural schemes, the synthesis parameters are determined by the given positions of the input and output links. Generation of the specified movements of output links (output objects) can be performed exactly and approximately. Exact reproduction of the required movements of a rigid body by linkage mechanisms is possible with a limited number of positions, depending on the structural scheme of the mechanism-generator, while the possibility of their approximate reproduction is not limited to the number of specified positions.

Exact methods for synthesis of mechanisms, or so-called geometric methods, are based on kinematic geometry. The fundamentals of kinematic geometry for finite positions of a rigid body in a plane motion were developed by Burmester and for finite-positions of a rigid body in space were developed by Shoenflies. Burmester in [17] developed the theory of a moving plane having four and five positions on circles. Shoenflies in [18] formulated theorems on the geometrical places of points of a rigid body having seven positions on a circle and three positions on a line. The graphical methods of Burmester and Schoenflies theories received an analytical interpretation [19–21], which is summarized in the monograph.

Geometric methods of mechanism synthesis are clear and simple. However, these methods are applicable only for a limited number of positions. Moreover, the algorithms for solving problems using these methods depend significantly on the number of specified positions, and their complexity increases with the number of positions. Approximation (algebraic) methods of mechanism synthesis are devoid of these disadvantages.

Problems of approximation synthesis of mechanisms were first formulated and solved in [22]. Least-square approximations are the most widely used in the approximation synthesis of mechanisms. For the development of this method, a new deviation function, a weighted difference with a parametric weight, proposed in [23], was important. In contrast to the actual deviation, the weighted difference can be reduced to linear forms (generalized polynomials). This makes it quite easy to apply linear approximation methods to the synthesis of mechanisms. This eliminates the limit on the maximum number of specified positions of the moving object.

Combining the main advantages of geometric and approximation methods, a new direction-approximation kinematic geometry of mechanism synthesis was formulated. It studies a special class of approximation problems related to the definition of points and lines of a rigid body describing the constraint of the synthesizing kinematic chains. In the works [24,25], the basics of approximation kinematic geometry of the plane and spatial movements are presented, where circular square points [24] and points with approximately spherical and coplanar trajectories [25] are defined, which correspond to binary links of the type RR, SS and SP_k. Further, in the works [26,27], the concept of discrete Chebyshev approximations was introduced for the kinematic synthesis of linkage mechanisms. Theorems characterizing the Chebyshev circle and straight line in plane motion [26] and the Chebyshev sphere and plane in spatial motion [27], as well as iterative algorithms for determining Chebyshev circular, spherical and other points based on minimizing the limit values of the weighted difference, are formulated.

Many approaches for kinematic analysis and synthesis of mechanisms and manipulators are based on the derivative of loop-closure equations using the vector, matrix and screw methods [28–42]. In this case, polynomials of high degrees are obtained [43,44].

In this paper, structural-parametric synthesis and kinematic analysis of the RoboMech class PM with two sliders is carried out on the base of modular approach, according to which, by the given laws of motions of the input and output links, the structural scheme and geometric parameters of links are determined from the separate simple structural modules [45]. The considered PM is formed by connecting two sliders using one passive and one negative closing kinematic chain (structural (modules) having zero and negative DOF, respectively. This PM can be used in a cold stamping technological—line as proposed in a case of study.

2. Structural Scheme of a Cold Stamping Technological Line

A scheme of a simple single-stream robotic cold stamping technological line [46] is shown in Figure 1, where TE is the main technological equipment, IR is an industrial robot and S is a piece-by-piece delivery store.

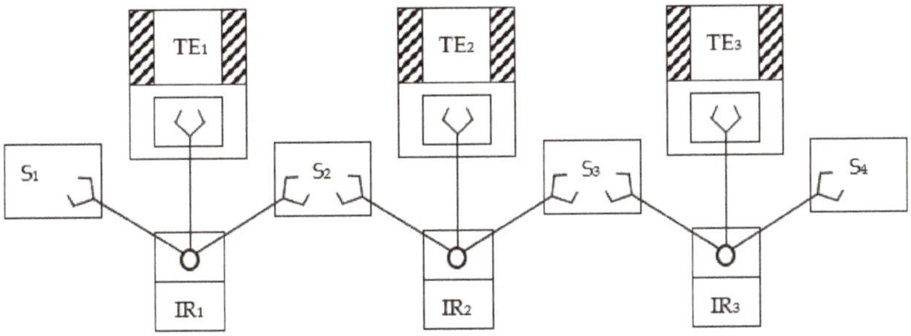

Figure 1. Scheme of a single-stream robotic cold stamping technological line.

This scheme is typical for technological processes with a small cycle of processing production items on technological equipment, in particular, in cold stamping process. In this scheme, there is no inter-operational transport system, and products (production items) are transferred from one piece of technological equipment to another directly by industrial robots.

The equipment of the presented scheme of the technological line operate in the following sequence: the first industrial robot IR_1 takes a workpiece in a certain position from the first store S_1 and delivers it to the first piece of technological equipment TE_1, where the workpiece is processed (stamped). After primary processing, the product is delivered by the same industrial robot IR_1 to the second store S_2, where the position of the product is changed for sequent processing. Then, the second industrial robot IR_2 delivers the product from the store S_2 to the second piece of technological equipment TE_2, where the second processing of the product is carried out. After this processing, the product is delivered to the store S_3 by the second industrial robot IR_2. Moreover, all equipment must operate in accordance with a given cyclogram of the technological line.

Thus, the considered technological line for processing the product with two changing positions of the product has two main pieces of technological equipment (hydraulic presses), I and II, four auxiliary pieces of equipment: a device III for feeding the workpiece, a device IV for removing the product after processing and two industrial robots V and VI (Figure 2). These devices in total have of eight DOF. It is known that the more DOF of equipment in technological lines for mass production of typical products, the lower their productivity and reliability.

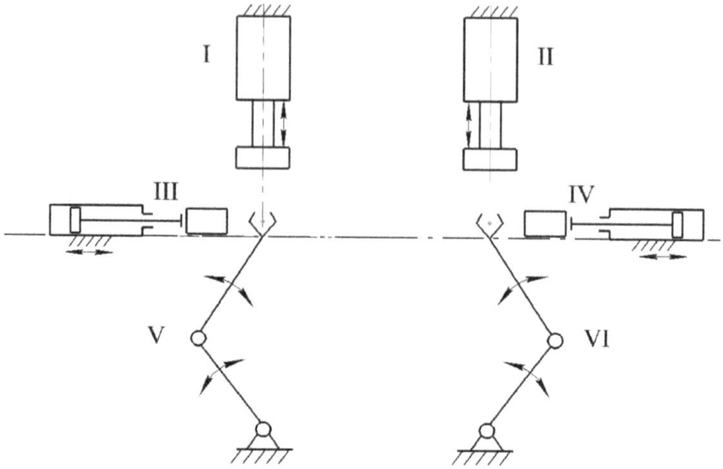

Figure 2. Scheme of the technological line with eight DOFs.

In order to eliminate the noted disadvantages of the technological line, we reduce its number of DOF, replacing the technological and auxiliary equipment with the RoboMech class PMs. According to the developed principle of forming the RoboMech class PMs [7], we combine the main technological equipment (hydraulic presses) I and II with devices for feeding the workpieces III and removing the workpieces IV, and also combine two industrial robots V and VI into one PM with two end-effectors.

Combination of the hydraulic presses I and II with the devices for feeding the workpieces III and removing the workpieces IV into PMs I′ and II′ with two sliders is carried out by connecting the punches Q' and Q'' of the hydraulic cylinders I and II with the sliders P' and P'' of the workpieces feeding and removing devices III and IV using passive CKCs $A'B'C'$ and $A''B''C''$, as well as negative CKCs $D'E'$ and $D''E''$, respectively (Figure 3).

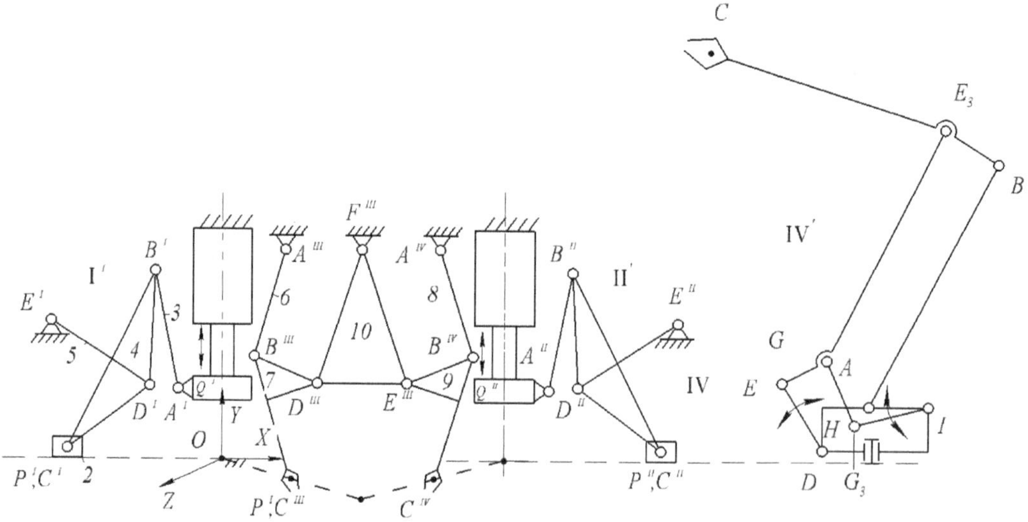

Figure 3. Structural scheme of the technological line with the RoboMech class PMs.

Combination of two industrial robots (serial manipulators) $A^{III}B^{III}C^{III}$ and $A^{IV}B^{IV}C^{IV}$ into one PM III′ with two output points is carried out by connecting the links $B^{III}C^{III}$ and $B^{IV}C^{IV}$ of the serial manipulators $A^{III}B^{III}C^{III}$ and A^{IV} $B^{IV}C^{IV}$ using negative CKC $D^{III}E^{III}F^{III}$.

As a result, we obtain a scheme of a technological line with the RoboMech class PMs with three DOFs, where the presses I′ and II′ have two DOF, the PM with two output point III′ have one DOFs. Hydraulic presses I′ and II′ with devices for feeding and removing workpieces work in the OXY plane, and the PM with two output points III′ works in the OXZ plane. Figure 3 also shows a scheme of the PM IV' operating in a cylindrical coordinate system. This PM is used to store finished products in bins.

The considered technological line with the RoboMech class PMs operates as follows. When feeding the workpiece for processing by the press I', the slider P' takes the right extreme position, and the punch Q' of the hydraulic cylinder takes the upper extreme position. When processing the workpiece, the punch Q' of this press takes the lower extreme position, and the slider P' returns to the left extreme position to deliver the next workpiece. At the moment of return of the punch Q' of the hydraulic cylinder I′ to the upper position, the first gripper C''' of the PM with two output points takes the extreme left position, captures the processed workpiece and delivers it to the store. At this moment, the second gripper C^{IV} delivers the previously processed workpiece to the press II′ for further processing, i.e., takes the extreme upper position. After the secondary processing of the workpiece, the finished product is delivered to the container by the slider P''. Then the cycle is repeated.

After accumulation of products in the container, it is stored in bins with the help of PM IV′. A gripper C^V of this PM reproduces the series of horizontal and vertical trajectories. In this case, the series of horizontal trajectories are reproduced by input link D^VE^V and the series of vertical trajectories are reproduced by input link I^VH^V drive. Rotation of the entire PM around the vertical axis provides a spatial movement of the gripper C^V in a cylindrical coordinate system. Structural-parametric synthesis of this PM is considered in [47]. Let us consider the structural-parametric synthesis of the PM with two sliders.

3. Structural-Parametric Synthesis of the PM with Two Sliders

The problem of structural-parametric synthesis of the PM with two sliders is to determine the structural scheme and geometric parameters of links, when the first slider Q (the punch of a hydraulic press) takes the lower extreme position with a stroke s_{Q_1}, the second slider P takes the left extreme position with a stroke s_{P_1} (Figure 4a) and also, when the first slider Q takes the upper extreme position with a stroke s_{P_N}, and the second slider P takes the right extreme position with a stroke s_{Q_N} (Figure 4b). For the convenience of reporting the strokes of the sliders, the absolute coordinate system OXY is located at the point of their intersection.

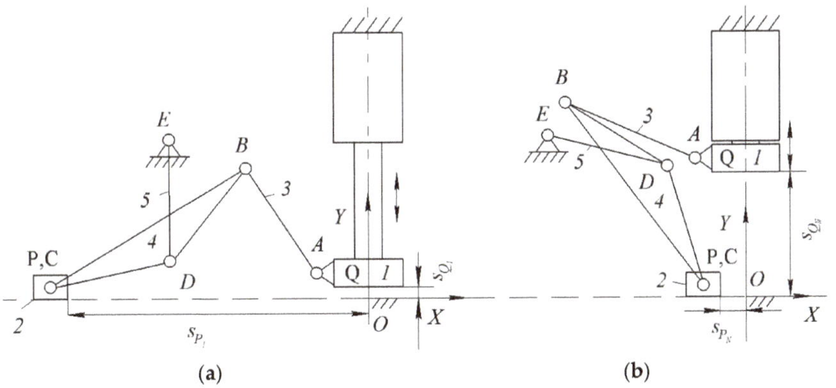

Figure 4. (**a**) lower extreme and (**b**) upper extreme positions of the slider Q.

As noted above, to form the PM with two sliders, providing their specified positions, we connect the punch Q of the hydraulic press and the slider P using the dyad *ABC* with revolute kinematic pairs. The dyad *ABC* has zero DOF and it is a passive CKC, which does not impose geometric constraints on the movements of the punch Q and the slider *P*. Therefore, the passive CKC *ABC* allows reproduction of the specified movements of the sliders Q and P. Then we connect the link *BC* of the dyad *ABC* with a base using a binary link *DE* with revolute kinematic pairs, which has one negative DOF, it is a negative CKC. Negative CKC *DE* imposes one geometric constraint on the movements of the sliders Q and P, and as a result, we obtain a structural scheme of the PM with structural formula I (0,1) → III (3,4,2,5), where the kinematic chain 3-4-2-5 represents the Assur group of the third class [48]. Figure 5 shows a block structure of the formed PM with two sliders.

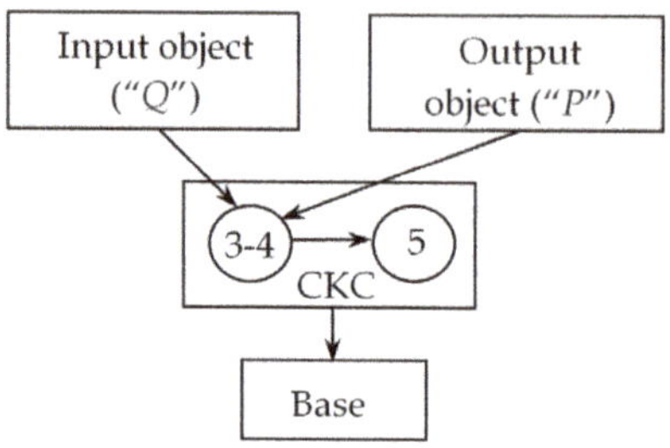

Figure 5. Block structure of the PM with two sliders.

For parametric synthesis of the PM with two sliders, let us consider its *i*-th intermediate position and attach the coordinate systems Qx_1y_1 and Px_2y_2 with the sliders (Figure 6), the axes Px_1 and Qx_2 which are directed parallel to the axis OX of the absolute coordinate system OXY. Then, the movements of the sliders are determined by the parameters s_{Q_i} and s_{P_i} of the coordinate systems Px_2y_2 and Qx_2y_2 movements, where *i* = 1, 2 ..., *N* (*N* is the number of given positions). Parametric synthesis of this PM with two sliders, according to its block structure (Figure 5), consists of a parametric synthesis of the passive CKC *ABC* and the negative CKC *DE*.

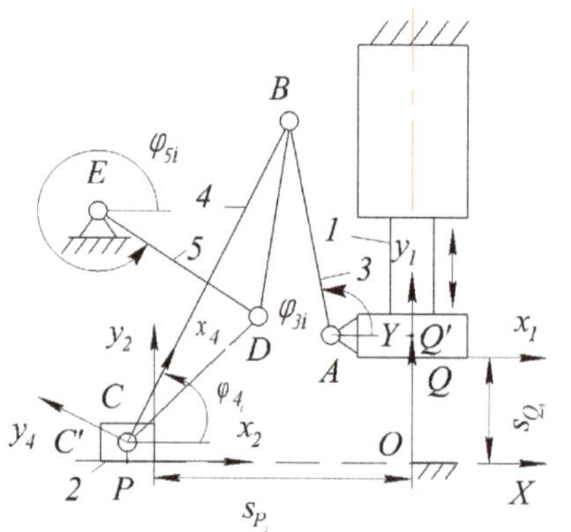

Figure 6. Intermediate position of the PM with two sliders.

The synthesis parameters (geometric parameters of the links) of the passive CKC *ABC* is a vector $r_1 = \left[x_A^{(1)}, y_A^{(1)}, x_C^{(2)}, y_C^{(2)}, l_{AB}, l_{BC}\right]^T$, where $x_A^{(1)}, y_A^{(1)}$ and $x_C^{(2)}, y_C^{(2)}$ are the coordinates of the joints A and C in the moving coordinate systems Qx_1y_1 and Px_2y_2 respectively, l_{AB} and l_{BC} are the lengths of the links AB and BC. Since the passive CKC ABC does not impose geometric constraint on the movements of the sliders Q and P, then its synthesis parameters are set, they are varied by the generator of LP_τ sequence [49] depending on the geometric constraint imposed by the negative CKC DE. Negative CKC DE imposes one geometric constraint on the movements of the links AB and BC of the passive CKC ABC; therefore, its synthesis parameters are determined.

Let us consider the parametric synthesis of the negative CKC *DE*. To do this, it is necessary to first determine the angle φ_{4i} by the expression

$$\varphi_{4_i} = \varphi_{(CA)_i} + \cos^{-1}\frac{l_{BC}^2 + l_{(CA)_i}^2 - l_{AB}^2}{l_{BC} l_{(CA)_i}}, \quad (1)$$

where

$$\varphi_{(CA)_i} = \mathrm{tg}^{-1}\frac{Y_{A_i} - Y_{C_i}}{X_{A_i} - X_{C_i}}, \quad (2)$$

$$l_{(CA)_i} = [(X_{A_i} - X_{C_i})^2 + (Y_{A_i} - Y_{C_i})^2]^{\frac{1}{2}}. \quad (3)$$

Coordinates X_{A_i}, Y_{A_i} and X_{C_i}, Y_{C_i} of the joints A and C in the absolute coordinate system OXY in Equations (2) and (3) are determined by the expressions

$$X_{A_i} = x_A^{(1)}, Y_{A_i} = s_{Q_i} + y_A^{(1)}, X_{C_i} = -s_{P_i} + x_C^{(2)}, Y_{Ci} = y_C^{(2)}. \quad (4)$$

With the link CB of the dyad ABC, we attach the coordinate system Cx_4y_4, the axis Cx_4 of which is directed along the link CB. Then, the synthesis parameters of the negative CKC DE are the vector $r_2 = \left[x_E^{(4)}, y_E^{(4)}, X_D, Y_D, l_{DE}\right]^T$, where $x_E^{(4)}, y_E^{(4)}$ and X_D, Y_D are the coordinates of the joints E and D in the coordinate systems Cx_4y_4 and OXY, respectively.

Let us consider the motion of the plane Cx_4y_4 in the absolute coordinate system OXY. In this case, the joint $E\left(x_E^{(4)}, y_E^{(4)}\right)$ moves along an arc of a circle with a center at the joint $D(X_D, Y_D)$ and a radius l_{DE}. Consequently, an equation of the geometric constraint imposed by the negative CKC DE of the RR type on the motion of the moving plane Cx_4y_4 is expressed as a weighted difference

$$\Delta q_i = (X_{E_i} - X_D)^2 + (Y_{E_i} - Y_D)^2 - l_{DE}^2, \quad (5)$$

where X_{E_i} and Y_{E_i} are the coordinates of the joint E in the absolute coordinate system OXY, which are determined by the expression

$$\begin{bmatrix} X_{E_i} \\ Y_{E_i} \end{bmatrix} = \begin{bmatrix} X_{C_i} \\ Y_{C_i} \end{bmatrix} + \begin{bmatrix} \cos\varphi_{4i} & -\sin\varphi_{4i} \\ \sin\varphi_{4i} & \cos\varphi_{4i} \end{bmatrix} \cdot \begin{bmatrix} x_E^{(2)} \\ y_E^{(2)} \end{bmatrix} \quad (6)$$

After substituting Equation (6) into Equation (5) and replacing the synthesis parameters of the form

$$\begin{bmatrix} p_1 \\ p_2 \end{bmatrix} = \begin{bmatrix} X_D \\ Y_D \end{bmatrix}, \begin{bmatrix} p_4 \\ p_5 \end{bmatrix} = \begin{bmatrix} x_E^{(4)} \\ y_E^{(4)} \end{bmatrix}, p_3 = \frac{1}{2}\left(X_D^2 + Y_D^2 + x_E^{(4)^2} + y_E^{(4)^2} - l_{ED}^2\right) \quad (7)$$

then Equation (5) is expressed linearly in two groups of synthesis parameters $\boldsymbol{p}_2^{(1)} = [p_1, p_2, p_3]^T$ and $\boldsymbol{p}_2^{(2)} = [p_4, p_5, p_3]^T$ in the form

$$\Delta q_i^{(k)}\left(\boldsymbol{p}_2^{(k)}\right) = 2\left(\boldsymbol{g}_i^{(k)^T} \cdot \boldsymbol{p}_2^{(k)} - g_{oi}^{(k)}\right), k = 1, 2, \quad (8)$$

where $g_i^{(k)}$ and $g_{oi}^{(k)}$ are the coefficients of the vectors $p_2^{(k)}$ and free terms depending on the remaining synthesis parameters, which have the form

$$\mathbf{g}_i^{(1)} = -\begin{bmatrix} X_{C_i} \\ Y_{C_i} \\ 1 \end{bmatrix} - \begin{bmatrix} \Gamma(\varphi_{4i}) & \begin{matrix} 0 \\ 0 \end{matrix} \\ \begin{matrix} 0 & 0 \end{matrix} & 1 \end{bmatrix} \cdot \begin{bmatrix} p_4 \\ p_5 \\ 0 \end{bmatrix} \tag{9}$$

$$\mathbf{g}_i^{(2)} = \begin{bmatrix} \Gamma(\varphi_{4i}) & \begin{matrix} 0 \\ 0 \end{matrix} \\ \begin{matrix} 0 & 0 \end{matrix} & 1 \end{bmatrix} \cdot \begin{bmatrix} X_{C_i} \\ Y_{C_i} \\ 1 \end{bmatrix} + \begin{bmatrix} \Gamma(\varphi_{4i}) & \begin{matrix} 0 \\ 0 \end{matrix} \\ \begin{matrix} 0 & 0 \end{matrix} & 1 \end{bmatrix} \cdot \begin{bmatrix} p_1 \\ p_2 \\ 0 \end{bmatrix} \tag{10}$$

$$g_{oi}^{(1)} = -\frac{1}{2}\left[X_{C_i}^2 + Y_{C_i}^2\right] + \left[X_{C_i}, Y_{C_i}\right] \cdot \Gamma(\varphi_{4i}) \begin{bmatrix} p_4 \\ p_5 \end{bmatrix} \tag{11}$$

$$g_{oi}^{(2)} = -\frac{1}{2}\left[X_{C_i}^2 + Y_{C_i}^2\right] - \left[X_{C_i}, Y_{C_i}\right] \cdot \begin{bmatrix} p_1 \\ p_2 \end{bmatrix}, \tag{12}$$

where $\Gamma(\alpha)$ is a rotation matrix of view

$$\Gamma(\alpha) = \begin{bmatrix} cos\alpha & -sin\alpha \\ sin\alpha & cos\alpha \end{bmatrix}. \tag{13}$$

The linear representability of Equation (5) allows one to formulate and solve the Chebyshev and least-square approximations for parametric synthesis [50]. In the Chebyshev approximation problem, the vectors of synthesis parameters are determined from the minimum of the functional

$$S^{(k)}\left(p_2^{(k)}\right) = \underset{i=\overline{1,N}}{max}\left|\Delta q_{ij}^{(k)}\left(p_2^{(k)}\right)\right|^2 \to \underset{p_2^{(k)}}{min} S^{(k)}\left(p_2^{(k)}\right) \tag{14}$$

In the least-square approximation problem, the synthesis parameters vectors are determined from the minimum of the functional

$$S^{(k)}\left(p_2^{(k)}\right) = \sum_{i=1}^{M}\left|\Delta q_i^{(k)}\left(p_2^{(k)}\right)\right|^2 \to \underset{p_2^{(k)}}{min} S^{(k)}\left(p_2^{(k)}\right). \tag{15}$$

The linear representability of Equation (5) allows the use of the kinematic inversion method, which is an iterative process, at each step of which one group of synthesis parameters $p_2^{(k)}$ is determined to solve the Chebyshev approximation problem (Equation (14)). In this case, the linear programming problem is solved [51]. To do this, we introduce a new variable $p' = \varepsilon$, where ε is the required approximation accuracy. Then the minimax problem (Equation (14)) leads to the following linear programming problem: determine the minimum of the sum

$$\sigma = \mathbf{c}^{\mathrm{T}} \cdot \mathbf{x} \to \underset{x}{min}\sigma, \tag{16}$$

with the following constraints

$$h_i'^T \cdot x + h_{0i} \geq 0, h''^T \cdot x - h_{0i} \geq 0, \tag{17}$$

where

$$c = [0, \ldots, 0, 1]^T, x = \left[p_2^{(k)}, p'\right]^T, h_i' = \left[-g_i^{(k)}, 0.5\right]^T, \tag{18}$$

$$h'' = \left[g_i^{(k)}, 0.5\right]^T, h_{0i} = g_{0i}. \tag{19}$$

The sequence of the obtained values of the function $S^{(k)}$ will decrease and have a limit as a sequence bounded below, because $S^{(k)}\left(p_2^{(k)}\right) \geq 0$ for any $p_2^{(k)}$.

Let consider the least-square approximation problem (Equation (15)) for the synthesis. The necessary conditions for the minimum of functions (Equation (15)) with respect to the parameters $p_2^{(k)}$

$$\frac{\partial S_i^{(k)}}{\partial p_2^{(k)}} = 0, \tag{20}$$

leading to the systems of linear equations

$$H^{(k)} \cdot p_2^{(k)} = h^{(k)}, \tag{21}$$

where

$$H^{(k)} = \sum_{i=1}^{N} \begin{bmatrix} g_{1i}^{(k)2} & g_{1i}^{(k)} g_{2i}^{(k)} & g_{1i}^{(k)2} \\ g_{1i}^{(k)} \cdot g_{2i}^{(k)} & g_{2i}^{(k)2} & g_{2i}^{(k)2} \\ g_{1i} & g_{2i}^{(k)2} & 1 \end{bmatrix}, \tag{22}$$

$$h^{(k)} = \sum_{i=1}^{N} \begin{bmatrix} g_{1i}^{(k)} \cdot g_{0i}^{(k)} \\ g_{2i}^{(k)} \cdot g_{0i}^{(k)} \\ g_{0i} \end{bmatrix}. \tag{23}$$

Solving these systems of linear equations for each group of synthesis parameters for given values of the remaining groups of synthesis parameters, we determine their values

$$r_2^{(k)} = H^{(k)^{-1}} \cdot h^{(k)}, \tag{24}$$

at $det(H^{(k)}) \neq 0$. If $det(H^{(k)}) = 0$, then the revolute kinematic pair is replaced by prismatic kinematic pair.

The matrix $H^{(k)}$ can be represented as a product $H_1 \cdot H_1{}^T$, where H_1 is a matrix with dimension $r \times N$ (in the considered case $r = 3$)

$$H_1 = \begin{bmatrix} g_{11}^{(k)} & g_{12}^{(k)} & \cdots & g_{1N}^{(k)} \\ g_{21}^{(k)} & g_{22}^{(k)} & \cdots & g_{2N}^{(k)} \\ \vdots & \vdots & \cdots & \vdots \\ g_{r1}^{(k)} & g_{r2}^{(k)} & \cdots & g_{rN}^{(k)} \end{bmatrix}. \tag{25}$$

According to the Binet-Cauchy formula [52], the determinant $detH^k$ becomes into the sum of the squares of all minors $H_{\alpha\beta\gamma}$ of order r (we assume that $N \geq r$) in the matrix H_1, compiled in ascending order of the column indices, i.e.,

$$detH^k = \sum_{1 \leq \alpha \leq \beta \leq \gamma} H^2_{\alpha\beta\gamma}. \tag{26}$$

Consequently, the determinant $detH^k$ is positive definite together with the principal minors and the solution of the set of linear Equation (21) corresponds to the minimum of the function S with respect to the parameters $p^{(k)}$. Hence, the least-square approximation problem (Equation (15)) can be solved by the linear iterations method, at each step of which one group of parameters $p^{(k)}$ is determined. The sequence of values of the function S will be decreasing and have a limit as a sequence bounded below.

4. Kinematic Analysis of the PM with Two Sliders

Given the synthesis parameters and the positions s_{Qi} of the hydraulic cylinder punch Q, it is necessary to find the kinematic parameters of the slider P.

This PM (Figure 6) has the structural formula

$$I(1) \rightarrow \text{III}(2,3,4,5), \tag{27}$$

i.e., the PM contains an Assur group of the third class with one external prismatic kinematic pair [41]. In the literature, there is no solution of kinematics of this type group.

4.1. Position Analysis

For position analysis of the considered PM, we use the method of conditional generalized coordinates [50]. According to this method, we remove the link 5 by disconnecting the elements of the joints D and E and select the slider P as a conditional input link due to the additional DOF that appears. Then, this PM of the third class is transformed into a mechanism of the second class with the structural formula

$$I(1) \rightarrow \mathrm{II}(3,4) \leftarrow I(2). \tag{28}$$

Derive the function

$$\Delta = l_{DE} - \widetilde{l}_{DE}, \tag{29}$$

where $\widetilde{l}_{DE}$ is a variable distance between the centers of the disconnected joints D and E, which is determined by the expression

$$\widetilde{l}_{DE} = \left[\left(Kh_{D_i} - Kh_E \right)^2 + \left(Y_{D_i} - Y_E \right)^2 \right]^{\frac{1}{2}}. \tag{30}$$

Coordinates Kh_{D_i} and Y_{D_i} of the joint D center in Equation (30) are determined by the equation

$$\begin{bmatrix} X_{D_i} \\ Y_{C_i} \end{bmatrix} = \begin{bmatrix} X_{C_i} \\ Y_{C_i} \end{bmatrix} + \begin{bmatrix} \cos\varphi_{4i} & -\sin\varphi_{4i} \\ \sin\varphi_{4i} & \cos\varphi_{4i} \end{bmatrix} \cdot \begin{bmatrix} x_D^{(4)} \\ y_D^{(4)} \end{bmatrix}, \tag{31}$$

where

$$\begin{bmatrix} X_{C_i} \\ Y_{C_i} \end{bmatrix} = \begin{bmatrix} -\left(s_{P_i} + x_C^{(2)} \right) \\ y_C^{(2)} \end{bmatrix}. \tag{32}$$

To determine the angle φ_{4_i} in Equation (31), we derive a vector ABC loop-closure equation

$$l_{AB}\boldsymbol{e}_{3_i} - l_{CDB}\boldsymbol{e}_{4_i} + l_{(CA)_i}\boldsymbol{e}_{(CA)_i} = 0, \tag{33}$$

$$l_{(CA)_i} = [(X_{A_i} - X_{C_i})^2 + (Y_{A_i} - Y_{C_i})^2]^{\frac{1}{2}}, \tag{34}$$

$$\varphi_{(CA)_i} = \mathrm{tg}^{-1}\frac{Y_{A_i} - Y_{C_i}}{X_{A_i} - X_{C_i}}, \tag{35}$$

$$\begin{bmatrix} X_{A_i} \\ Y_{A_i} \end{bmatrix} = \begin{bmatrix} -x_A^{(1)} \\ s_{Q_i} + y_A^{(1)} \end{bmatrix}. \tag{36}$$

In Equation (33) $\boldsymbol{e}$ denotes the unit vector.

We transfer $l_{AB}\boldsymbol{e}_{3_i}$ to the right side of Equation (33) and square both sides. As a result, we obtain

$$\varphi_{4i} = \varphi_{(CA)_i} + \cos^{-1}\frac{l_{CB}^2 + l_{(CA)_i}^2 - l_{AB}^2}{2l_{CB}l_{(CA)_i}}, \tag{37}$$

Next, we define

$$\begin{bmatrix} X_{B_i} \\ Y_{B_i} \end{bmatrix} = \begin{bmatrix} X_{C_i} \\ Y_{C_i} \end{bmatrix} + l_{CB}\begin{bmatrix} \cos\varphi_{4i} \\ \sin\varphi_{4i} \end{bmatrix}, \tag{38}$$

$$\varphi_{3i} = \mathrm{tg}^{-1}\frac{Y_{B_i} - Y_{A_i}}{X_{B_i} - X_{A_i}}. \tag{39}$$

Thus, Equation (29) is a function of one variable: the conditional generalized coordinate s_{P_i}, for the given values of the real generalized coordinate s_{Q_i}. Minimizing

Equation (29) with respect to a variable s_{P_i} by the bisection method [53], we determine its values for given values s_{Q_i}. In this case, the angles φ_{3i} and φ_{4i} are simultaneously determined. The angle φ_{5i} is determined by the expression

$$\varphi_{5i} = \text{tg}^{-1}\frac{Y_{D_i} - Y_E}{X_{D_i} - X_E}. \tag{40}$$

4.2. Analogues of Velocities and Accelerations

To solve the problems of analogues of velocities and accelerations of the PM with two sliders, we select its independent vector contours, the number of which is equal to half the number of links of the Assur group, i.e., it is equal to two. As independent vector contours, we choose the contours $OQ'ABCC'O$ and $OQ'ABDEO$, the vector loop-closure equations of which have the forms

$$\left.\begin{array}{c} l_{OQ'}\mathbf{e}_{OQ'} + l_{Q'A}\mathbf{e}_{Q'A} + l_{AB}\mathbf{e}_{3i} - l_{CB}\mathbf{e}_{4i} + l_{CC'}\mathbf{e}_{CC'} - l_{(OC')_i}\mathbf{e}_{OC'} = 0 \\ l_{OQ'}\mathbf{e}_{OQ'} + l_{Q'A}\mathbf{e}_{Q'A} + l_{AB}\mathbf{e}_{3i} - l_{DB}\mathbf{e}_{(DB)_i} - l_{ED}\mathbf{e}_{5i} - l_{OE}\mathbf{e}_{OE} = 0 \end{array}\right\} \tag{41}$$

Project the system of Equation (41) on the axes OX and OY of the absolute coordinate system OXY

$$\left.\begin{array}{c} l_{Q'A} + l_{AB}\cos\varphi_{3i} - l_{CB}\cos\varphi_{4i} - l_{(OC')_i} = 0 \\ l_{OQ'} + l_{AB}\sin\varphi_{3i} - l_{CB}\sin\varphi_{4i} + l_{CC'} = 0 \\ l_{Q'A} + l_{AB}\cos\varphi_{3i} - l_{DB}\cos(\varphi_{4i} - \alpha_4) - l_{ED}\cos\varphi_{5i} - l_{OE}\cos\varphi_{OE} = 0 \\ l_{OQ'} + l_{AB}\sin\varphi_{3i} - l_{DB}\sin(\varphi_{4i} - \alpha_4) - l_{ED}\sin\varphi_{5i} - l_{OE}\sin\varphi_{OE} = 0 \end{array}\right\} \tag{42}$$

Differentiate the system of Equation (42) with respect to the generalized coordinate s_{Q_i}

$$\left.\begin{array}{c} -l_{AB}\sin\varphi_{3i}\cdot\varphi'_{3i} - l_{CB}\sin\varphi_{4i}\cdot\varphi'_{4i} - u_{P_i} = 0 \\ 1 + l_{AB}\cos\varphi_{3i}\cdot\varphi'_{3i} - l_{CB}\cos\varphi_{4i}\cdot\varphi'_{4i} = 0 \\ -l_{AB}\sin\varphi_{3i}\cdot\varphi'_{3i} + l_{DB}\sin(\varphi_{4i} - \alpha_4)\cdot\varphi'_{4i} + l_{ED}\sin\varphi_{5i}\cdot\varphi'_{5i} = 0 \\ 1 + l_{AB}\cos\varphi_{3i}\cdot\varphi'_{3i} - l_{DB}\cos(\varphi_{4i} - \alpha_4)\cdot\varphi'_{4i} - l_{ED}\cos\varphi_{5i}\cdot\varphi'_{5i} = 0 \end{array}\right\} \tag{43}$$

From the system of Equation (43) we determine the analogues of velocities

$$\mathbf{u} = \mathbf{A}^{-1}\cdot\mathbf{b}, \tag{44}$$

where

$$\mathbf{A} = \begin{bmatrix} Y_{A_i} - Y_{B_i} & Y_{B_i} - Y_{C_i} & 0 & -1 \\ X_{B_i} - X_{A_i} & X_{C_i} - X_{B_i} & 0 & 0 \\ Y_{A_i} - Y_{B_i} & Y_{B_i} - Y_{D_i} & Y_{D_i} - Y_E & 0 \\ X_{B_i} - X_{A_i} & X_{D_i} - X_{B_i} & X_E - X_{D_i} & 0 \end{bmatrix} \tag{45}$$

$$\mathbf{u} = \begin{bmatrix} \varphi'_{3_i} \\ \varphi'_{4_i} \\ \varphi'_{5_i} \\ u_{P_i} \end{bmatrix}, \mathbf{b} = \begin{bmatrix} 0 \\ -1 \\ 0 \\ 0 \end{bmatrix}. \tag{46}$$

Differentiate the system of Equation (43) with respect to the generalized coordinate s_{Q_i}

$$\left.\begin{array}{c} -l_{AB}\cos\varphi_{3i}\cdot\varphi'^2_{3i} - l_{AB}\sin\varphi_{3i}\cdot\varphi''_{3i} + l_{CB}\cos\varphi_{4i}\cdot\varphi'^2_{4i} + l_{CB}\sin\varphi_{4i}\cdot\varphi''_{4i} - w''_{P_i} = 0 \\ -l_{AB}\sin\varphi_{3i}\cdot\varphi'^2_{3i} + l_{AB}\cos\varphi_{3i}\cdot\varphi''_{3i} + l_{CB}\sin\varphi_{4i}\cdot\varphi'^2_{4i} + l_{CB}\cos\varphi_{4i}\cdot\varphi''_{4i} = 0 \\ -l_{AB}\cos\varphi_{3i}\cdot\varphi'^2_{3i} - l_{AB}\sin\varphi_{3i}\cdot\varphi''_{3i} + l_{DB}\cos(\varphi_{4i} - \alpha_4)\cdot\varphi'^2_{4i} + \\ +l_{DB}\sin(\varphi_{4i} - \alpha_4)\cdot\varphi''_{4i} + l_{ED}\cos\varphi_{5i}\cdot\varphi'^2_{5i} + l_{ED}\sin\varphi_{5i}\cdot\varphi''_{5i} = 0 \\ -l_{AB}\sin\varphi_{3i}\cdot\varphi'^2_{3i} - l_{AB}\cos\varphi_{3i}\cdot\varphi''_{3i} + l_{DB}\sin(\varphi_{4i} - \alpha_4)\cdot\varphi'^2_{4i} - \\ -l_{DB}\cos(\varphi_{4i} - \alpha_4)\cdot\varphi''_{4i} + l_{ED}\sin\varphi_{5i}\cdot\varphi'^2_{5i} - l_{ED}\cos\varphi_{5i}\cdot\varphi''_{5i} = 0 \end{array}\right\} \tag{47}$$

Then we obtain the analogues of accelerations

$$w = A^{-1} \cdot c, \tag{48}$$

where

$$w = \begin{bmatrix} \varphi''_{3_i} \\ \varphi''_{4_i} \\ \varphi''_{5_i} \\ w_{P_i} \end{bmatrix}, \; c = \begin{bmatrix} (X_{B_i} - X_{A_i}) \cdot \varphi'^2_{3_i} + (X_{C_i} - X_{B_i}) \cdot \varphi'^2_{4_i} \\ (Y_{B_i} - Y_{A_i}) \cdot \varphi'^2_{3_i} + (Y_{C_i} - Y_{B_i}) \cdot \varphi'^2_{4_i} \\ (X_{B_i} - X_{A_i}) \cdot \varphi'^2_{3_i} + (X_{D_i} - X_{B_i}) \cdot \varphi'^2_{4_i} + (X_{E_i} - X_{D_i}) \cdot \varphi'^2_{5_i} \\ (Y_{B_i} - Y_{A_i}) \cdot \varphi'^2_{3_i} + (Y_{D_i} - Y_{B_i}) \cdot \varphi'^2_{4_i} + (Y_{E_i} - Y_{D_i}) \cdot \varphi'^2_{5_i} \end{bmatrix} \tag{49}$$

5. Numerical Results and Prototyping

$N = 11$ positions s_{Q_i} and s_{P_i} of the input and output sliders of the PM with two sliders are shown in Table 1.

Table 1. Positions of the input and output sliders.

i	1	2	3	4	5	6	7	8	9	10	11
s_{Q_i}, mm	0	6.0	12.0	18.0	24.0	30.0	36.0	42.0	48.0	54.0	60
s_{P_i}, mm	−97.88	−87.86	−80.29	−73.46	−66.8	−59.97	−52.63	−44.31	−34.18	−19.99	0

Tables 2 and 3 show the obtained values of the synthesis parameters of the passive CKC *ABC* and negative CKC *DE*, respectively.

Table 2. Synthesis parameters of the passive CKC *ABC*.

$x_A^{(1)}$,mm	$y_A^{(1)}$,mm	$x_C^{(2)}$,mm	$y_C^{(2)}$,mm	l_{AB},mm	l_{BC},mm
−7.5012	2.0817	−2.5335	1.7562	60.0174	100.0207

Table 3. Synthesis parameters of the negative CKC *DE*.

$x_D^{(4)}$,mm	$y_D^{(4)}$,mm	X_E,mm	Y_E,mm	l_{ED},mm
50.0628	−20.0408	−69.5361	67.9353	60.7365

3D CAD model of the synthesised PM with two sliders is shown in Figure 7.

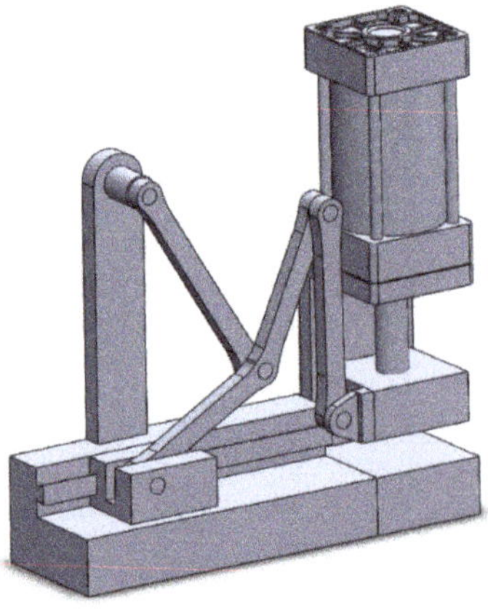

Figure 7. 3D CAD model of the PM with two sliders.

Table 4 shows the obtained values of the positions s_{P_i} and analogues of the linear velocities u_{P_i} and linear accelerations w_{P_i} of the output slider P.

Table 4. Positions and anologues of the linear velocities and accelerations of the output slider P.

i	1	2	3	4	5	6	7	8	9	10	11
s_{P_i}, mm	−97.88	−87.86	−80.29	−73.46	−66.8	−59.97	−52.63	−44.31	−34.18	−19.99	0
u_{P_i},	29.0364	24.8392	20.6181	16.0497	10.9754	5.2438	−1.3316	−9.0244	−18.3301	−30.4705	−44.6353
w_{P_i}, mm^{-1}	1.8788	1.4345	1.1417	08.1868	07.3624	1.02159	1.4427	2.2557	3.5165	3.6692	1.9001

Graphics of the parameters s_{P_i}, u_{P_i}, w_{P_i} are shown in Figure 8.

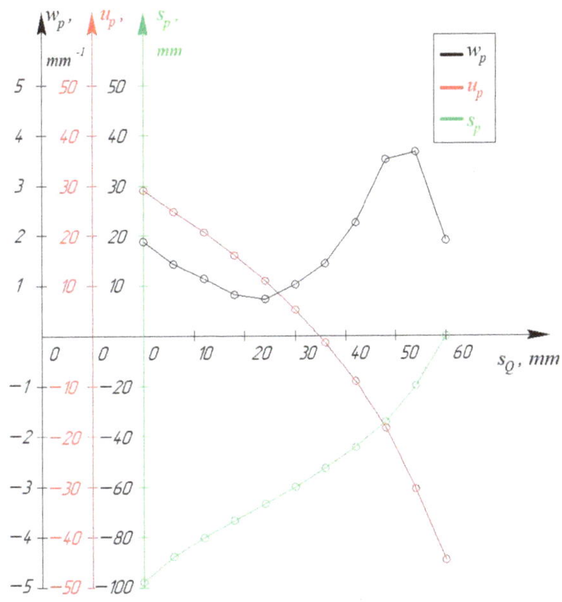

Figure 8. Graphics of the parameters s_{P_i},u_{P_i},w_{P_i}.

A prototype of the PM with two sliders, and a block scheme of its characteristics are shown in Figures 9 and 10, respectively.

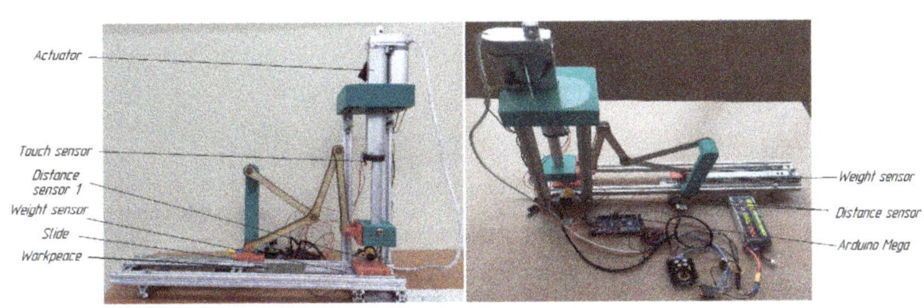

(a) (b)

Figure 9. Prototype of the PM with two sliders: (**a**) the first position (**b**) the second position.

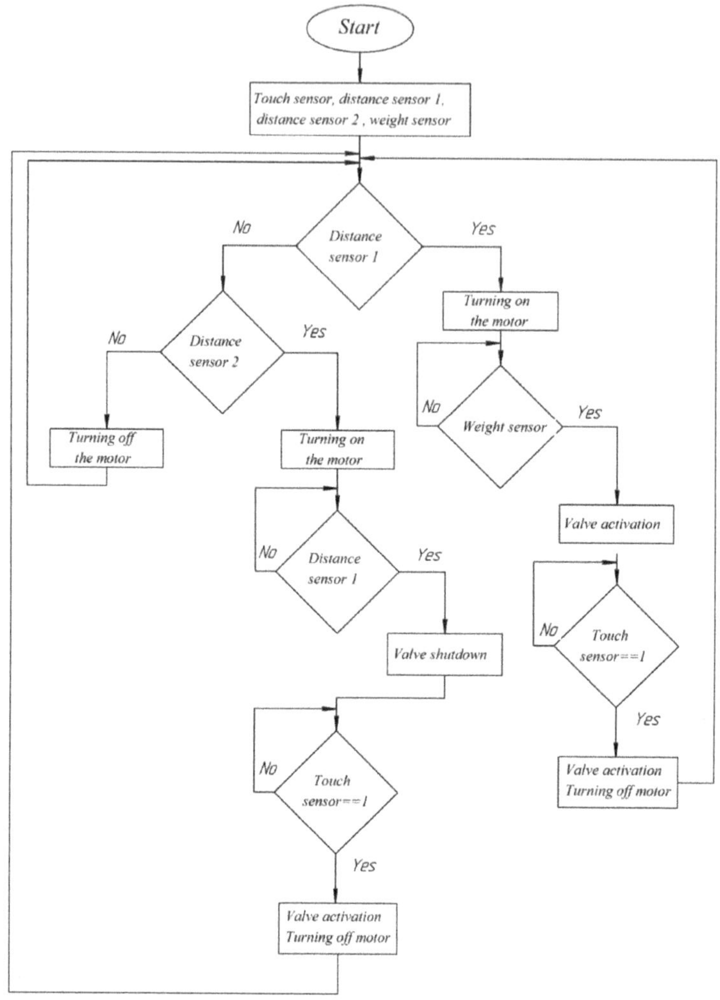

Figure 10. Block scheme of the PM with two sliders work.

At the beginning, the hydraulic cylinder punch is located at the upper extreme position, and the distance sensor 1 checks a presence of the workpiece for stamping in the die. If there is no workpiece, then the presence of the workpiece in the store is checked using the distance sensor 2. If there is no workpiece in the store, the motor does not turn on. If there is the workpiece in the store, the motor turns on and the punch moves to the lower working position. At this time, the PM slider moves to the left extreme position for the next workpiece.

The first stroke of the punch will be idle. After reaching the lower extreme position of the punch, the distance sensor 1 gives the command to the hydraulic cylinder valve to switch and the punch rises until reaching the touch sensor. In this case, the hydraulic cylinder valve switches, the hydraulic cylinder motor is turned off, and the PM delivers the workpiece to the die (working area). Further, the distance sensor 1 checks the presence of the workpiece in the die.

After delivering the workpiece to the die, the motor turns on and the punch goes down to the lower working position, where the weight sensor is located, which regulates the press force for high-quality stamping. The hydraulic cylinder valve switches after punching, and the punch rises until it reaches the touch sensor. In this case, the hydraulic cylinder valve is switched and its motor is turned off. This cycle is repeated until the end of the workpieces in the store. The connection scheme of the sensors and motor is shown in Figure 11.

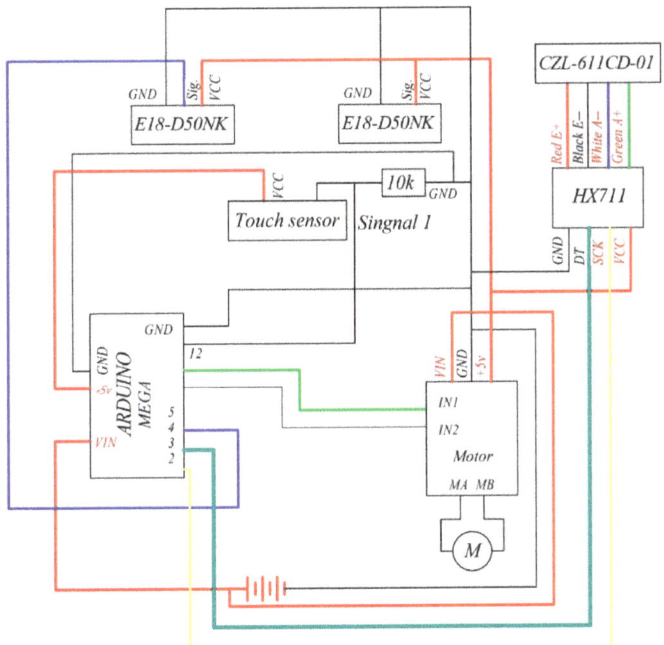

Figure 11. Connection scheme of the sensors and motor.

6. Conclusions

The scheme of a cold-stamping technological line with the use of RoboMech class PMs has been developed. This technological line uses three RoboMech class PMs: a PM with two sliders, a PM with two end-effectors, and a PM working in a cylindrical coordinate system. The PM with two sliders is formed by connecting two sliders (input and output objects) and a base using one passive and one negative CKC. The formed PM with two sliders contains an Assur group of the third class with one external prismatic kinematic pair. Geometric parameters of the negative CKC are determined on the basis of the Chebyshev and least-square approximations. The problem of the positions of the PM with two sliders is solved using the method of conditional generalized coordinates. The 3D CAD model and prototype of the PM with two sliders have been made.

Author Contributions: Z.B., M.A.L., G.C. developed methods of parametric synthesis and kinematic analysis, A.M., B.A., Y.Z. performed numerical calculations and prototyping of the RoboMech class PM with two sliders. All authors have read and agreed to the published version of the manuscript.

Funding: This research is funded by the Science Committee of the Ministry of Education and Science of Kazakhstan (Grant No AP08857522).

Institutional Review Board Statement: Not applicable.

Informed Consent Statement: Not applicable.

Data Availability Statement: The data presented in the study are available on request from the corresponding author.

Conflicts of Interest: The authors no conflict of interest.

References

1. Fu, K.S.; Gonzalez, Z.C.; Lee, C.S.G. *Robotics: Control, Sensing, Vision and Intelligence*; McGraw–Hill Book Company: London, UK, 1987.
2. Yoshikava, T. *Foundation of Robotics: Analysis and Control*; MIT Press: Cambridge, MA, USA, 1990.
3. Gupta, K.C. *Mechanics and Control of Robots*; Springer: New York, NY, USA, 1997.
4. Spong, M.W.; Hutchinson, S.; Vidyasagar, M. *Robot Modeling and Control*; Hobekon, N.J., Ed.; John Wiley & Sons, Inc.: Hoboken, NJ, USA, 2006.
5. Song, S.M.; Lee, J.K.; Waldron, K.J. Motion Study of Two—and Three–Dimensional Panthograph Mechanisms. *Mech. Mach. Theory* **1987**, *22*, 321–331. [CrossRef]
6. Waldron, K.J.; Kinzel, G.L. The Relation between Actuator Geometry and Mechanical Efficiency in Robots. In Proceedings of the 4th CISM–IFToMM Symposium on Theory and Practices of Manipulators, Zaborow, Poland, 8–12 September 1981; pp. 366–374.
7. Baigunchekov, Z.; Kalimoldayev, M.; Ibrayev, S.; Izmambetov, M.; Baigunchekov, T.; Naurushev, B.; Aisa, N. Parallel Manipulator of a Class RoboMech. *Lect. Notes Electr. Eng.* **2017**, *408*, 547–557.
8. Assur, L.V. Investigation of plane hinged mechanism with lower pairs from the point of view of their structure and classification (in Russian). *Bull. Petrograd Polytech. Inst.* **1914**, *21*, 187–283.
9. Hunt, K.H. *Kinematic Geometry of Mechanisms*; Clarendon Press: Oxford, UK, 1998.
10. Kong, X.; Gosselin, C.M. *Type Synthesis of Parallel Mechanisms*; Springer: Berlin/Heidelberg, Germany, 2007.
11. Sunkari, R.P.; Schmidt, L.C. Structural Synthesis of Planar Kinematic Chains by Adapting McKay–Type Algorithm. *Mech. Mach. Theory* **2006**, *41*, 1021–1030. [CrossRef]
12. Gogu, G. *Structural Synthesis of Parallel Robots. Part 3: Topologies with Planar Motion of the Moving Platform*; Springer: Dordrecht, The Netherlands, 2010.
13. Notash, L.; Zhang, J. *Structural Synthesis of Kinematic Chains and Mechanisms. Integrated Design and Manufacturing in Mechanical Engineering*; Springer: Dordrecht, The Netherlands, 2002; pp. 391–398.
14. Huang, P.; Ding, H. Structural Synthesis of Assur Groups with up to 12 Links and Creation of Their Classified Databases. *Mech. Mach. Theory* **2020**, *145*, 103668. [CrossRef]
15. Liu, Y.; Li, Y.; Yao, Y.A.; Kong, X. Type Synthesis of Multi-mode Mobile Parallel Mechanisms Based on Refined Virtual Chain Approach. *Mech. Mach. Theory* **2020**, *152*, 103908. [CrossRef]
16. Meng, X.; Gao, F.; Wu, S.; Ge, Q.J. Type synthesis of Parallel Robotic Mechanisms: Framwork and Brief Review. *Mech. Mach. Theory* **2014**, *78*, 177–186. [CrossRef]
17. Burmester, L. *Lehrbuch der Kinematik*; A. Felix Verlag: Leipzig, Germany, 1988.
18. Schoenflies, A. *Geometric der Bewegung in Synthetischer Darstellung*; BG Teubner: Leipzig, Germany, 1886.
19. Bottema, O.; Roth, B. *Theoretical Kinematics*; North-Holland Publishing Company: Amsterdam, The Netherlands, 1979.
20. Bai, S.; Li, Z.; Li, R. Exact Synthesis and Input–Output Analysis of 1-DOF Planar Linkages for Visiting 10 Poses. *Mech. Mach. Theory* **2020**, *143*, 103625. [CrossRef]
21. Hamida, I.B.; Laribi, M.A.; Mlika, A.; Romdhane, L.; Zeghloul, S. Dimensional Synthesis and Performance Evaluation of Four Translational Parallel Manipulators. *Robotica* **2021**, *39*, 233–249. [CrossRef]
22. Chebyshev, P.L. Sur Les Parallélogrammes Composés de Trois Éléments Quelconques. *Mémoires l'Académie Sci. St.-Pétersbourg* **1879**, *36*, 1–16.
23. Levitskii, N.I. *Design of Plane Mechanisms with Lower Pairs*; Izd. Akad. Nauk USSR: Moscow/Leningrad, Russia, 1950.
24. Sarkissyan, Y.L.; Gupta, K.C.; Roth, B. Kinematic Geometry Associated with the Least-Square Approximation of a Given Motion. *Trans. ASME Ser. B* **1973**, *95*, 503–510. [CrossRef]
25. Sarkissyan, Y.L.; Gupta, K.C.; Roth, B. Spatial Least-Square Approximation of a Motion. In Proceedings of the IFToMM International Symposium on Linkages and Computer Design Methods, Bucharest, Romania, 7–13 June 1973.
26. Sarkissyan, Y.L.; Gupta, K.C.; Roth, B. Chebyshev Approximations of Finite Point Sets with Application to Planar Kinematic Synthesis. *Trans. ASME J. Mech. Des.* **1979**, *101*, 32–40.
27. Sarkissyan, Y.L.; Gupta, K.C.; Roth, B. Chebyshev Approximations of Spatial Point Sets Using Spheres and Planes. *Trans. ASME J. Mech. Des.* **1979**, *101*, 499–503.
28. Paul, B. *Kinematics and Dynamics of Planar Machinery*; Prentice-Hall: Upper Saddle River, NJ, USA, 1979.
29. Hartenberg, R.S.; Denavit, J. *Kinematic Synthesis of Linkages*; McGraw-Hill: New York, NY, USA, 1964.
30. Rodhavan, M.; Roth, B. Solving Polynomial Systems for the Kinematic Analysis and Synthesis of Mechanisms and Robot Manipulators. *J. Mech. Des.* **1995**, *117*, 71–79.
31. Erdman, A.G.; Sandor, G.N.; Kota, S. *Mechanism Design: Analysis and Synthesis Lecture Notes*; McGill University: Montreal, QC, Canada, 2001.
32. Uicker, J.J.; Pennack, G.R.; Shigley, J.E. *Theory of Machines and Mechanisms*, 4th ed.; Oxford University Press: New York, NY, USA, 2001.
33. Hernandez, A.; Petya, V. Position Analysis of Planar Mechanisms with R-pairs Using Geometrical-Iterative Method. *Mech. Mach. Theory* **2004**, *39*, 50–61. [CrossRef]

34. Pennock, G.R.; Israr, A. Kinematic Analysis and Synthesis of an Adjustable Six-Bar Linkage. *Mech. Mach. Theory* **2009**, *44*, 306–323. [CrossRef]
35. McCarthy, J.M.; Soh, G.M. *Geometric Design of Linkages*, 2nd ed.; Springer Science & Business Media: Berlin/Heidelberg, Germany, 2010.
36. Angeles, J.; Bai, S. *Kinematic Synthesis, Lecture Notes*; McGill University: Montreal, QC, Canada, 2016.
37. Arikawa, K. Kinematic Analysis of Mechanisms Based on Parametric Polynomial System: Basic Concept of a Method Using Gröbner Cover and Its Application to Planar Mechanisms. *J. Mech. Robot.* **2019**, *11*, 020906. [CrossRef]
38. Baskar, A.; Plecnik, M. Synthesis of Watt-Type Timed Curve Generators and Selection from Continuous Cognate Spaces. *J. Mech. Robot.* **2021**, *13*, 051003. [CrossRef]
39. Fomin, A.; Antonov, A.; Glazunov, V.; Carbone, G. Dimensional (Parametric) Synthesis of the Hexapod-Type Parallel Mechanism with Reconfigurable Design. *Machines* **2021**, *9*, 117. [CrossRef]
40. Duffy, J. *Analysis of Mechanisms and Robot Manipulator*; Edward Arnold: London, UK, 1980.
41. Tsai, L.W. *Robot Analysis: The Mechanics of Serial and Parallel Manipulators*; John Wiley and Sons, Inc.: New York, NY, USA, 1999.
42. Glazunov, V. Design of Decoupled Parallel Manipulators by Means of the Theory of Screws. *Mech. Mach. Theory* **2010**, *45*, 209–230. [CrossRef]
43. McCarthy, J.M. Kinematics, Polynomials, and Computers-A Brief History. *J. Mech. Robot.* **2011**, *3*, 010201–010203. [CrossRef]
44. McCarthy, J.M. 21st Kinematics Synthesis, Compliance, and Tensegrity. *J. Mech. Robot.* **2011**, *3*, 020201–020203. [CrossRef]
45. Baigunchekov, Z.; Ibrayev, S.; Izmambetov, M.; Baigunchekov, T.; Naurushev, B.; Mustafa, A. Synthesis of Cartesian Manipulator of a Class RoboMech. In *Mechanisms and Machine Science*; Springer: Berlin/Heidelberg, Germany, 2019; Volume 66, pp. 69–76.
46. Grabchenko, I.P. *Introduction to Mechatronics*; HPI: Harkov, Ukraine, 2017.
47. Baigunchekov, Z.; Mustafa, A.; Tarek, S.; Patel, S.; Utenov, M. A Robomech Class Parallel Manipulator with Three Degrees of Freedom. *East.-Eur. J. Enterp. Technol.* **2020**, *3*, 44–56. [CrossRef]
48. Artobolevskiy, I.I. *Theory of Mechanisms and Machines*; Nauka: Moscow, Russia, 1988.
49. Sobol, I.M.; Statnikov, R.B. *Choice of Optimal Parameters in Problems with Many Criteria*; Nauka: Moscow, Russia, 1981.
50. Baigunchekov, Z.; Naurushev, B.; Zhumasheva, Z.; Mustafa, A.; Kairov, R.; Amanov, B. Structurally Parametric Synthesis and Position Analysis of a RoboMech Class Parallel Manipulator with Two End-Effectors. *IAENG Int. J. Appl. Math.* **2020**, *5*, 1–11.
51. Gill, P.E.; Murray, W.; Wright, M.H. *Practical Optimisation*; Academic Press: London, UK; New York, NY, USA; Toronto, ON, Canada; San Francisco, CA, USA, 1981.
52. Korn, G.A.; Korn, T.M. *Mathematical Handbook for Scientifists and Engineers*; Dover Publications. Inc.: Mineola, NY, USA, 2000.
53. Burden, R.L.; Faires, J.D. *Numerical Analisys*, 3rd ed.; PWS Publishers: Boston, MA, USA, 1985.

applied sciences

MDPI

Article

A Parallel Robot with Torque Monitoring for Brachial Monoparesis Rehabilitation Tasks

Doina Pisla [1], Daniela Tarnita [2], Paul Tucan [1,*], Nicoleta Tohanean [3], Calin Vaida [1] Ionut Daniel Geonea [2], Gherman Bogdan [1], Cristian Abrudan [3], Giuseppe Carbone [4] and Nicolae Plitea [1]

1 CESTER, Technical University of Cluj-Napoca, 400641 Cluj-Napoca, Romania; doina.pisla@mep.utcluj.ro (D.P.); calin.vaida@mep.utcluj.ro (C.V.); bogdan.gherman@mep.utcluj.ro (G.B.); nicolae.plitea@mep.utcluj.ro (N.P.)
2 Department of Applied Mechanics and Civil Construction, Faculty of Mechanics, University of Craiova, 200585 Craiova, Romania; daniela.tarnita@edu.ucv.ro (D.T.); ionut.geonea@edu.ucv.ro (I.D.G.)
3 Department of Neurology, University of Medicine and Pharmacy "Iuliu Hatieganu", 400012 Cluj-Napoca, Romania; nicoleta_alexa@yahoo.com (N.T.); cristian.abrudan@yahoo.fr (C.A.)
4 DIMEG, University of Calabria, 87036 Cosenza, Italy; giuseppe.carbone@unical.it
* Correspondence: paul.tucan@mep.utcluj.ro

Abstract: Robots for rehabilitation tasks require a high degree of safety for the interaction with both the patients and for the operators. In particular, high safety is a stable and intuitive control of the moving elements of the system combined with an external system of sensors able to monitor the position of every aspect of the rehabilitation system (operator, robot, and patient) and overcome in a certain measure all the events that may occur during the robotic rehabilitation procedure. This paper presents the development of an internal torque monitoring system for ASPIRE. This is a parallel robot designed for shoulder rehabilitation, which enables the use of strategies towards developing a HRI (human–robot interaction) system for the therapy. A complete analysis regarding the components of the robotic system is carried out with the purpose of determining the dynamic behavior of the system. Next, the proposed torque monitoring system is developed with respect to the previously obtained data. Several experimental tests are performed using healthy subjects being equipped with a series of biomedical sensors with the purpose of validating the proposed torque monitoring strategy and, at the same time, to satisfy the degree of safety that is requested by the medical procedure.

Keywords: robotic rehabilitation; brachial monoparesis; torque monitoring; human–robot interaction

Citation: Pisla, D.; Tarnita, D.; Tucan, P.; Tohanean, N.; Vaida, C.; Geonea, I.D.; Bogdan, G.; Abrudan, C.; Carbone, G.; Plitea, N. A Parallel Robot with Torque Monitoring for Brachial Monoparesis Rehabilitation Tasks. *Appl. Sci.* **2021**, *11*, 9932. https://doi.org/10.3390/app11219932

Academic Editor: Manuel Armada

Received: 13 September 2021
Accepted: 20 October 2021
Published: 24 October 2021

Publisher's Note: MDPI stays neutral with regard to jurisdictional claims in published maps and institutional affiliations.

1. Introduction

Monoplegia is a type of paralysis that affects one of the limbs (arm or leg), permanently or temporarily [1].When monoplegia is located on one of the upper limbs it is referred to as brachial monoplegia and if it is located on one of the lower limbs it is referred to as crural monoplegia [2].There are several neurological disorders that may cause monoplegia, most common are: stroke, cerebral palsy, Guillain–Barre syndrome, peripheral neuropathy, Lou Gehrig's disease, head or spinal trauma, and neuropathy. With respect to the neurological disease that caused the monoplegia, sometimes it can evolve to paraplegia or hemiplegia. A less severe form of monoplegia is called monoparesis and monoparesis of the upper limb is referred as brachial monoparesis. Pure brachial monoparesis is usually caused by stroke [3] and its effects are weakness, spasticity, numbness, paralysis, pain, and headaches. Stroke may occur in any age group, but it is more frequent in ages of 55 to 85. In 2017 United Nations released a series of highlights regarding trending in the ageing of the worldwide population [4]. According to these statistics in 2017 population aged over 60 years numbered 962 million worldwide. The same study reveals that by 2030 the number will rise to 1.41 billion and by 2050 it will surpass 2 billion.

There is no cure for brachial monoparesis but with the help of physical therapy performed under the supervision and help of a physiotherapist, some muscular tonus and functionalities may be regained [5]. Physical therapy of brachial monoparesis implies repetitive motion of the upper limb with motor deficit. With respect to the affected area (shoulder, elbow, wrist, fingers) different rehabilitation motions are used, targeting the specific motion of each articulation, but each motion must be done by special medical personnel.

The capability of the NHS (National Health System) to provide medical care for each person in need has been seriously stressed lately. Previous statistics preconized that in 2030 the NHS will be unable to provide medical care for the people in need due to personal shortage [5]; momentarily there is a crisis regarding staff recruiting in the NHS [6] so this prevision seams quite close. All the above data raises a question: Who will provide care for the ones in need in the near future? There is no simple and straight forward answer to this question, but there are research domains trying to overcome the predicted personnel crisis and one of these domains is medical robotics. Developing robotic solutions for the medical care should not imply substituting the medical personnel from the operation/therapy room but providing means for the health careers to deliver a more precise and effective treatment with less effort and allowing them to serve more efficiently a larger number of patients. The effectiveness of the physical therapy when discussing rehabilitation of stroke or other neurological disease survivors is largely dependent on how fast the treatment is delivered to the patient. With the help of robotic devices this treatment may be delivered in the acute phase, immediately after patient stabilization, following a stroke incident.

Robotic devices for brachial monoparesis should be able to mold the impaired upper limb and mimic the rehabilitation motion of the targeted anatomical articulation. Several solutions for upper limb rehabilitation have been developed; one can classify these solutions in exoskeletons and end-effectors systems [7].

Gull et al. [8] developed an adaptive 4-DOF exoskeleton for upper limb rehabilitation, mountable on a wheelchair, able to carry the patient throughout daily activities like eating and drinking. The exoskeleton uses an open-chain structure to mimic the anatomy of the upper limb joints and it is divided into three modules: shoulder module, elbow module, and wrist module. The exoskeleton uses a proportional-derivative control method implemented using ROS (Robotic Operating System) and it uses the dynamic model of the mechanism and the weights of the upper limb segments to hold the arm in position during daily tasks. The system has yet to undergo clinical trials and user acceptance.

Studies regarding the development of upper limb exoskeletons were also published in [9–12], some of them being based on the study of the dynamic behavior of the exoskeletons and the possibility to control the robotic solutions to maximize the efficiency of the rehabilitation.

Paolucci [13] presents MOTORE (Mobile Robot for upper limb neuro-ortho rehabilitation), an end-effector based robotic system for upper limb rehabilitation able to offer different modes of operation ranging from offering support for functional patients to assisted movement for patients with severe hemiplegia. The device comes with a visual feedback system that helps in performing daily tasks such reaching and carrying with the goal of rehabilitating the shoulder and the elbow. The device underwent clinical trials resulting in a good efficiency of the rehabilitation, reduction of spasticity, and improvements in the upper limb functions and at the same time reduction of the pain. The authors also stated a series of limitations of the study like neuropsychological limitations, absence of instrumental evaluation to assess the correlation between the variation of spasticity and muscle strength, and lack of analysis at aphasia level. There are several studies regarding the development of end-effector based rehabilitation devices [14–17], some of them developed until the stage of clinical trials.

All the above studies have in common use of several tools to achieve the safety required by the medical procedure. The safety may be achieved through an effective correlation between the kinematics of the medical robot [18–23] and the control system [24–31] used to actuate the robotic structure with respect to human–robot interactions [32–35].

Regarding rehabilitation robots where human–robot interaction is required, the safety of the patient during the rehabilitation procedure should be of utmost importance. For achieving a high level of safety, the dynamic behavior of the robotic structure should be carefully and completely analyzed, and the control system should be able to perform the rehabilitation procedure with respect to the data obtained after the dynamic behavior analysis and at the same time to be able to interact with the patient during the procedure in order to react to any deviation from the initial rehabilitation plan.

This paper focuses on the development of an internal torque monitoring system for a rehabilitation system. Section 2.1 of the paper presents the ASPIRE robotic structure. The mechanical structure passed through a few optimization phases to improve the behavior of the structure during the rehabilitation procedure. The dynamic behavior of each rehabilitation module is studied in Section 2.2. to obtain the nominal working torques and the torque monitoring system is developed with respect to the obtained results regarding the dynamic behavior in Section 2.3 followed by Section 2.4 where the torque monitoring system is experimentally tested and validated. The third section of the paper consists of results of the research, followed by discussions and conclusions.

2. Materials and Methods

2.1. Short Description of ASPIRE

ASPIRE is a parallel robot for shoulder rehabilitation, whose design is outlined in [36]. The robotic structure is based on a spherical architecture, and it is capable of performing flexion, extension, adduction, and abduction of the shoulder. The system is also able to perform the pronation and supination of the forearm. The kinematic scheme of ASPIRE robotic structure is presented in Figure 1 [37]. The robot is composed of two modules: the first module is the one containing tools for performing the shoulder rehabilitation procedure (one axis for each rehabilitation motion). The OXYZ reference system indicates the center of shoulder joint, and all rehabilitation motions are described using this point. The adduction-abduction motion is performed using the q_1 axis that rotates a circular guide (G_1) attached to circular guide G_2 using a passive revolute joint (r_f). The flexion-extension motion is performed by q_2 axis that rotates the circular guide G_2. The interconnected circular guides impose a spherical trajectory of O′X′Y′Z′ reference system with respect to the shoulder joint center. The pronation-supination motion is performed using axis q_3 that reorients the O′X′Y′Z′ reference system around X′ axis.

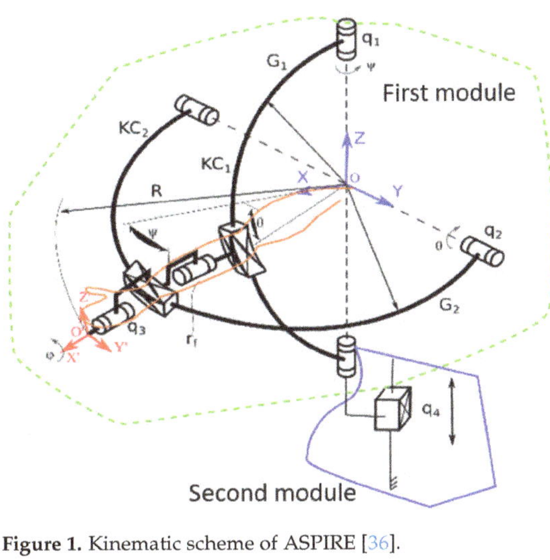

Figure 1. Kinematic scheme of ASPIRE [36].

The second module is used for adjusting the rehabilitation mechanisms based on patient height and it contains one axis (q_4) that displaces vertically the OXYZ reference system of the structure. The motion amplitudes for each rehabilitation motions are

±80 degrees [38] and the limb positions during the rehabilitation may be seen in Figure 2a. During the rehabilitation procedure, the patient is seated, and his right shoulder is pressed against the shoulder plate of the robot; the forearm is attached within the forearm support. His palm is placed on the pronation-supination mechanism and held in place by an elastic band as it can be seen in Figure 2b where the experimental model of the ASPIRE is presented.

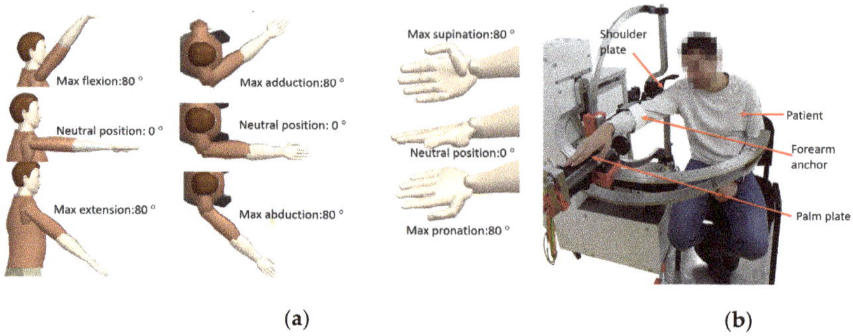

Figure 2. Arm position during rehabilitation and motion amplitudes (**a**); patient position during rehabilitation, experimental model (**b**).

At the end of 2019, the robotic structure was enrolled for first set of clinical trials within a rehabilitation hospital in Cluj-Napoca. During this clinical trial, among validating the functionality of the system, a series of risk factors were identified and analyzed. The identified risk factors along with their associated risk and overcome measures are given in Table 1. Torque monitoring was revealed to be part of solution to some of the identified risks and the next step was to design the manner that torque should influence the rehabilitation procedure.

Table 1. Risk factors identified during the physical rehabilitation.

Risk Factor	Associated Risk	Risk Overcome Measure
Tremor	The patient may detach from the rehabilitation device. Creates supplementary jerk in the system.	By monitoring the torque, the rehabilitation procedure may be performed with respect to the tremor frequency of the patient
Spasticity	May harm the patient arm. May break soft parts in the rehabilitation system.	By monitoring the torque in the system, the spasticity may be identified. In addition, using torque monitoring the patient arm may be slowly stretched over the spasticity limit
Unable to maintain seating position	The patient may fall and detach from the rehabilitation device.	The patient must be strapped by the rehabilitation chair. The rehabilitation chair must be a sturdy one and allow harness fixtures.
Spasms Muscular pain	May harm the patient arm. May break soft parts in the rehabilitation system.	By monitoring the torque in the system, the spasms may be identified. When spasms occur, the system requires medical intervention and the motion is stopped until further instructions.
Headaches/Migraine, Depression	The patient cannot concentrate to the rehabilitation process.	Here, multimodal simulation and interactive procedures can be of help. By keeping the patient busy, he/she will focus more on the tasks. In this case, torque-based control allows multiple exercises, including AR ones.
Seizures	May harm the patient arm. May break soft parts in the rehabilitation system.	By monitoring the torque in the system, the seizures may be identified. When seizures occur, the system require medical intervention and the motion is stopped until further instructions.

Table 1. *Cont.*

Risk Factor	Associated Risk	Risk Overcome Measure
Wounds Skin diseases	Cannulas or wounds in anchor points may create discomfort during the rehabilitation	When the patient presents wounds over the anchor zones, bandage must be applied before attaching the rehabilitation device
Overall poor health state	Pulmonary, respiratory, cardiovascular diseases may influence the effects of the rehabilitation.	When enrolling the physical rehabilitation, the patient must not have diseases that may obstruct the physical therapy.

2.2. The Dynamic Analysis of ASPIRE

As stated in the first section of the paper, a robotic system for physical rehabilitation should undergo a dynamic analysis to determine the behavior of the robotic structure during the rehabilitation procedure, hence this section extends the dynamic analysis previously presented in [37], with a series of properties regarding the components of the robotic structure. Each motion module of ASPIRE (the module for flexion/extension, the one for adduction/abduction and the one for pronation supination) is analyzed with respect to the composing components. By studying the dynamic behavior of each module using virtual simulation the required torque to perform different rehabilitation scenarios can be computed. Figure 3 presents the ASPIRE structure divided into rehabilitation modules. The mass of each module was computed using Siemens NX; the adduction-abduction module weighs 25.39 kg, the flexion-extension module weighs 18.25 kg, and the pronation-supination module weighs 1.29 kg.

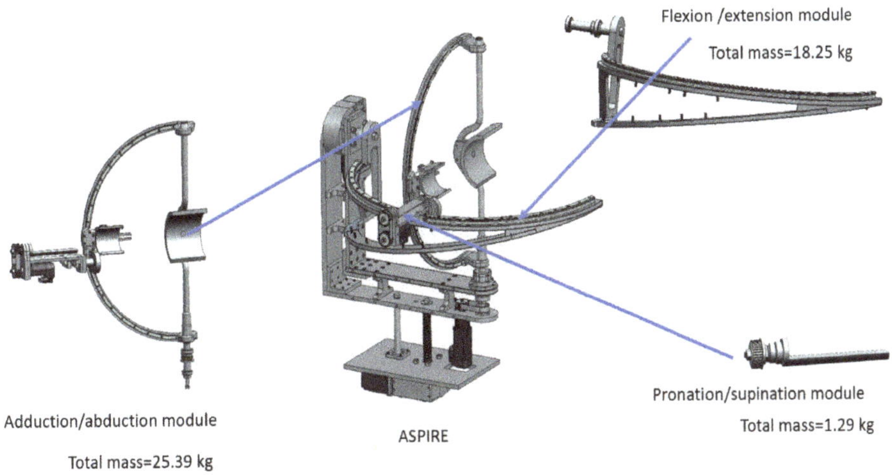

Figure 3. ASPIRE rehabilitation modules.

To analyze the dynamic behavior of each rehabilitation module a harmonic motion was used as input for each active revolute joint (Amplitude: 50°, Frequency: 100 °/s) and the recorded torques of each joint are graphically represented. To achieve more accurate results, a simulated manikin was used during the simulation. For computing the biomechanical model of the patient Plagenhoef model [38] was used which defines standard man and women limb segments weights. The values for the upper limb extracted from [38] are given in Table 2.

Table 2. Weight of upper-limb segment according to Plagenhoef [39].

Segment	Segment Weight	
	Men	Woman
Hand	0.526 kg	0.306 kg
Forearm	1.519 kg	0.965 kg
Upper arm	2.633 kg	1.777 kg

For the virtual simulation max value of segments weights was used for each upper-limb segment (0.526 kg for hand weight, 1.519 kg for forearm segment, and 2.633 kg for upper arm segment); the virtual setup is displayed in Figure 4.

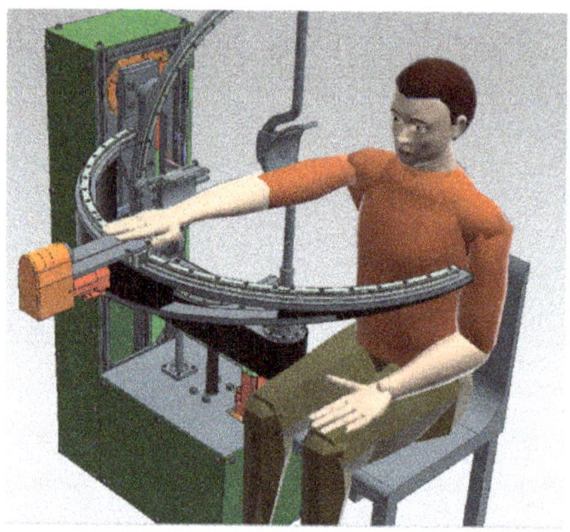

Figure 4. Virtual setup for simulation.

The virtual manikin was defined using the above data and the position of the patient during the procedure was simulated.

For the flexion-extension module, a 25-s simulation was carried on and the max torque obtained was 127.202 Nm while the minimum value was −13.231 Nm. Figure 5 represents the time-based torque variation during the simulation.

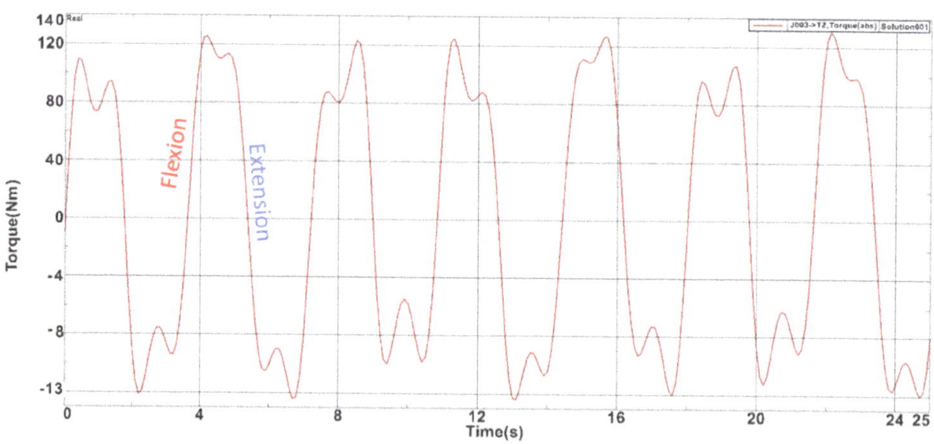

Figure 5. Torques recorded during the flexion-extension module simulation.

For the adduction-abduction module also a 25-s simulation was carried on and the max torque obtained was 5.040 Nm while the minimum value was −4.81 Nm. Figure 6 represents the time-based torque variation during the simulation. For the pronation-supination module the max torque obtained was 1.91 Nm while the minimum value was −1.89 Nm. Figure 7 represents the time-based torque variation during the simulation.

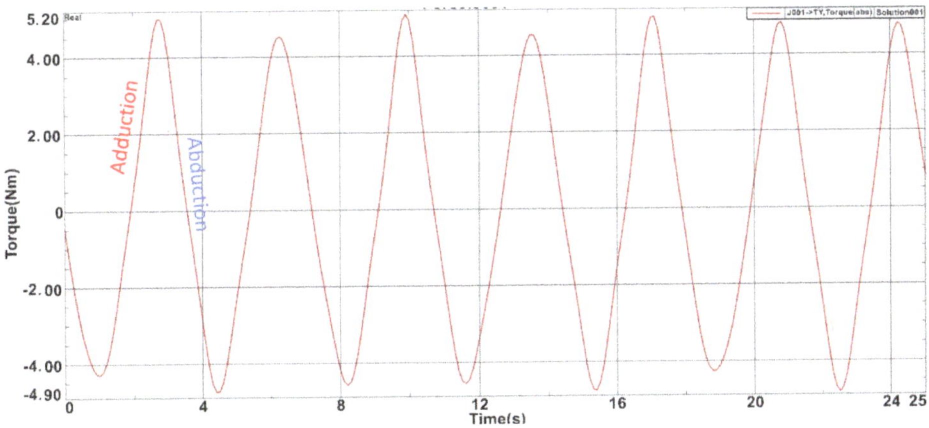

Figure 6. Torques recorded during the adduction-abduction module simulation.

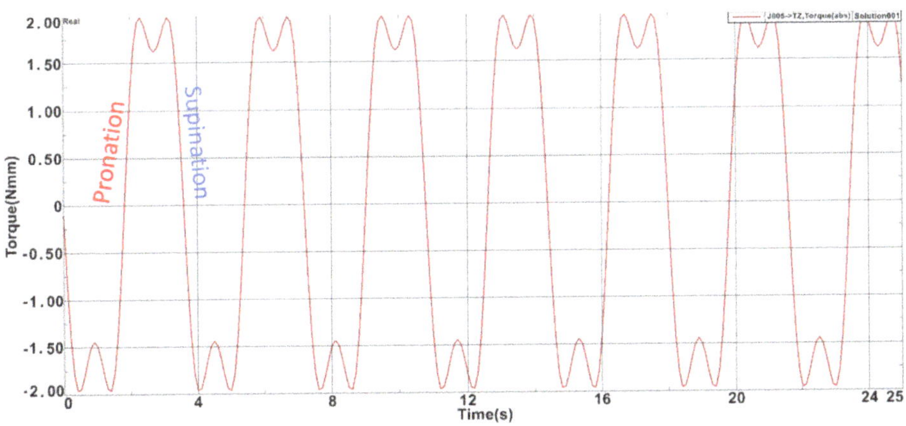

Figure 7. Torques recorded during the pronation-supination module simulation.

The results obtained during the virtual simulation imply the use of quite big motors to perform the rehabilitation motion; to overcome this problem, gearboxes were used. For the flexion-extension mechanism a worm gearbox gear ratio I = 45:1 was used, for the adduction-abduction mechanism a planetary gearbox with gear ratio I = 100:1 was used, and for the pronation-supination mechanism a planetary gearbox with gear ration I = 11:1 was used.

After the proposed torque-controlled strategy has been implemented, a series of experimental runs are performed, and the values obtained are checked against the ones obtained via virtual simulation.

2.3. The Development of Torque Monitoring System

The need to monitor torque during the procedure came as a result of in-hospital tests performed on patients [39–42], where the ASPIRE succeeded in performing the required rehabilitation motion, but there were times when the operator had to manually stop the robotic system because the patient arm was unable to achieve the desired amplitude due to its spasticity.

Figure 8 presents the dynamic control loop of the rehabilitation system. The functioning of the system is based on *Normal Run* sequence. In *Normal Run*, motion parameters (motion amplitudes, velocity, and number of repetitions) are given using the graphical user interface. Using inverse kinematics, data required for reaching the imposed amplitudes are transferred to the PLC which sends as inputs to the drivers the positions, velocities, accelerations, and a torque limit for the imposed motion (P.V.A.T). The drivers send the data to the motors and in exchange they read the state parameters of each motor (P.V.A.T.) and transmit the data to the PLC which provides dynamic feedback to the user. If there are differences between the data imposed via the Normal Run mode and those provided by the dynamic feedback, the system enters another state named *Correction Run*.

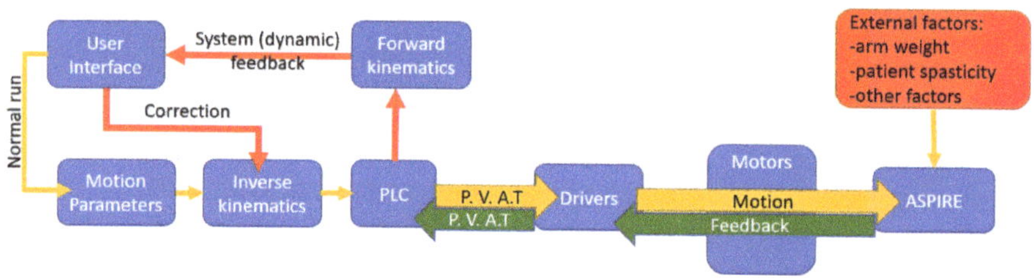

Figure 8. Dynamic control loop (P.V.A.T. = Positions, Velocities, Accelerations, Torque).

The *correction run* is activated when torque monitor system identifies some irregularities in torque variation created by external factors. In this case the system signals the event and requires the physiotherapist interventions in fixing the problem.

When *correction run* is activated, the robotic system identified a disturbing external factor that affects the rehabilitation procedure. Different scenarios are required for this correction run based on the external factor type and the observation of the physiotherapist. After analyzing the risk factors from above, three scenarios have been developed (Figure 8):

- ✓ The first scenario (**Red scenario**) implies spasm seizures or other critical situations that may endanger the patient. In this case the operator must press the emergency button and the patient is detached from the rehabilitation robot and primary care is administrated. The procedure may be resumed when the patient is in stable condition. This scenario is represented with red arrows and red blocks in Figure 9, and the output of this scenario is the end of the rehabilitation procedure due to unplanned events.
- ✓ The second scenario (**Yellow scenario**) includes reaching a spasticity limit or striving muscular pain. In this case, the system may return to the normal run after reversing direction. The system may re-enter the correction run mode when/if the spasticity limit is reached in the opposite direction. This scenario is represented with yellow arrows and yellow blocks in Figure 9 and the output of this scenario is Normal Run, the system that resumes the rehabilitation procedure until the next event.
- ✓ The third scenario (**Green scenario**) is related to some non-critical situations (the patient rearranged his position, tremor, etc.). In this case the physiotherapist may resume the normal run. This scenario is represented with green arrows and green blocks in Figure 9, and the output of the scenario is also Normal Run, the system that resumes the rehabilitation motion until the next event.

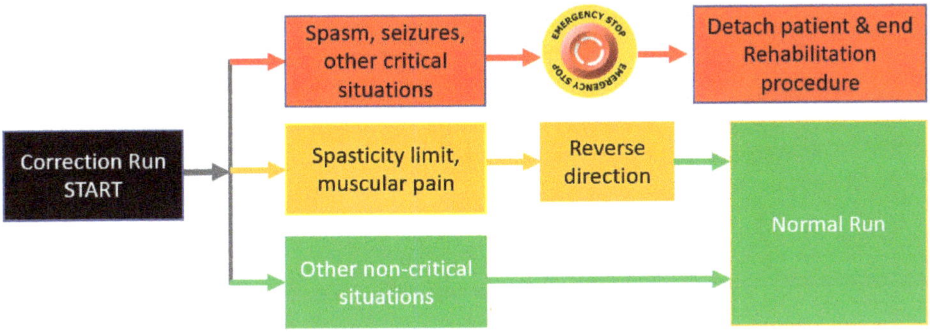

Figure 9. Actions following correction run event.

After developing the torque-monitoring strategy, the hardware architecture of the robotic system was reviewed, and the communication protocol was modified (old system used Virtual Network Connection while the new torque-monitoring system was based on Modbus protocol-required by the programming unit. In addition, the old interface was implemented in the PLC with the obvious limitations regarding graphical quality and extra functions (such as a database and exercise logging). The hardware configuration of the system is presented in Figure 10.

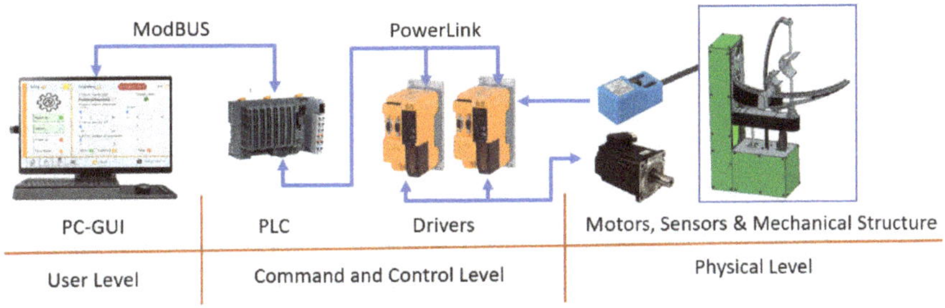

Figure 10. Hardware configuration of the ASPIRE robot control system.

The components of the robotic system are divided into 3 levels. First level is the User Level, and it contains the graphical user interface (GUI) running on a PC and containing all the tools needed for inputting the rehabilitation parameters. The second level is the Command-and-Control Level, and it contains the Industrial PLC of the robotic system and the drivers required for controlling the motors of the robotic structure. The third level is the Physical Level, and it contains the motors required to drive the mechanical structure, the sensors needed for system initialization, and the mechanical structure of ASPIRE. The connection between the User Level and the Command-and-Control Level is made using Ethernet Cable and ModBus protocol, while the connection between the Command-and-Control Level and the Physical Level is made using special connector compatible with the drivers, motors, and sensors.

The entire process is coordinated using GUI that communicates with a PLC via ModBUS protocol. The PLC controls two drivers and each driver communicates with two servomotors and two proximity sensors. The communication between the drivers and servomotors is bidirectional; the driver communicates the required position and, at the same time, monitors the real-time position, velocity, acceleration, and torque recorded at the motor spindle. The communication between the sensors and the drivers is unidirectional from sensors to drivers. The sensors are used for system initialization. The process diagram for initialization is given in Figure 11.

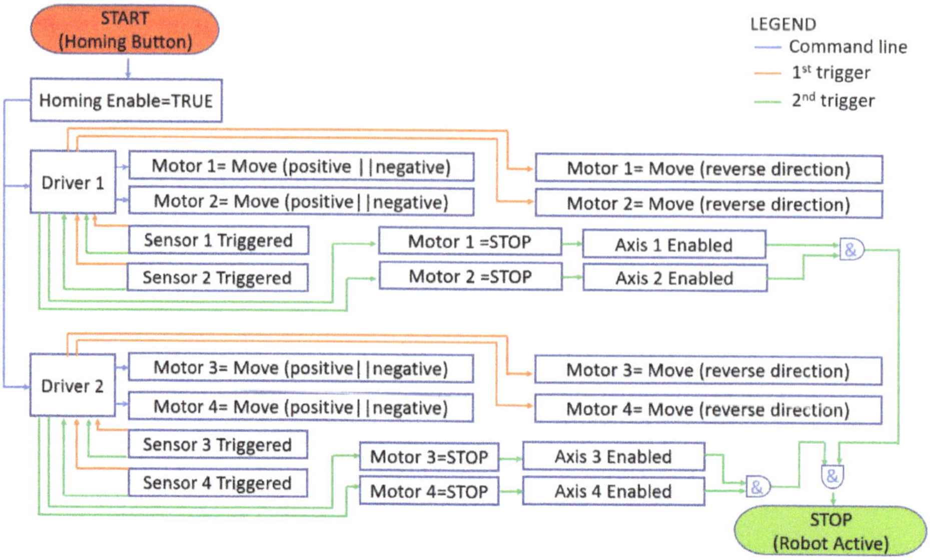

Figure 11. Homing process flowchart.

After receiving the initialization command (Homing Button and blue arrows in Figure 11) each driver starts the assigned motors in positive or negative direction until the proximity sensors are triggered. When a sensor is triggered for the first time after the robot start-up (orange arrow in Figure 11), after receiving the rising edge signal the driver reverses the moving directions (also orange arrows) of the motor until the sensor is triggered again (green arrow in Figure 11). After the second rising edge signal the motion of the motor is stopped, and the axis is considered initialized (Axis is enabled). The operator may use the robotic system only after all axes ($1 \div 4$) are enabled. If the initialization of all axes is complete the system will signal the operator using a green LED placed in the GUI, and if there is an error and the initialization could not be completed, the LED color will turn red, and the error message is prompted for the user.

For implementing the torque monitoring system, Automation Studio [43] was used. The programming environment allows easy assembly of instructions for controlling the servo motors of the robotic structure using embedded functions tested in industrial environment, proving the stability of the control system. For easy integration of torque monitoring, Mapp Motion is used, a tool provided by B&R [43] that allows easy access to the parameters of the axes and at the same time enables the torque control for the motors.

2.4. Experimental Setup and Torque Control Validation

After the development of the torque-monitoring system, a series of in-lab tests were performed to validate the improved control system after receiving ethical approval to carry out the experiments. Ten healthy subjects were selected for the experiments from the research center staff and the experimental setup was defined.

Figure 12 presents the ASPIRE robot placed in CESTER research center within Technical University of Cluj-Napoca for performing experimental tests. The robotic structure is physically attached to the control unit and because the participants in the experiment were healthy subjects and the main objective of the experiment was to validate the torque-monitoring system, a regular chair was used (during the previous tests performed in the hospital the patients were carried to the rehabilitation system using a wheelchair).

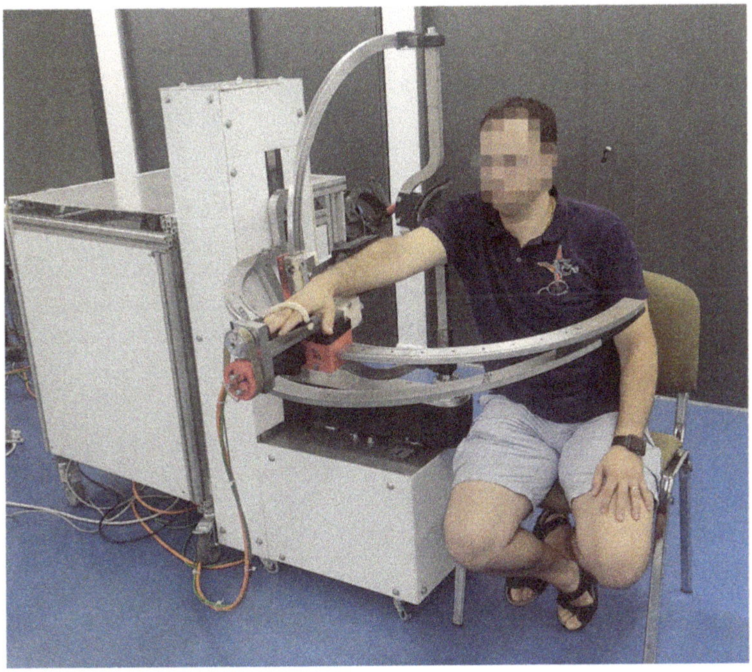

Figure 12. ASPIRE experimental model.

Besides the robotic system, a sensor system was used (Biometrics [44]). The sensor system uses a series of goniometers to simultaneously measure the angles in up to two planes of movement. During the experimental tests two goniometers were used. One was placed under the armpit of the subject to measure angles in two directions (flexion/extensions and adduction/abduction of the shoulder) and the other one was placed on the forearm of the subject to measure the rotation of the wrist with respect to the forearm (pronation/supination of the forearm). The decision to place the goniometers in this position came after a series of tests that revealed the optimal position to obtain most accurate reading during the rehabilitation motion. An image regarding this aspect of the experimental test is shown in Figure 13 along with a figure of the subject performing shoulder rehabilitation using the ASPIRE system. The amplitudes recorded by the biomedical sensors were not passed to the control system of the robot; the sensor system uses a separate graphical interface to calibrate each goniometer, and, after the calibration, the time-based variation of the amplitudes is graphically represented. The calibration of each goniometer is performed before placing the sensors on the patient by placing both ends of the goniometers on the same plane surface at the same time and pressing the interface "Calibrate" button. The entire calibration procedure for each goniometer takes up to 15 s. Each goniometer can read values between ±180 degrees with an accuracy of ±2 degrees measured over a range of 90 degrees and a repeatability of ±1 degree measured over a range of 90 degrees. The advantage of the system is given by the 12 h lasting battery and the wireless range up to 30 m. An example of the output of the sensor system for the pronation/supination motion is presented in Figure 14. The graph shows the rotation amplitudes recorded during a 20-repetitions session using a 50 degrees amplitude for pronation and 50 degrees amplitude for supination.

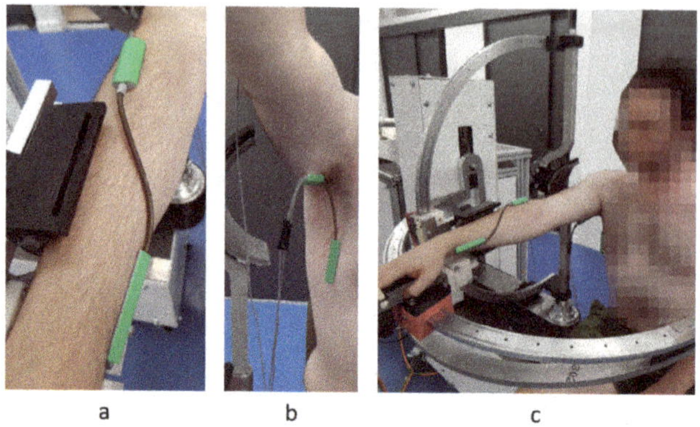

Figure 13. Goniometer used for measuring the pronation supination (**a**) Torsiometer used for measuring the pronation/supination of the forearm (**b**) Goniometer used for measuring adduction/abduction/flexion/extension for the shoulder; (**c**) the subject's position during the in-lab experimental tests.

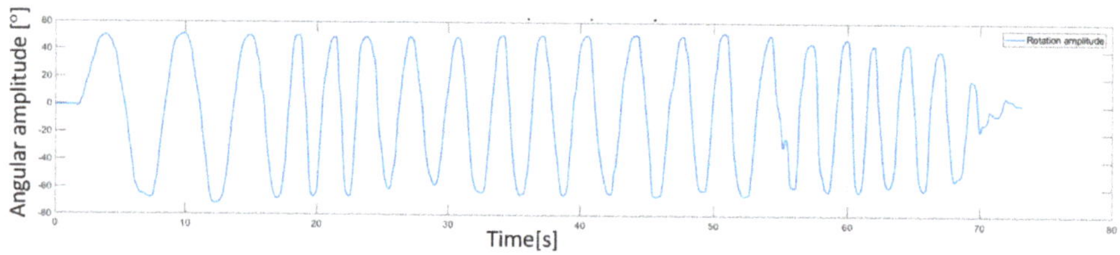

Figure 14. Amplitudes recorded using Biometrics sensor system for a pronation/supination rehabilitation session.

Classical rehabilitation protocol, previously used for the tests performed in the hospital, was also used for performing the in-lab experimental test. The subject was sitting on the rehabilitation chair with the shoulder resting in the shoulder plate. The forearm was attached to the forearm anchor device, and his palm was supported by the palm support. Using the GUI of the robotic system, motion amplitudes were introduced for each rehabilitation motion. While the robotic system was performing the motion, a background thread was used to monitor the torque. During this stage of the experimental test, the goniometers were used only for validating the motion amplitudes given within the user interface. Each of 10 healthy subjects received the same treatment, meaning the same motion amplitudes (maximum amplitudes), the same velocity during the rehabilitation, and the same number of repetitions. In addition, each subject performed the rehabilitation two times, the first time letting the robotic system perform the entire motion and the second time trying to move against the rehabilitation motion to create excessive torque in the system. The obtained results are furthermore analyzed in the following sections.

3. Results

After performing the experimental tests with the healthy subjects, a supplementary dry run was made in order to establish the torques in the system when it is moving without any patient performing the rehabilitation. Figure 15 graphically represents the data recorded during the experimental test, which are then statistically analyzed. Median values from all runs were extracted and graphically represented using MatLab [45]. The first column of plots from Figure 15 represents the dry run test performed at the end of the experimental tests. These results should be similar to the ones obtained during the virtual simulation performed in the previous section of the paper and, as it can be seen, there are inconsistent differences between the results of the virtual simulation and the ones obtained using the experimental model. Table 3 summarizes the comparison between these data. The second column of plots represents the median values obtained during the experimental test, when the subject was creating no extra forces in the system. The third column represents the median values obtained when the subject of the experiment was opposing the rehabilitation motion. As it can be observed, there are some differences in the pronation/supination module. The maximum absolute value obtained is 1.10 Nm for the dry run test, 1.62 Nm for normal run test, and 2.45 Nm for the resisting test. There are also a series of differences in the flexion/extension module but it is not easy to identify them on the graphical representation. The maximum value obtained during the dry run test was 122 Nm, 131.39 Nm during the normal run, and 147.37 Nm during the resisting tests. Regarding the adduction/abduction module the maximum obtained value is 9.20 Nm for the dry run test, 10.47 Nm for the normal run, and 14.40 Nm for the resisting test.

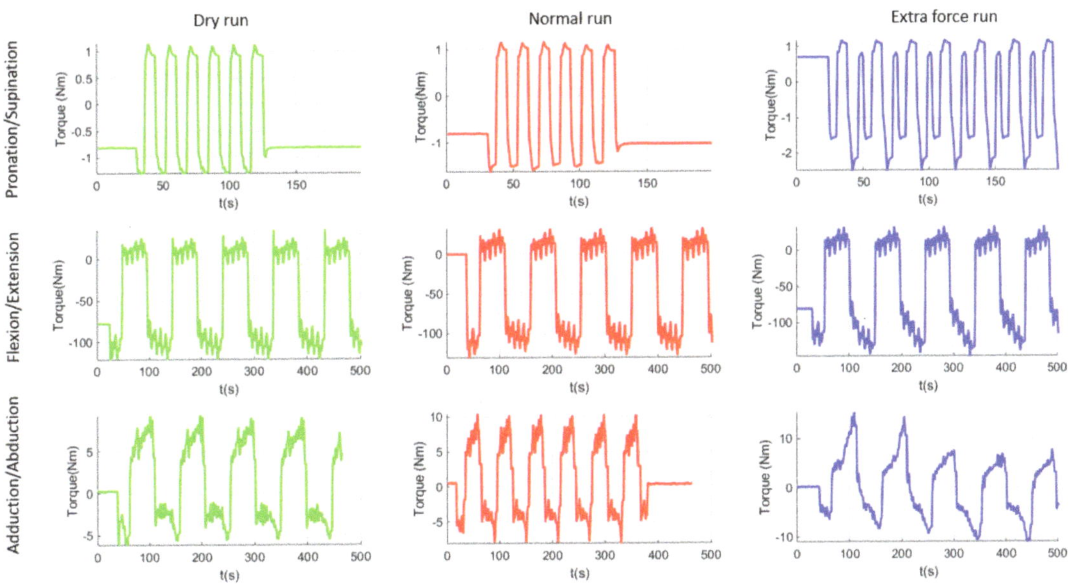

Figure 15. Graphical representation of the experimental tests' results.

Table 3. Torque analysis results.

Rehabilitation Motion	**Maxim Torque Recorded (Nm)**							
	Virtual Sim.	SD	Dry run	SD	Normal run	SD	Resisting test	SD
Pronation/Supination	1.91	0.6985	1.10	0.0759	1.62	0.0850	2.45	0.1227
Flexion/Extension	127.2	1.1354	122	1.2097	131.39	1.2378	147.37	1.3663
Adduction/Abduction	5.04	0.0495	9.20	0.0490	10.47	0.0463	14.40	0.0591
Percentual difference between different running modes (between max values)								
Pronation/Supination		−43%		+47%		+51%		
Flexion/Extension		−4%		+7%		+12%		
Adduction/Abduction		+82%		+13%		+37%		

The values that resulted from the torque analysis are quite similar. There are some differences between different running modes but torque variation between different functioning modes is quite visible. The values obtained during the virtual simulation are just for checking the similarity of the constructive model with the experimental model. The differences between these two simulations are given by the imperfect homogeneity of the material, the manufacturing imperfections, the machining accuracy, the elasticity of material, and other material properties, but overall the values are correlated. Regarding the values obtained using the experimental model, a 47% torque difference was recorded between the dry run and normal run and 51% between the dry run and resisting test in the case of pronation -supination motion. For the flexion-extension motion the percentual difference was 7% between dry run and normal run and 12% between normal run and resisting test. Given the length of the moving element (almost 500 mm radius from origin) the force arm is quite big, and any load can create supplementary torques for the motor. For the adduction/abduction motion there was a 13% difference between the dry run and normal run and 37% difference between the normal run and resisting test, the difference also given by the geometry of the moving element, making it easy to increase the force arm applied on the motor. The trigger for the correction run may be set by the operator either to a given torque or to a certain percentual overpass.

In previous clinical trials feedback coming from the physiotherapists has helped the authors in optimizing the robotic system in terms of mechanical structure (aesthetics and functionality) and of control of the robotic system. All the optimizations in the end have the great target to increase the safety of the patient and the acceptance of the robotic system by the patient and by the personal operating the robotic system. As a result, the robotic system must comply with a series of requests to fulfill the needs of the clinician and to increase the trust of the patient (and physiotherapist) in the rehabilitation system. By unifying the control system of the robot with the biomedical sensor system (after proofing the feasibility of the sensors system also with the help of the physiotherapists), a complete system can be obtained to help the physiotherapist (the system allows user data registration and with the help of the goniometers the baseline assessment may be performed easily; the manual evaluation of the patient is time consuming, while using biomedical sensors can be performed in a few minutes). Using a user database to store all data for a specific user, a personalized rehabilitation procedure may be provided for each patient by evaluating the evolution or regression of the patient through user database reports delivered at the end of each rehabilitation session.

4. Discussion

After analyzing the results obtained at the end of the experimental tests, the torque-monitoring system was validated. The system performed accordingly, and no events were reported during the experimental runs. The system was able to record the torques during the entire rehabilitation procedure and a special section was implemented in the user interface to show torque variation during the rehabilitation using a code of colors (green for normal run and yellow for when the torque is not normal, but it is in allowable limits, and red for when the torque in the system exceeds a certain imposed value). The allowable torque for each patient is established by the physiotherapist after the first general evaluation of the patient using a series of dynamometers to test the spasticity of the patient and the ability to perform a certain motion.

Using the Biometrics sensors, the rehabilitation system may be also used as an evaluation device. The robotic system is able to perform "blind" rehabilitation. Using this feature, after the general health assessment of the patient by the physiotherapist, ASPIRE may be used to measure the spasticity level of the patient. The goniometers are placed on the patient body in the previously identified areas (Section 2.4). After calibrating the sensors, the patient is placed in the rehabilitation system and prepared for the rehabilitation procedure. In the user interface maximum amplitudes for rehabilitation motion are introduced and the torque monitor system is started. The speed of the rehabilitation motion is set to lowest, and the number of repetitions is set to 1. After the robotic system starts to perform the rehabilitation motion, the sensor system records the real-time amplitude of the motion performed, and the torque is monitored within the allowable limits. If the system succeeds in performing the entire motion without triggering the torque monitor, then the patient has no spasticity, but if the torque monitor is triggered then the spasticity degree of the patient is computed using the value given by the sensor measuring system. This value can be used as a starting point for the future rehabilitation sessions and after each session this value can be tested to see if there are some improvements.

Torque monitor can be a useful tool to test some of the effects of some neurological diseases. For example, if after rehabilitation, graphical results show regular variation that can be interpreted as noise, further analysis may reveal that the noise is caused by the patient's tremor.

Based on the success of the experimental tests, and on the feedback received during the experimental tests from both the subject and from the robotic system, the functionality of the improved control system was validated. The system is able to accurately monitor the torques from the system using only the encoder of the servomotors without using any additional sensors for torque monitoring. The obtained values were checked against the values that resulted from the virtual simulation of the system and the differences were within an allowable range, the differences caused by the structure of the material, density, and even the manufacturing of some components.

Future work targets the implementation of interactive user interface using human–robot interaction modalities in order to prepare the robotic system for another session of clinical trials for the fall of 2021.

5. Conclusions

Torque monitoring system was successfully implemented in the control of the ASPIRE robotic system. The development of the system started from the requests identified during the clinical trials performed in the fall of 2019 in Cluj-Napoca. A series of effects that may influence the effectiveness of the rehabilitation were presented and torque control proved to be a solution in identifying and overcoming some of these effects. After the development of the torque control system the performance was tested using healthy subjects chosen among the staff of the research center. The results have validated the functionality of the proposed improved control system and pushed the development of the experimental model towards the next level of maturity, preparing it for further clinical trials. At the same time the torque-based control adds to patient safety and the possibility to improve the HRI characteristics of the robotic system with high levels of safety and user-friendliness.

Author Contributions: Conceptualization, D.P., P.T., C.V., G.B. and G.C.; Data curation, D.T., N.T., I.D.G. and C.A.; Formal analysis, N.P.; Investigation, D.T., P.T., N.T., C.V., I.D.G., G.B., C.A. and G.C.; Methodology, I.D.G., G.B., G.C. and N.P.; Project administration, D.P.; Resources, D.T.; Software, G.B.; Validation, C.V.; Writing—original draft, D.P. and P.T.; Writing—review and editing, P.T. All authors have read and agreed to the published version of the manuscript.

Funding: This work was supported by a grant from the Romanian Ministry of Research and Innovation, CCCDI—UEFISCDI, project number PN-III-P2-2.1-PED-2019-3022/546PED/2020 (NeuroAssist) within PNCDI III, and by the project POCU/380/6/13/123927—ANTREDOC, "Entrepreneurial competencies and excellence research in doctoral and postdoctoral studies programs", co-funded from the European Social Fund through the Human Capital Operational Program 2014–2020.

Institutional Review Board Statement: Ethical review and approval were waived for this study, due to the fact that the validation tests were intended only to assess the feasibility and range of motion of ASPIRE for the rehabilitation of the upper limb, being achieved in laboratory conditions, which required the participation of volunteers, healthy subjects. Furthermore, the volunteers have given their written consent after being informed about the test circumstances and the robotic system's operation, as well as any hazards or safety concerns that may arise during the tests.

Informed Consent Statement: Informed consent was obtained from all subjects involved in the study.

Data Availability Statement: The data presented in this study are openly available in reference number [37,39–41].

Conflicts of Interest: The authors declare no conflict of interest.

References

1. Healthline. Available online: https://www.healthline.com/health/monoplegia (accessed on 15 July 2021).
2. Garcia, R.V. The Brachial and Crural Indices of Modern North American Population. Master's Thesis, Texas State University, San Marcos, TX, USA, 2015, unpublished.
3. Iqbal, J.; Bruno, A.; Berger, M. Stroke causing pure brachial monoparesis. *J. Stroke Cerebrovasc. Dis.* **1995**, *5*, 88–90. [CrossRef]
4. Department of Economic and Social Affairs. *World Population Ageing 2017 Highlights*; United Nations: New York, NY, USA, 2017.
5. Eyigor, S.; Durmaz, B.; Karapolat, H. Monoparesis with complex regional pain syndrome-like symptoms due to brachial plexopathy caused by varicella zoster virus: A case report. *Arch. Phys. Med. Rehabil.* **2006**, *87*, 1653–1655. [CrossRef] [PubMed]
6. NHS Funding. Available online: https://nhsfunding.info/ (accessed on 15 July 2021).
7. Baniqued, P.D.E.; Stanyer, E.C.; Awais, M.; Alazmani, A.; Jackson, A.E.; Mon-Williams, M.A.; Mushtaq, F.; Holt, R.J. Brain–computer interface robotics for hand rehabilitation after stroke: A systematic review. *J. Neuroeng. Rehabil.* **2021**, *18*, 1–25. [CrossRef] [PubMed]
8. Gull, M.A.; Thoegersen, M.; Bengtson, S.H.; Mohamadi, M.; Struijk, L.N.S.A.; Moeslund, T.B.; Bak, T.; Bai, S. A 4-DOF upper limb exoskeleton for physical assistance: Design, modeling, control and performance evaluation. *Appl. Sci.* **2021**, *11*, 5856. [CrossRef]
9. Chaparro-Rico, B.D.M.; Cafolla, D.; Castillo-Castaneda, E.; Ceccarelli, M. Design of arm exercises for rehabilitation assistance. *J. Eng. Res.* **2020**, *8*, 204–218. [CrossRef]
10. Li, J.; Cao, Q.; Dong, M.; Zhang, C. Compatibility evaluation of a 4-DOF ergonomic exoskeleton for upper limb rehabilitation. *Mech. Mach. Theory* **2021**, *156*, 104146. [CrossRef]
11. Kim, B.; Deshpande, A.D. An upper-body rehabilitation exoskeleton Harmony with an anatomical shoulder mechanism: Design, modeling, control, and performance evaluation. *Int. J. Robot. Res.* **2017**, *36*, 414–435. [CrossRef]
12. Brahmi, B.; Driscoll, M.; El Bojairami, I.K.; Saad, M.; Brahmi, A. Novel adaptive impedance control for exoskeleton robt rehabilitation using a nonlinear time-delay disturbance observer. *ISA Trans.* **2021**, *108*, 381–392. [CrossRef]
13. Paolucci, T.; Agostini, F.; Mangone, M.; Berneti, A.; Pezzi, L.; Liotti, V.; Recubini, E.; Cantarella, C.; Bellomo, R.G.; D'Azurio, C.; et al. Robotic rehabilitation for end-effector device and botulinum toxin in upper limb rehabilitation in chronic post-stroke patients: An integrated rehabilitative approach. *Neurol. Sci.* **2021**. [CrossRef]
14. DeBoon, B.M.; Foley, R.C.A.; Nokleby, S.B.; La Delfa, N.J.; Rossa, C. Nine degree-of-freedom kinematic modeling of the upper-limb complex for constrained workspace evaluation. *J. Biomech. Eng.* **2021**, *143*, 1–13. [CrossRef]
15. Vaida, C.; Carbone, G.; Major, K.; Major, Z.; Plitea, N.; Pisla, D. On human robot interaction modalities in the upper limb rehabilitation after stroke. *Acta Tech. Napoc. Ser. Appl. Math. Mech. Eng.* **2017**, *60*, 91–102.
16. Motos, A.B.; Blanco, A.; Badesa, F.J.; Barios, J.A.; Zollo, L.; Garcia-Aracil, N. Human arm joints reconstruction algorithm in rehabilitation therapies assisted bt end-effector robotic devices. *J. Neuroeng. Rehabil.* **2018**. [CrossRef]
17. Zhang, L.; Guo, S.; Sun, Q. Development and assist-As-needed control of an end-effector upper limb rehabilitation robot. *Appl. Sci.* **2020**, *10*, 6684. [CrossRef]
18. Vaida, C.; Pisla, D.; Schadlbauer, J.; Husty, M.; Plitea, N. Kinematic analysis of an innovative medical parallel robot using study parameters. In *New Trends in Medical and Service Robots. Mechanisms and Machine Science*; Wenger, P., Chevallereau, C., Pisla, D., Bleuler, H., Rodić, A., Eds.; Springer: Cham, Switzerland, 2016; Volume 39. Available online: https://doi.org/10.1007/978-3-319-30674-2_7 (accessed on 10 July 2021). [CrossRef]
19. Vaida, C.; Plitea, N.; Gherman, B.; Szilaghyi, A.; Galdau, B.; Cocorean, D.; Covaciu, F.; Pisla, D. Structural analysis and synthesis of parallel robots for brachytherapy. In *New Trends in Medical and Service Robots. Mechanisms and Machine Science*; Pisla, D., Bleuler, H., Rodic, A., Vaida, C., Pisla, A., Eds.; Springer: Cham, Switzerland, 2014; Volume 16. Available online: https://doi.org/10.1007/978-3-319-01592-7_14 (accessed on 10 July 2021). [CrossRef]
20. Pisla, D.; Plitea, N.; Gherman, B.G.; Vaida, C.; Pisla, A.; Suciu, M. Kinematics and design of a 5-DOF parallel robot used in minimally invasive surgery. In *Advances in Robot Kinematics: Motion in Man and Machine*; Lenarcic, J., Stanisic, M., Eds.; Springer: Dordrecht, The Netherlands, 2010. Available online: https://doi.org/10.1007/978-90-481-9262-5_11 (accessed on 10 July 2021). [CrossRef]
21. Schonstein, C. Considerations about matrix exponentials in geometrical modeling of the robots. *Acta Tech. Napoc. Ser. Appl. Math. Mech. Eng.* **2019**, *62*, 337–342.
22. Schonstein, C. Kinematic control functions for a serial robot structure based on the time derivative Jacobian matrix. *Acta Tech. Napoc. Ser. Appl. Math. Mech. Eng.* **2018**, *61*, 219–224.
23. Nedezki, C.M. Position problem for 3rpr 3-mobile plan parallel manipulator. *Acta Tech. Napoc. Ser. Appl. Math. Mech.* **2013**, *56*, 317–322.
24. Zhao, F.; Yang, Z.; Li, Z.; Gua, D.; Li, H. Extract executable action sequences from natural language instructions based on DQN for medical service robots. *Int. J. Comput. Commun. Control* **2021**, *16*. [CrossRef]
25. Escalante, F.M.; Jutinico, A.L.; Jaimes, J.C.; Terra, M.H.; Siqueira, A.A.G. Markovian robust filtering and control applied to rehabilitation robotics. *IEEE/ASME Trans. Mechatron.* **2021**, *26*. [CrossRef]
26. Kim, D.; Lee, J. Robust control of a system with pneumatic spring. *J. Frankl. Inst.* **2021**, *358*, 555–574. [CrossRef]

27. Masengo, G.; Zhang, X.D.; Yin, G.; Alhassan, A.B.; Dong, R.L.; Orban, M.; Mudaheranwa, E. A design of lower limb rehabilitation robot and its control for passive training. In Proceedings of the 10th Institute of Electrical and Electronics Engineers International Conference on Cyber Technology in Automation, Control, and Intelligent Systems (IEEE-CYBER 2020), Xi'an, China, 10–13 October 2020; pp. 152–157.
28. Antal, T.A. Addendum modification of spur gears with equalized efficiency at the points where the meshing starts and ends. *Mechanika* **2016**, *6*, 480–485.
29. Antal, T.A.; Antal, A. The use of genetic algorithms for the design of mechatronic transmission with improved operating conditions. In Proceedings of the 3rd International Conference on Human System Interaction, Rzeszow, Poland, 13–15 May 2010. [CrossRef]
30. Aliabadi, M.; Mashayekhifard, J.; Moha, B. Intelligent and classic control of rehabilitation robot with robust PID and Fuzzy methods. *Majlesi J. Mechatron. Syst.* **2020**, *9*, 31–36.
31. Bouteraa, Y.; Abdallah, I.B.; ElMogy, A.; Tariq, U.; Ahmad, T. A fuzzy logic architecture for rehabilitation robotic systems. *Int. J. Comput. Commun. Control.* **2020**, *15*, 1–17. [CrossRef]
32. Lyu, S.; Cheah, C.C. Human-robot interaction control based on a general energy shaping method. *IEEE Trans. Control Syst. Technol.* **2020**, *28*, 2445–2460. [CrossRef]
33. Shi, D.; Zhang, W.X.; Zhang, W.; Ju, L.H.; Ding, X.L. Human-centered adaptive control of lower limb rehabilitation robot based on human-robot interaction dynamic. *Mech. Mach. Theory* **2021**, *162*. [CrossRef]
34. Bouteraa, Y.; Ben Abdallah, I.; Alnowaiser, K.; Ibrahim, A. Smart solution for pain detection in remote rehabilitation. *Alex. Eng. J.* **2021**, *60*, 3485–3500. [CrossRef]
35. Erwin, A.; McDonald, C.G.; Moser, N.; O'Malley, M.K. The SE-AssessWrist for robot-aided assessment of wrist stiffness and range of motion: Development and experimental validation. *J. Rehabil. Assist. Technol. Eng.* **2021**, *8*. [CrossRef]
36. Vaida, C.; Plitea, N.; Pîslă, D.; Carbone, G.; Gherman, B.; Ulinici, I.; Pislă, A. Spherical Robot for Medical Recovery of Upper Limb Proximal Area. Patent RO132233-A0, 29 November 2017.
37. Tucan, P.; Plitea, N.; Vaida, C.; Gherman, B.; Carbone, G.; Luchian, I.; Pisla, D. Inverse dynamics and simulation of a parallel robot used in shoulder rehabilitation. In *New Trends in Mechanism and Machine Science. EuCoMeS 2020. Mechanisms and Machine Science*; Pisla, D., Corves, B., Vaida, C., Eds.; Springer: Cham, Switzerland, 2020; Volume 89.
38. Plagenhoef, S.; Gaunor, E.V.; Abdelnour, T. Anatomical data for analyzinghuman motion. *J. Res. Q. Exerc. Sport* **1983**, *54*, 169–178. [CrossRef]
39. Tucan, P.; Vaida, C.; Plitea, N.; Pisla, A.; Carbone, G.; Pisla, D. Risk-based assessment engineering of a parallel robot used in post-stroke upper limb rehabilitation. *Sustainability* **2019**, *11*, 2893. [CrossRef]
40. Major, Z.Z.; Vaida, C.; Major, K.A.; Tucan, P.; Simori, G.; Banica, A.; Brusturean, E.; Burz, A.; Craciunas, R.; Ulinici, I.; et al. The impact of robotic rehabilitation on the motor system in neurological diseases. A multimodal neurophysiological approach. *Int. J. Environ. Res. Public Health* **2020**, *17*, 6557. [CrossRef]
41. Tucan, P.; Vaida, C.; Ulinici, I.; Banica, A.; Burz, A.; Pop, N.; Birlescu, I.; Gherman, B.; Plitea, N.; Antal, T.; et al. Optimization of the ASPIRE spherical parallel rehabilitation robot based on its clinical evaluation. *Int. J. Environ. Res. Public Health* **2021**, *18*, 3281. [CrossRef]
42. Vaida, C.; Ulinici, I.; Banica, A.; Burz, A.; Gherman, B.; Tucan, P.; Pisla, A.; Carbone, G.; Pisla, D. First clinical evaluation of a spherical robotic system for shoulder rehabilitation. In *New Trends in Medical and Service Robotics; MESROB 2020. Mechanisms and Machine, Science*; Rauter, G., Cattin, P.C., Zam, A., Riener, R., Carbone, G., Pisla, D., Eds.; Springer: Cham, Switzerland, 2020; Volume 93. Available online: https://doi.org/10.1007/978-3-030-58104-6_8 (accessed on 10 July 2021). [CrossRef]
43. B&R Industrial Automation GmbH. Available online: https://www.br-automation.com/de/ (accessed on 21 July 2021).
44. Biometrics Ltd. Available online: https://www.biometricsltd.com/ (accessed on 21 July 2021).
45. MatLab. Available online: https://www.mathworks.com/ (accessed on 21 July 2021).

applied sciences

MDPI

Article

Innovative Collaborative Method for Interaction between a Human Operator and Robotic Manipulator Using Pointing Gestures

Marek Čorňák *, Michal Tölgyessy and Peter Hubinský

Faculty of Electrical Engineering and Information Technology, Institute of Robotics and Cybernetics, Slovak University of Technology in Bratislava, Ilkovičova 3, 812 19 Bratislava, Slovakia; michal.tolgyessy@stuba.sk (M.T.); peter.hubinsky@stuba.sk (P.H.)
* Correspondence: marek.cornak@stuba.sk

Abstract: The concept of "Industry 4.0" relies heavily on the utilization of collaborative robotic applications. As a result, the need for an effective, natural, and ergonomic interface arises, as more workers will be required to work with robots. Designing and implementing natural forms of human–robot interaction (HRI) is key to ensuring efficient and productive collaboration between humans and robots. This paper presents a gestural framework for controlling a collaborative robotic manipulator using pointing gestures. The core principle lies in the ability of the user to send the robot's end effector to the location towards, which he points to by his hand. The main idea is derived from the concept of so-called "linear HRI". The framework utilizes a collaborative robotic arm UR5e and the state-of-the-art human body tracking sensor Leap Motion. The user is not required to wear any equipment. The paper describes the overview of the framework's core method and provides the necessary mathematical background. An experimental evaluation of the method is provided, and the main influencing factors are identified. A unique robotic collaborative workspace called Complex Collaborative HRI Workplace (COCOHRIP) was designed around the gestural framework to evaluate the method and provide the basis for the future development of HRI applications.

Keywords: HRI (human–robot interaction); Leap Motion; UR5; hand gesture recognition; pointing gestures; collaborative robotics

Citation: Čorňák, M.; Tölgyessy, M.; Hubinský, P. Innovative Collaborative Method for Interaction between a Human Operator and Robotic Manipulator Using Pointing Gestures. *Appl. Sci.* **2022**, *12*, 258. https://doi.org/10.3390/app12010258

Academic Editors: Giuseppe Carbone and Med Amine Laribi

Received: 10 November 2021
Accepted: 20 December 2021
Published: 28 December 2021

Publisher's Note: MDPI stays neutral with regard to jurisdictional claims in published maps and institutional affiliations.

1. Introduction

Collaborative robotic applications are, nowadays, well-established and commonly used concepts of technology. The main goal of such applications is to combine the strengths of both humans and robotic systems to achieve maximum effectiveness in completing a specified task while minimizing the risks imposed on the human worker. In the manufacturing process, these applications are crucial in enabling the concept referred to as "Industry 4.0". Industry 4.0 focuses on developing so-called cyber-physical systems or CPS for short, aiming to create highly flexible production systems capable of fast and easy changes, addressing the need for individualized mass production of the current markets [1].

Collaborative applications are becoming increasingly popular on the factory floors due to their various benefits, such as lower deployment costs, compact sizes, and easier repurposing, than standard robotic systems [2]. As a result, more factory workers will be required to work in close contact with robotic systems. This, however, introduces new arduous challenges to overcome, primarily how to secure the safety of a human worker, while ensuring high work effectiveness. The former problem focuses mainly on minimizing potential risks and avoiding accidents from collaborative work between the human worker and a robot, not considering the work efficiency. The latter aims to find the methods capable of maximizing the overall productivity of such collaboration. Albeit the safety of the human worker is an absolute priority, also being the most currently researched topic [3], the economic benchmarks are vital in determining further deployment of these applications. Thus, it is necessary to pay more attention to the interaction between the human and robotic systems, also referred to as HRI, researching the impact and influence of the whole human–robot relationship, developing new approaches capable of meeting today's production demands.

In today's conventional robotic application, usually involving robust industrial manipulators, the robot works in a cell that must be wholly separated from the workers' environment by a physical barrier [4]. The human only comes in contact with the robot during maintenance and programming, carried out by highly qualified personnel. All of this is not only expensive, but also time-consuming. Alternatively, in collaborative applications, factory workers may often get into contact with parts of the robot and are even encouraged to influence the robot's actions based on the current circumstances to secure maximum work flexibility. This could pose a severe problem, where technically unskilled workers may suffer great difficulties communicating with the robotic system or conducting some form of control. A solution to this problem lies in developing such HRI methods, which would enable natural and ergonomic human–robot communication without the need for prior technical knowledge of the robotic systems. Hence, the workers could solely focus on the task at hand and not waste time figuring out the details of the application's interface.

For humans, the most natural way of communication is through speech, body, or facial gestures. Implementing these types of interfaces into collaborative applications could yield many benefits, which lead to increased productivity and overall worker contentment [5].

This paper presents a gesture-based collaborative framework for controlling a robotic manipulator with flexible potential deployment in the manufacturing process, academic field, or even a commercial sector. The underlying basis of the proposed framework is based on the previous work of Tölgyessy et al. [6], which introduced the concept of "linear HRI," initially designed for mobile robotic applications.This paper further develops this concept, widening its usability from mobile robotic applications to robotic arm manipulation while utilizing a new state-of-the-art visual-based sensory system. The main goal of the proposed framework is to provide systematic foundations for gesture-based robotic control, which would support a wide variety of potential use-cases, could be applied universally regardless of the specific software or robotic platform while providing a basis on which other more complex applications could be built.

2. Related Work

Gestural HRI is a widely researched topic across the whole spectrum of robotics. Allowing the robot to detect and recognize human movements and act accordingly is a powerful ability enabling the robot and the human worker to combine their strengths and achieve various challenging tasks. In mobile robotics, gestures can be used to direct and influence the robot's movement. Cho and Chung [7] used a mobile robotic platform, Pioneer 3-DX, and a Kinect sensor to recognize a human body and follow the operator's movement. Tölgyessy et al. [6] proposed a method for controlling a mobile robotic platform, iRobot Create, equipped with a Kinect sensor via pointing gestures. Chen et al. [8] used a Leap Motion Controller to control the movement of the mobile robotic platform with a robotic arm. Both the chassis and the robotic arm could be controlled via hand gestures. Gesture-based control was also used in projects RECO [9] and TeMoto [10], which focused on intuitive multimodal control of mobile robotic platforms. Besides mobile robotic platforms, force-compliant robotic manipulators are ideal for implementing various forms of gestural interaction, potentially deployable in multiple scenarios, such as object handling, assembly, assistance, etc. A significant part of the research is focused on teleoperation, in which the robot mirrors the human operator's movements. Hernoux et al. [11] used a Leap Motion sensor and a collaborative manipulator UR10 to reproduce the movement of the operator's hands. G. Du et al. [12] proposed a similar approach using Leap Motion to control a dual-arm robot with both hands. They used the interval Kalman filter and improved particle filter methods to improve hand tracking. Kruse et al. [13] proposed a gesture control dual-arm telerobotic system, in which the operator controlled the position of an object held by the robotic system. Microsoft Kinect was used to track the human body. Many other researchers used Microsoft Kinect or a Leap Motion controller to control the robotic manipulator [14–20]. Tang and Webb [5] studied the feasibility of gesture control to replace conventional means of direct control through teach-pendant. Zhang et al. [21] proposed a gesture control for the delta architecture robot. Cipolla and Hollinghurst [22] investigated a gestural interface for teleoperating a robotic manipulator based on pointing gestures. Using stereo cameras, collineations, and an active contour model, they were able to pinpoint a precise location in the 40 cm workspace of the robot at which the user would point at with his finger. Object picking is another promising application area. A human hand can pick different objects of various sizes and shapes. Therefore, several works were conducted to mimic the human hand, to grasp objects [23–25]. An interesting concept was proposed by Razjigaev et al. [26], the authors proposed a gestural control for a concentric tube robot. The work focused on the potential use of such an interface in noninvasive surgical procedures. A gestural interface may be a promising concept in the control of UAV drones; some works [27,28] show potential for future development. Social robotics is another vast research area where gestural interfaces could yield many benefits. Natural interfaces could be especially advantageous in interaction with humanoid robots. Yu et al. [29] proposed a gestural control of the NAO humanoid robot. Cheng et al. [30] used this robot and a Kinect sensor to facilitate a gestural interaction between humans and the humanoid robot.

In Table 1, there is a comparison of key works related to the interaction method designed by us. The vast majority focus mainly on teleoperation of the robotic manipulator used, while the user has visual feedback of the resulting robot motion. Only three approaches present some form of direct interaction of the operator with the manipulator's workspace and two of these allow the user to point and select objects present in the workspace. The major novelty and contribution of our design is that the operator can select objects on the planar surface; furthermore, she or he can point to any spot of the workspace and subsequently navigate the end effector to the desired destination.

Table 1. Comparison of key related works.

Lit.	Robotic System	Sensor	Visual Tracking	Weareables	Tele-Operation	Workspace Interaction	Pointing to Objects
[11]	UR10	Leap Motion	Yes	-	Yes	-	-
[17]	UR10 simulation	Leap Motion	Yes	-	Yes	-	-
[16]	EPSON SCARA LS3-401x	Leap Motion	Yes	-	Yes	-	-
[21]	Custom Delta architecture	Leap Motion	Yes	-	Yes	-	-
[5]	UR5	Leap Motion	Yes	-	Yes	-	-
[12]	GOOGOL GRB3016	Leap Motion	Yes	-	Yes	-	-
[31]	UR5	Leap Motion	Yes	-	Yes	-	-
[20]	ABB IRB120	Kinect V2	Yes	-	Yes	-	-
[32]	Baxter	MYO Armband	-	Yes	Yes	Yes	-
[33]	UR5	Microsoft HoloLens	Yes	Yes	Yes	Yes	Yes
[19]	Custom 7DOF robot	Kinect V2	Yes	-	Yes	-	-
[18]	UR5	Kinect V2	Yes	-	Yes	-	-
[13]	Motoman SDA 10	Kinect V1	Yes	-	Yes	-	-
[22]	Scorbot ER-7	Stereo CCD cameras	Yes	-	Yes	Yes	Yes

3. Our Approach

In a conventional manufacturing process, a worker usually performs repetitive manual and often unergonomic tasks. Workers do most of these tasks mainly using their hands, such as assembly, machine tending, object handling, material processing, etc. Manual work, however, has many limitations which consequently influence the whole process efficiency.

Integrating a collaborative robotic solution could significantly improve efficiency while, at the same time, it can alleviate a human worker's physical and mental workload.

Our proposed concept for the gestural HRI framework for the collaborative robotic arm aims to create such interaction, where the human worker could interact with the robot as naturally and conveniently as possible. Secondly, besides ergonomics, the framework focuses on flexibility, providing the user with functionality that could be used in numerous application scenarios. Lastly, the framework lays basic foundations for further application development, pushing the flexibility even further.

The fundamental principles of the proposed framework are based on the so-called "linear HRI" concept introduced by Tölgyessy et al. [6], which formulates three simple laws for HRI that state:

1. Every pair of two joints of a human sensed by a robot form a line.
2. Every line defined by the first law intersects with the robot's environment in one or two places.
3. Every intersection defined by the second law is a potential navigation target or a potential object of reference for the robot.

The proposed framework was developed under the laws of linear HRI, where the core functionality lies in the ability to send the end effector of the robotic arm to a specific location on a horizontal/vertical plane using the operator's hand pointing gestures.

Humans naturally use pointing gestures to direct someone's attention to the precise location of something in space. Combined with speech, it can be a powerful communication tool. Pointing gestures have many great use-cases among people to signify the importance of a particular object ("That's the pen I was looking for" (pointing to the specific one)), express intention ("I'm going there!" (pointing to the place)), or specify the exact location of something ("Can you hand me that screwdriver please?" (pointing to the specific location in space)). Pointing gestures are incredibly efficient when the traditional speech is insufficient or impossible due to circumstances, such as loud and noisy environments.

Let us imagine a collaborative application scenario where the worker performs a delicate assembly task that requires specific knowledge. A collaborative robot would fulfill the role of an intelligent assistant/co-worker. The worker could point to the necessary tools and parts out of his reach, which the robot would then bring to him; thus, he could specifically focus on the task while keeping his workspace clutter-free. Due to the natural character of the whole interaction, the worker could control the robot's behavior conveniently without any prior technical skills. Such application would be accessible and easy to use across the entire worker spectrum, with various technical backgrounds. Our gestural framework aims to enable the described interaction in real-world conditions using the principles of linear HRI and state-of-the-art hardware. However, first, the following main challenges need to be addressed:

- Choosing suitable technology and a method of detecting the position and orientation of human joints.
- Calculating the exact position of the intersection between the formed line and the environment surface.
- Selecting the pair of joints to form and calculate the line following the first two principles of linear HRI.

Solving these challenges is crucial for ensuring efficient, reliable, and natural interaction. Additionally, we aimed to make the framework universal, not dependent on the specific robotic platform. For that reason, we chose the Robot Operating System or ROS as our software environment. ROS supports a wide variety of robotic platforms and is considered standard for programming robots in general. For the robot's control and trajectory planning, the ROS package MoveIt was used.

In summary, the primary objectives of our approach lie in natural interaction, application flexibility, and scalability, while following the underlying concept of so-called linear HRI.

4. The Human Body Tracking and Joint Recognition

Precise recognition and localization of joints are vital in gestural interaction. Several technologies and approaches exist, providing the user with tracking capabilities. According to Zhou and Hu [34], human motion tracking technologies can be divided into non-visual, visual, and robot-aided tracking.

In non-visual tracking, the human motion and joints are mapped via various sensors placed onto the human body; typically, MEMS IMU or IMMU devices are used. These sensors are usually a part of a suit or so-called "data glove", which must be worn by the user [35–39]. Although these systems are precise and relatively cheap, they suffer from many shortcomings, such as the need for the user to wear the sensory suit, which may be uncomfortable and often calibrated for the specific user. The sensors on the suit may require extensive cabling, which can limit the user's range of motion. The sensors themselves may suffer from several issues, such as null bias error, scale factor error, noise, or interference. Due to this, the general application focus shifted towards the use of visual-based human body tracking.

This approach uses optical sensors, such as RGB, IR, or depth cameras to extract the user's spatial position. Complex image processing algorithms are then applied to pinpoint the precise position and orientation of human joints. These sensors do not require any equipment that the user needs to wear, thus not limiting him in any way. They can be easily calibrated to the specific user, which greatly improves their flexibility. Today's camera technologies and image processing algorithms make them fast and accurate and relatively cheap. These attributes made these sensors popular and widely used in various commercial applications, mainly in HCI and video-gaming industry. However, visual-based sensors still have multiple drawbacks, such as the sensitivity to the lighting conditions and worsened tracking capabilities when the human body is occluded. The most widely recognizable sensor for visual body tracking is the Microsoft Kinect, released in 2010, initially intended for video gaming. However, due to its capabilities, the sensor became widely used in other applications. The Kinect's principle of body tracking relies on capturing the depth data from the depth sensors and then applying their image processing methods to produce a so-called "skeletal stream" representing the human figure. The first iteration of the sensors used an IR projection pattern to acquire the depth data. The latter versions used the so-called "time-of-flight" technology or TOF. According to Shotton et al. [40], body part classification (BPC) and offset joint regression (OJR) algorithms, specially developed for Microsoft Kinect, are used to determine the parts of the human body. Following the success of Microsoft Kinect, other similar sensors were released, such as Intel RealSense or ASUS Xtion.

Another widely popular vision-based motion tracking sensor is the Leap Motion controller. This compact device was specially designed to track the human hands and arms at very high precision and accuracy. Alongside gaming and VR applications, the controller's design was initially meant to replace conventional peripheral devices, such as a mouse and keyboard, and provide a more sophisticated and natural HCI. The Leap Motion controller uses two IR cameras, capturing emitted light from three LEDs with a wavelength of 850 nm. The depth data are acquired by applying unique algorithms on the raw data, consisting of infrared brightness values and the calibration data [41]. According to the manufacturer, the accuracy of position estimation is around ± 0.01 mm. However, several studies show [42,43] that this is highly dependent on the conditions.

Sensor Choice for the Proposed Concept

Due to the advantages of vision-based body tracking, we decided to use this technology in our concept. We believe that the proper HRI should not rely on "human-dependent devices" as data gloves or body-mounted sensors, but on the perceptive ability of the robot itself, as this, in our opinion, most accurately represents natural and ergonomic interaction.

As for the specific sensor, the Leap Motion controller was picked as the most suitable option. The main reason is the sensor's application focus. The core of the proposed concept centers around the hand gesture interaction between the robot and the human worker, as most manufacturing tasks are done by hand. Furthermore, the whole gestural framework is built on pointing gestures, performed by the arrangement of individual fingers. Leap Motion provides accurate and precise human hand tracking, explicitly focusing on the position and orientation of fingers. Other commercially available sensors on the market are not yet capable of such precision and generally focus on tracking the whole human body. The controller depicted in Figure 1 can track hands within a 3D interactive zone that extends up to 60 cm (24″) or more, extending from the device in a 140 × 120° typical field of view, which is illustrated in Figure 2. The controller produces an output in the form of gray-scale images captured by the IR cameras and the skeletal representation of the human hand. The data are shown in Figure 3. Leap Motion's software can differentiate between 27 distinct hand elements, such as bones and joints, and track them even when other hand parts or objects obscure them. The positions of joints are represented in the controller's coordinate system depicted in Figure 4. The connection with the PC is facilitated via USB 2.0 or USB 3.0 connection. The manufacturer also provides a software development kit or SDK for the Leap Motion controller, allowing the developers to create custom applications.

Figure 1. The Leap Motion controller (LMC).

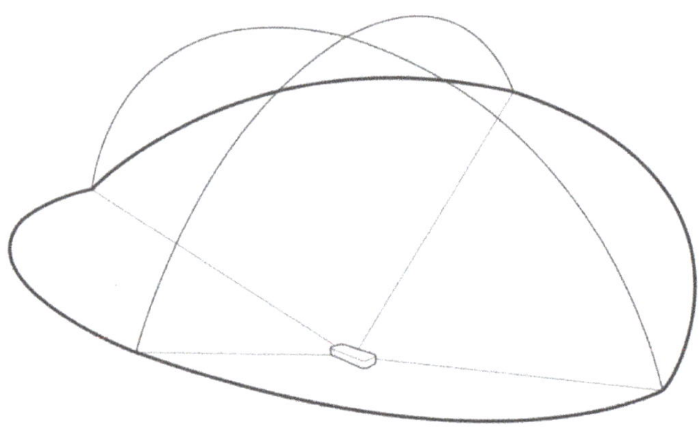

Figure 2. Approximate workspace of the LMC.

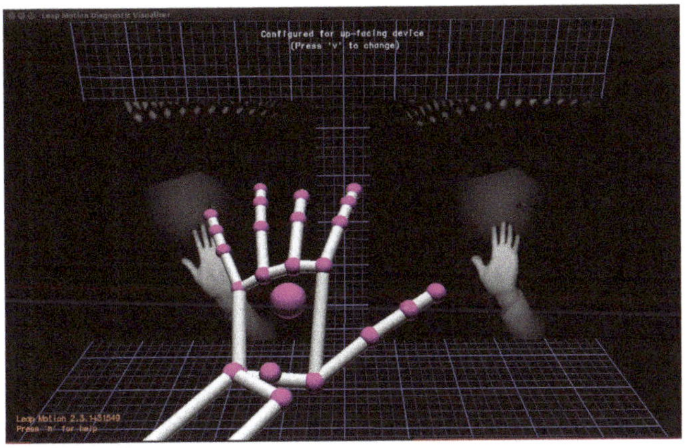

Figure 3. Visualization of detected hand joints.

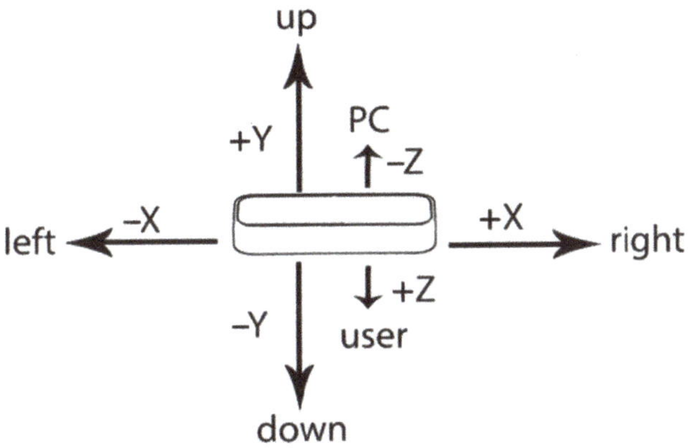

Figure 4. Coordinate system of the LMC.

5. Method Design

The core functionality of the proposed framework follows specific consecutive steps. The operator first performs a pointing gesture, pointing to the particular location in the (planar) workspace of the robot. Then, the robot computes the specified location defined by the planar surface and the half-line formed by the operator's joints. When the operator is satisfied with the pointed place, he performs a command gesture, triggering the signal to move the robot's end effector to the desired location. The whole process is illustrated in Figure 5. The half-line is defined by the direction of the index finger, as it is the most common way to represent a pointing gesture. The following command gesture was specifically designed to be performed naturally, fluently connecting with the previous gesture while not significantly influencing the pointing precision. The command gesture is achieved by extending the thumb, forming a so-called "pistol" gesture. Both gestures are depicted in Figure 6. Custom ROS packages and nodes were created for gesture recognition and intersection computation. The ROS-enabled machine connected to the robotic arm manages these nodes, ensuring proper communication between the individual parts of the whole application. The method's architecture in ROS framework is depicted in Figure 7. The flowchart of the method's process is in Figure 8. The most prominent geometrical features for the mathematical description of the concept are illustrated in Figure 9. The global coordinate system G defines the planar ground surface π.

Figure 5. Process of controlling the robot via pointing gestures. 1—The operator points to a location. 2—The operator performs "pistol" gesture. 3—The robot's TCP moves to the appointed location.

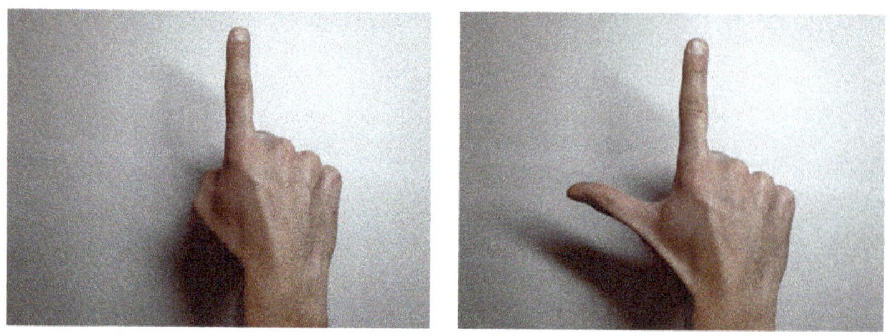

(**a**) Pointing gesture (**b**) Pistol gesture

Figure 6. Two basic gestures used to interact with the system.

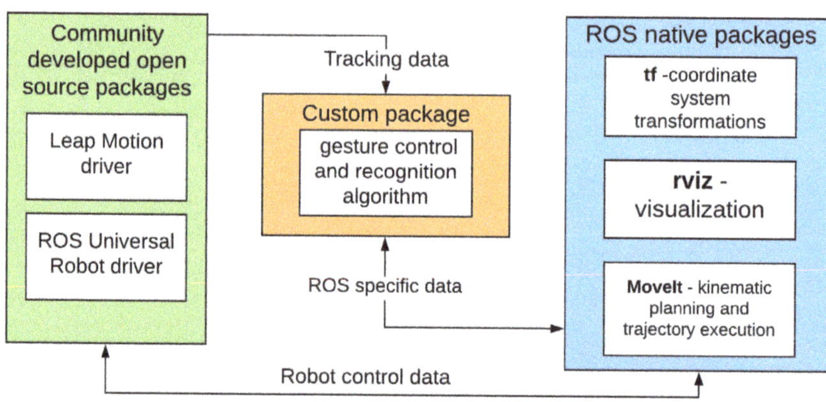

Figure 7. The package architecture of the method in ROS.

The Leap Motion sensor determines the position of the joints of the human hand in its coordinate system L; furthermore, the robotic manipulator operates in its coordinate system R. The unification of those coordinate systems is vital to obtain points A and B coordinates defining the half-line p, which ultimately determines the coordinates of intersection I.

All of the calculations are done in the global coordinate system G. Hence, the following homogeneous transformations between R and G, L, and G are defined:

$$\begin{pmatrix} X_G \\ Y_G \\ Z_G \\ 1 \end{pmatrix} = \begin{pmatrix} 1 & 0 & 0 & d_x \\ 0 & \cos\alpha & -\sin\alpha & d_y \\ 0 & \sin\alpha & \cos\alpha & d_z \\ 0 & 0 & 0 & 1 \end{pmatrix} \cdot \begin{pmatrix} X_L \\ Y_L \\ Z_L \\ 1 \end{pmatrix}$$
$$\begin{pmatrix} X_G \\ Y_G \\ Z_G \\ 1 \end{pmatrix} = \begin{pmatrix} 0 & \cos\alpha & -\sin\alpha & d_x \\ 0 & \sin\alpha & \cos\alpha & d_y \\ 0 & 0 & 1 & 0 \\ 0 & 0 & 0 & 1 \end{pmatrix} \cdot \begin{pmatrix} X_R \\ Y_R \\ Z_R \\ 1 \end{pmatrix} \quad (1)$$

Analytical geometry and vector algebra are used to calculate the desired coordinates. The data acquired from the Leap Motion sensor are transformed based on Equation (1). The points and vectors used in the calculation are illustrated in Figure 10.

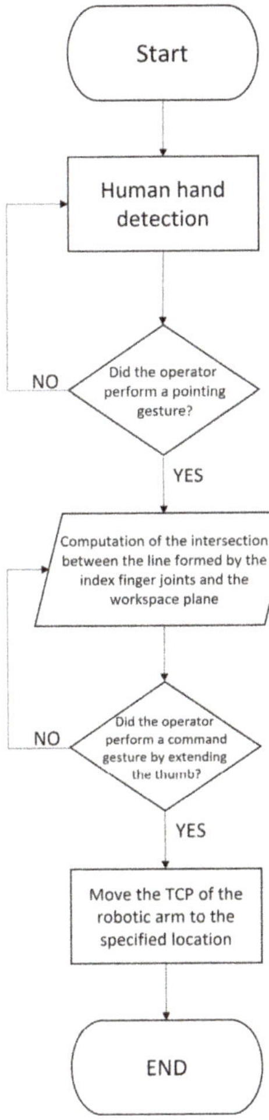

Figure 8. State machine flowchart of the method.

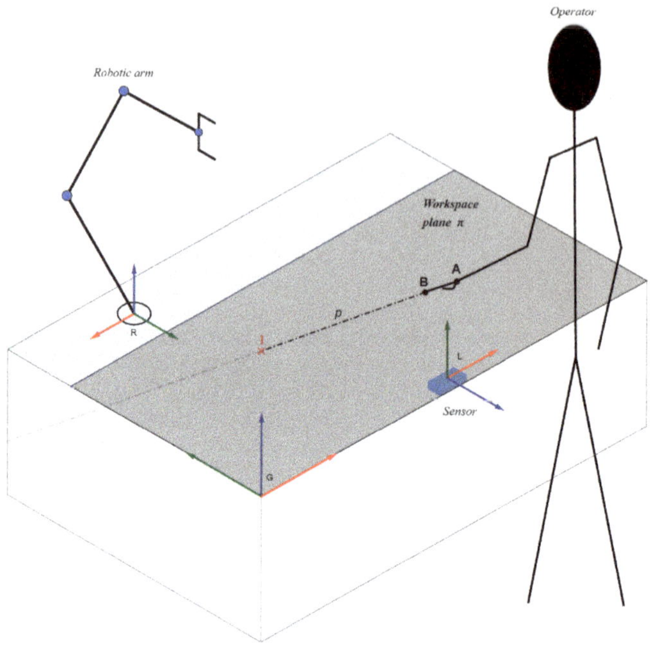

Figure 9. The abstract overview of the method.

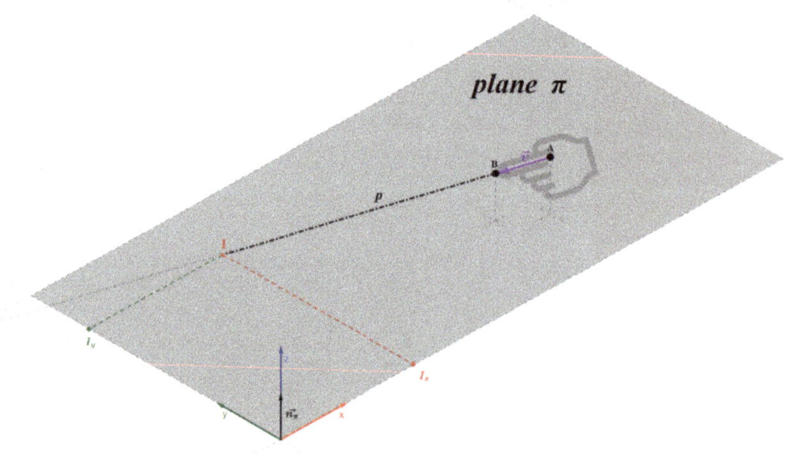

Figure 10. Extraction of fundamental geometrical features.

First, the direction of the vector $\vec{v}$ defined by points A and B needs to be found. Vector $\vec{v}$ is calculated by subtracting coordinates of point A from point B (positions of index finger joints).

$$\vec{v} = B - A \tag{2}$$

The direction of vector $\vec{v}$ is calculated by dividing the coordinates of vector $\vec{v}$ with its magnitude.

$$\hat{v} = \frac{\vec{v}}{||\vec{v}||} \tag{3}$$

Now, it is possible to define the half-line p by parametric equations.

$$\begin{aligned} x &= x_0 + \hat{v_1} \times t \\ y &= y_0 + \hat{v_2} \times t \\ z &= z_0 + \hat{v_3} \times t \end{aligned} \tag{4}$$

where t is the parameter of p. The plane π is defined by the x and y axes with an arbitrary normal vector $\vec{n}$. Equation (5) can mathematically describe plane π.

$$z = 0 \tag{5}$$

By substituting Equation (5) into Equation (4), the parameter t can be calculated as follows.

$$t = -\frac{z_0}{\hat{v_3}} \tag{6}$$

Now, using substitution, the X and Ycoordinates of the intersection $I(X, Y)$ can be calculated by the following equations.

$$\begin{aligned} I_X &= A_X + \hat{v_1} \times (-\frac{z_0}{\hat{v_3}}) \\ I_Y &= A_Y + \hat{v_2} \times (-\frac{z_0}{\hat{v_3}}) \end{aligned} \tag{7}$$

Respectively, it is possible to calculate the intersection on the plane ω perpendicular to π mathematically defined by the Equation (7).

$$y = d, \tag{8}$$

where d is the distance of the plane on y axis. The scenario is illustrated in Figure 11.

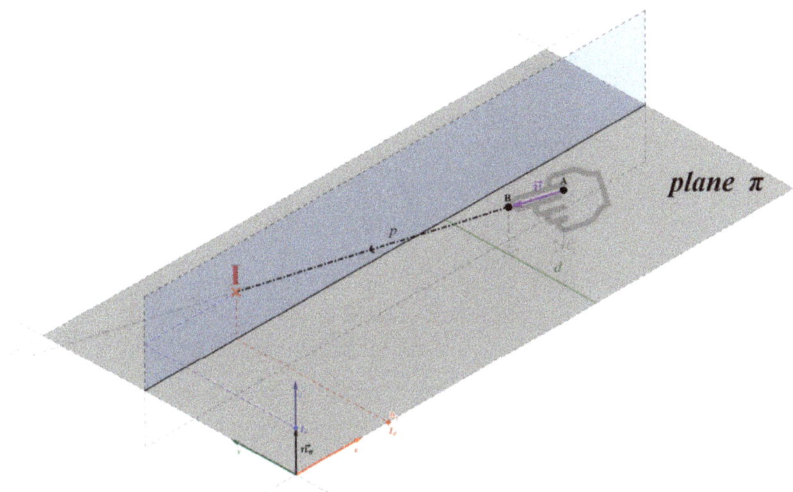

Figure 11. Abstraction of pointing to the perpendicular plane.

6. Method Implementation

The main application focus of the proposed concept is to create a natural HRI workplace where humans and robots can work together efficiently. For this reason, a specialized robotic workplace was built around the core concept's functionality, supporting the ergonomy of the whole interaction between human and the robot, trying to maximize the efficiency and convenience for the worker. Furthermore, it also acts as a modular foundation for implementing, testing, and evaluating other HRI concepts.

The whole workplace, called COCOHRIP, an abbreviation for Complex Collaborative HRI WorkPlace, is depicted in Figure 12. The COCOHRIP consists of three main parts, the sensors, the visual feedback, and the robotic manipulator. The sensory part contains the various sensors that gather data about the human worker and the environment. The visual feedback part consists of two LED monitors, providing the user with visual feedback from the sensor. The robotic part comprises the force-compliant robotic manipulator. All these parts are connected through the ROS machine, which manages all the communication and logic of the application. Transformation matrices between coordinate systems were:

$$T_{LG} = \begin{pmatrix} 1 & 0 & 0 & 670 \\ 0 & \cos(0.707) & -\sin(0.707) & 15 \\ 0 & \sin(0.707) & \cos(0.707) & 11 \\ 0 & 0 & 0 & 1 \end{pmatrix}$$
$$T_{RG} = \begin{pmatrix} 0 & \cos(3.14) & -\sin(3.14) & 1050 \\ 0 & \sin(3.14) & \cos(3.14) & 693.3 \\ 0 & 0 & 1 & 103 \\ 0 & 0 & 0 & 1 \end{pmatrix} \quad (9)$$

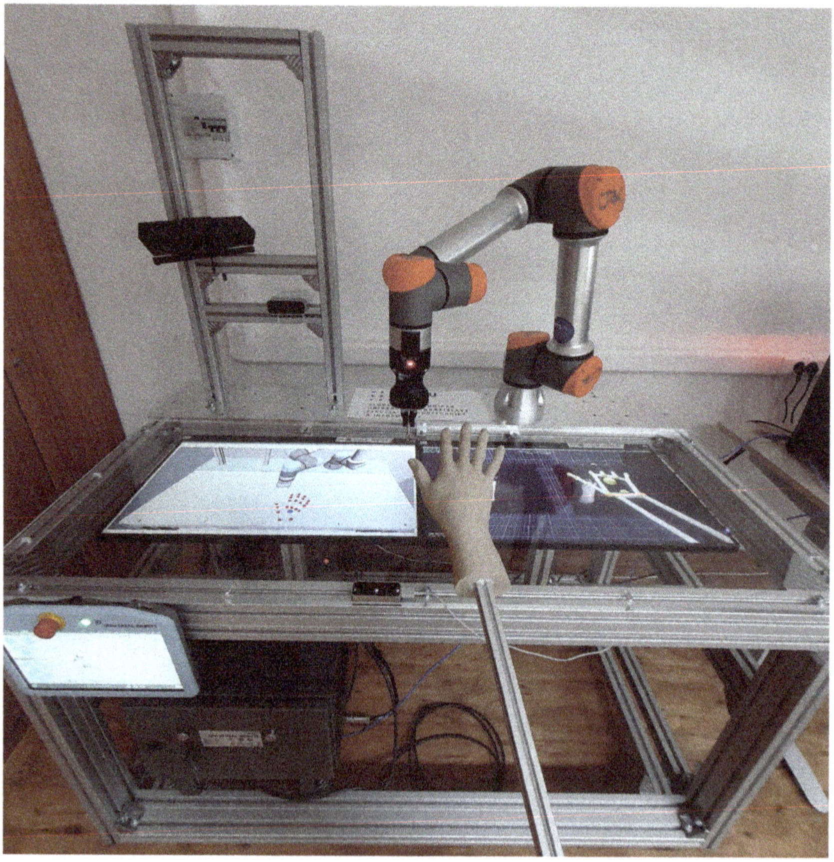

Figure 12. The COCOHRIP workplace.

7. Experimental Evaluation

In this section, we present two experimental scenarios to evaluate the basic concept principles, which quantify the overall usability of the concept and serve as a baseline for further development. The scenarios of each of the two experiments are derived from the proposed gesture pointing method, in which the operator points to a specific location on the workspace plane and performs a command gesture, sending the robot to the desired location. Each of the experiment scenarios has two variants. In the first scenario, the operator points to the designated horizontally placed markers; in the second scenario, the operator points at vertically placed markers. In the first variant, the operator has no visual feedback about where she or he is pointing. His task is to rely solely on his best guess. In the second variant, the user receives visual feedback from the interactive monitors showing him the exact place he is pointing. The illustration of the experimental setup and the precise positions of the markers in the global coordinate system is depicted in Figures 13 and 14. The position values are in millimeters. The process of obtaining one measurement is as follows:

1. The operator's hand is calibrated for optimal gesture recognition.
2. The operator points to the first point, and when he is satisfied with his estimation, he performs a command gesture.
3. When the operator performs a command gesture, the spatial data are logged, and the robot's TCP moves to the pointed location.
4. The operator repeats these previous two steps, iterating through each target point in order, as illustrated in Figures 13 and 14.

In total, the operator points ten times to each of the target markers for the current experiment. During the pointing, the only requirement on the user is to keep the hand approximately 200–250 mm above the Leap Motion sensor, as this is believed to be the optimal distance according to [42] and the empirical data acquired during the method pre-testing. The user is encouraged to perform the pointing gestures as naturally as possible, allowing him to position his body and arm as he sees fit. Seven male participants of different height, physical constitution, and height executed the experiment scenarios.

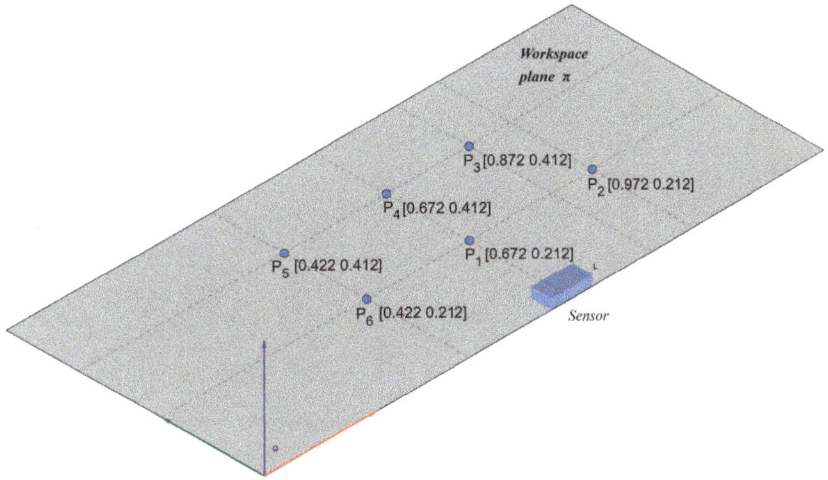

Figure 13. Scenario of experiment no. 1.

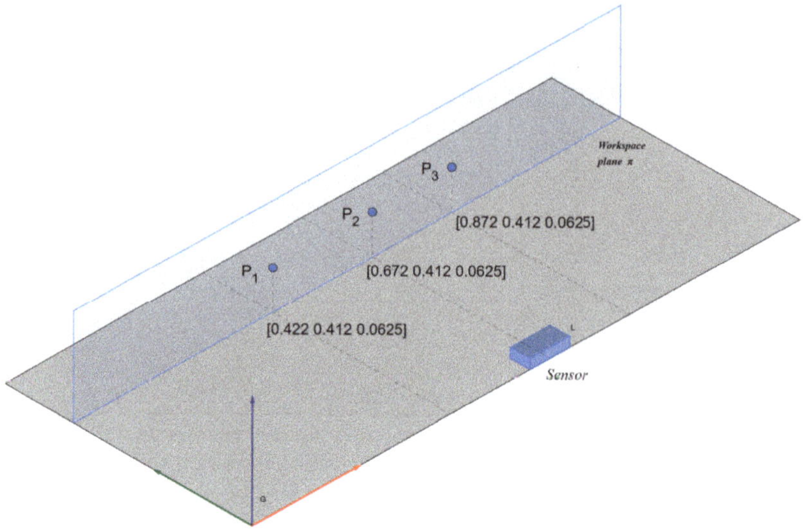

Figure 14. Scenario of experiment no. 2.

8. Results

The measured coordinates for each target marker relative to the desired position are depicted in Figures 15 and 16. The blue markers represent the first variant, where the user had no visual feedback. The red markers represent the second variant of each experiment. The euclidean distance between each measured point and the desired position was computed to quantify the precision of pointing for each experiment. Figures 17 and 18 show the boxplots of the euclidean distance data for each experiment scenario. The mean deviation in position and the overall deviation are shown in Tables 2 and 3.

Additionally, the dispersion of measured points in each axis was calculated to evaluate the repeat-ability of pointing. The results are shown in Tables 4 and 5. All the data are represented in millimeters.

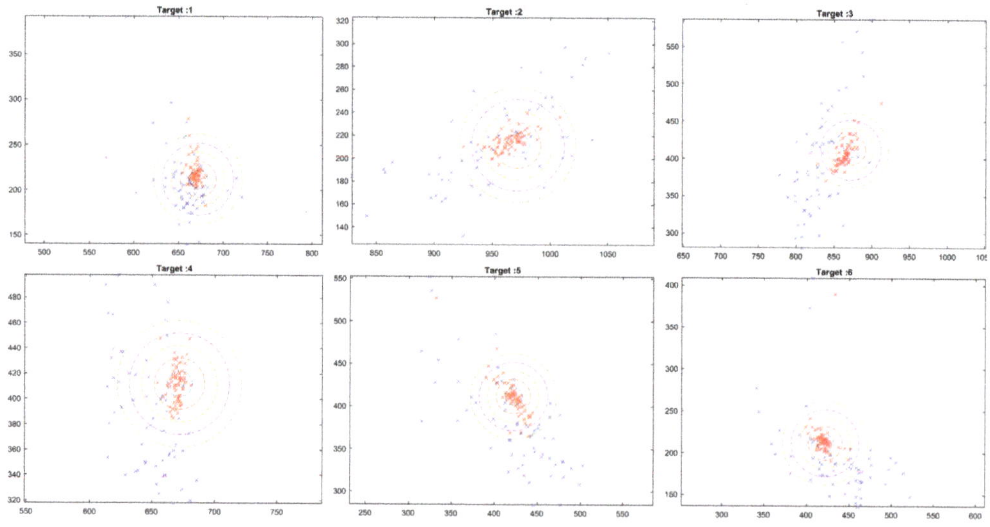

Figure 15. The acquired position data for experiment no. 1.

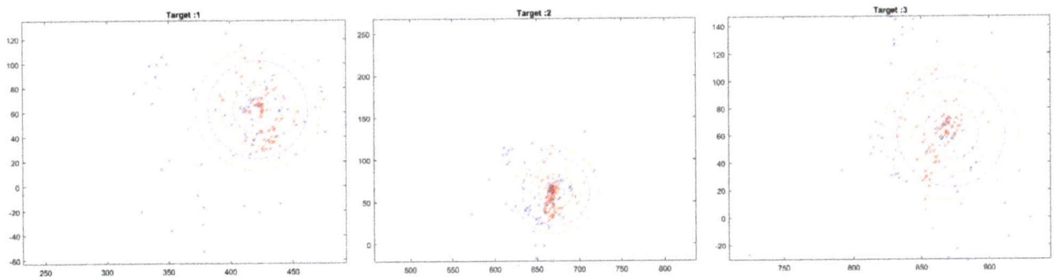

Figure 16. The acquired position data for experiment no. 2.

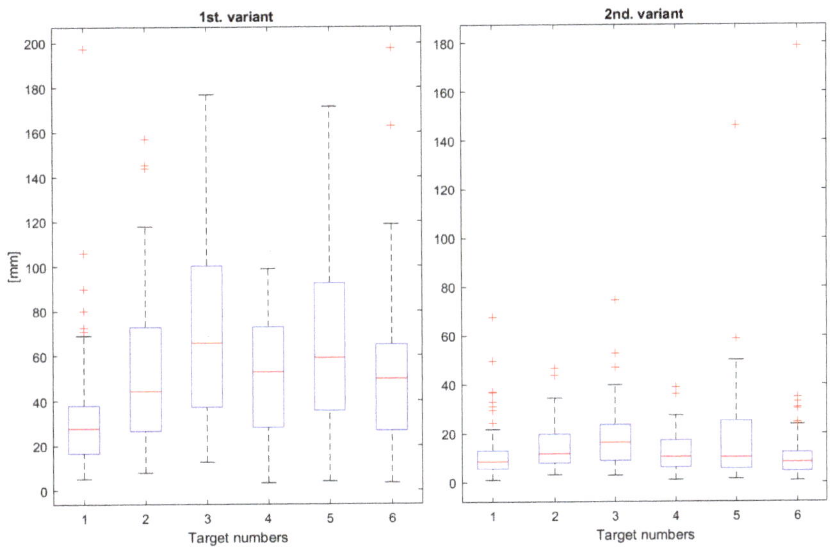

Figure 17. Boxplots of euclidean distance data for experiment no. 1.

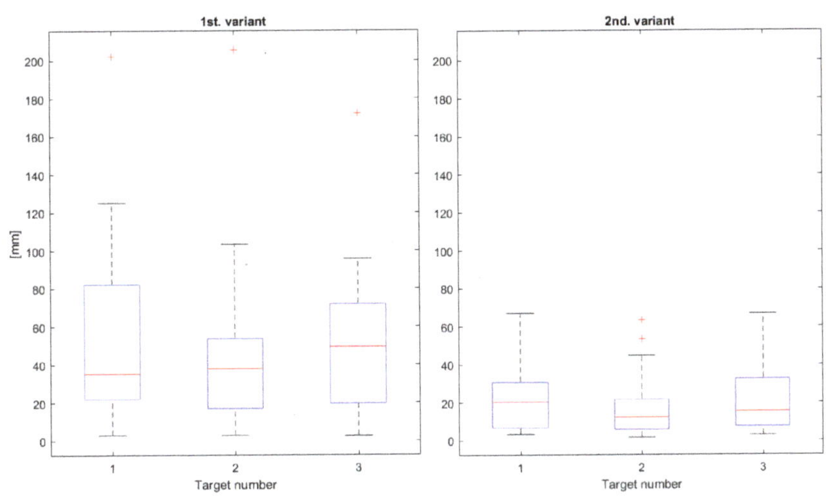

Figure 18. Boxplots of euclidean distance data for experiment no. 2.

Table 2. The mean of euclidean distances between the desired position and measured data for each horizontally placed target marker.

Target Marker No.	1st. Variant (mm)	2nd. Variant (mm)
1	33.42	12.16
2	54.18	14.81
3	70.71	18.68
4	50.88	12.60
5	63.99	17.64
6	51.45	12.54
overall	54.11	14.74

Table 3. The mean of euclidean distances between the desired position and measured data for each vertically placed target marker.

Target Marker No.	1st. Variant (mm)	2nd. Variant (mm)
1	49.77	21.17
2	40.56	15.62
3	47.25	20.21
overall	45.86	19.0

Table 4. The dispersion of measured points for each target in the first experimental scenario.

	1st. Variant (mm)		2nd. Variant (mm)	
	x Axis	y Axis	x Axis	y Axis
1	22.42	35.32	4.94	14.07
2	50.14	36.93	13.28	8.35
3	26.92	65.54	11.24	17.64
4	20.08	44.55	4.91	13.99
5	44.31	52.0	15.46	22.34
6	38.71	43.99	9.06	22.82
overall	33.77	46.39	9.81	16.53

Table 5. The dispersion of measured points for each target in the second experiment scenario.

	1st. (mm)		2nd. Variant (mm)	
	x Axis	y Axis	x Axis	y Axis
1	46.10	37.44	14.47	20.79
2	27.74	40.20	5.61	17.93
3	32.88	43.96	12.66	19.40
overall	35.57	40.53	10.91	19.37

9. Discussion

The data show that the error of pointing to the specific location, relying solely on the user's estimation, is approximately 50 mm. The user's consistency of pointing is around 30–50 mm, depending on the user's distance to the target. The results also show significant improvement in both accuracy and consistency when visual feedback is provided. These results outline the potential application use regarding the possible size of manipulated objects as mentioned previously in a real-life scenario. They also highlight the importance of some feedback that should be implemented to potential real use-case applications. However, the data are not yet sufficient to derive general conclusions. Further, more thorough analysis is needed. During the experiments, several factors were identified that could significantly influence the accuracy of pointing. The first factor is the posture and position of the human body and the distortion caused by the user's point of view. The second factor is the tracking capabilities and precision of the LMC sensor. During experiments, numerous glitches of

LMC controller were reported. The LMC occasionally misinterpreted the position of joints and fingers, which led to inaccuracies in the measurements. Another factor influencing the results is the joints selected to represent the half-line intersecting with the working plane. The last identified factor is the influence of the command pistol gesture on pointing accuracy, as extending the thumb may change the intended pointing location. Further evaluation of the method, considering all of the mentioned factors, must be done in the future to determine the full potential of the proposed gestural framework. In all, we believe that the experiments and the data served their purpose. The main factors influencing the accuracy were identified, and the obtained data can be used as the base for further experiments and evaluation of the method.

10. Conclusions

In this work, we proposed and successfully implemented a framework for gestural HRI using pointing gestures. The core concept of the method was based on the previous work by Tölgyessy et al. [6], particularly on the idea of linear HRI. In the work, they proposed a similar concept of controlling a mobile robotic platform via pointing gestures. This article further improves the work by extending the overall potential of this innovative concept from mobile platforms to other fields of robotics. Additionally, it promotes the universal nature of linear HRI as the concept uses a different sensory system and a different type of robotic system. The work contains a general overview as well as the mathematical background for the core method. A concept of the collaborative robotic workplace, also known as COCOHRIP, was built around the proposed framework. Practical experiments in a laboratory environment were carried out to prove the described method. The obtained data provided the basis for further development of the whole gestural framework.

The main contribution of this work is a novel approach to the control of the collaborative robotic arm using pointing gestures. In the majority of the cited work, the researchers focus on teleoperation of the robotic arm either through movement mirroring or by binding the control of the robot to a specific gesture. The operator is not directly interacting with the (shared) workspace. The robot is controlled directly by the user, and its movements are entirely subordinate to the operator's movements. In our approach, the operator interacts with the robot through the shared workspace, thus not teleoperating the robot directly. The robot is in the role of an assistant rather than a subordinate. Consequently, this dynamic could facilitate more effective collaboration between humans and the robotic systems. Cippolla and Hollinghurst [22] presented a gesture-based interface method that is most comparable to our approach. They similarly use pointing gestures to direct the robot's movement. In their work, hand tracking is achieved using stereo monochromatic CCD cameras. However, their method suffers shortcomings such as the need for high contrast background and specific light conditions. Additionally, the method's workspace is much smaller compared to our method. Our proposed concept has far more robust gesture recognition and tracking performance, additionally being able to recognize multiple gestures. Other cited works in which the user interacts with a workspace require some sort of wearable equipment. We believe the future of HRI lies in the utilization of multiple interconnected visual-based sensors with the ultimate purpose to rid humans of any additional restrictive equipment, allowing them to interact with robots as naturally and ergonomically as possible, without the need for superior technical skills. Furthermore, the core mathematical method in the proposed framework is universally applicable, not relying on the specific hardware or software.

The practical utilization of the proposed framework was outlined in the form of a collaborative robotic assistant that could potentially be used in the manufacturing process, academic field, or research and development. This paper further develops this concept, widening its usability from mobile robotic applications to robotic arm manipulation while utilizing a new state-of-the-art visual-based sensory system. Our framework, along with the built COCOHRIP workplace, could serve as a basis for more complex applications, such as naturally controlled object picking and manipulation, robot-assisted assembly, or material processing. Our future research will focus on improving the system's accuracy based on the identified factors, and designing and implementing more complex algorithms and methods to further increase the usability and flexibility of the proposed gestural framework.

Author Contributions: Conceptualization, M.T. and M.Č.; methodology, M.T.; software, M.Č.; validation, M.Č.; formal analysis, M.T. and M.Č.; investigation, M.T. and M.Č.; resources, M.T. and M.Č.; data curation, M.Č.; writing—original draft preparation, M.Č.; writing—review and editing, M.T. and M.Č.; visualization, M.Č.; supervision, M.T.; project administration, M.T.; funding acquisition, P.H. All authors have read and agreed to the published version of the manuscript.

Funding: This work was funded by APVV-17-0214.

Data Availability Statement: All data presented in this paper are available at: https://drive.google.com/drive/folders/1V4cUwvD7barr6BBPcmY9KFzPz3vc1Ee8?usp=sharing.

Conflicts of Interest: The authors declare no conflict of interest.

References

1. Tang, N. *Securing the Future of German Manufacturing Industry Recommendations for Implementing the Strategic Initiative INDUSTRIE 4.0 Final Report of the Industrie 4.0 Working Group*; Acatech: Munich, Germany, 2013.
2. Matheson, E.; Minto, R.; Zampieri, E.G.G.; Faccio, M.; Rosati, G. Human—Robot Collaboration in Manufacturing Applications: A Review. *Robotics* **2019**, *8*, 100. [CrossRef]
3. Gualtieri, L.; Rauch, E.; Vidoni, R. Emerging research fields in safety and ergonomics in industrial collaborative robotics: A systematic literature review. *Robot. Comput.-Integr. Manuf.* **2021**, *67*, 101998. [CrossRef]
4. Thoben, K.; Wiesner, S.; Wuest, T. "Industrie 4.0" and Smart Manufacturing—A Review of Research Issues and Application Examples. *Int. J. Autom. Technol.* **2017**, *11*, 4–16. [CrossRef]
5. Tang, G.; Webb, P. The Design and Evaluation of an Ergonomic Contactless Gesture Control System for Industrial Robots. *J. Robot.* **2018**, *2018*, 9791286. [CrossRef]
6. Tölgyessy, M.; Dekan, M.; Duchoň, F.; Rodina, J.; Hubinský, P.; Chovanec, L. Foundations of Visual Linear Human–Robot Interaction via Pointing Gesture Navigation. *Int. J. Soc. Robot.* **2017**, *9*, 509–523. [CrossRef]
7. Cho, H.; Chung, W. Preliminary research on robust leg-tracking indoor mobile robots by combining the Kinect and the laser range finder information. In Proceedings of the 12th International Conference on Ubiquitous Robots and Ambient Intelligence (URAI), Goyangi, Korea, 28–30 October 2015; [CrossRef]
8. Chen, M.; Liu, C.; Du, G. A human—Robot interface for mobile manipulator. *Intell. Serv. Robot.* **2018**, *11*, 269–278. [CrossRef]
9. Burger, B.; Ferrané, I.; Lerasle, F.; Infantes, G. Two-handed gesture recognition and fusion with speech to command a robot. *Auton. Robot.* **2012**, *32*, 129–147. [CrossRef]
10. Valner, R.; Kruusamäe, K.; Pryor, M. TeMoto: Intuitive Multi-Range Telerobotic System with Natural Gestural and Verbal Instruction Interface. *Robotics* **2018**, *7*, 9. [CrossRef]
11. Hernoux, F.; Béarée, R.; Gibaru, O. Investigation of dynamic 3D hand motion reproduction by a robot using a Leap Motion. In Proceedings of the 2015 Virtual Reality International Conference. Association for Computing Machinery, VRIC '15, Laval, France, 8–10 April 2015; pp. 1–10. [CrossRef]
12. Du, G.; Zhang, P.; Liu, X. Markerless Human–Manipulator Interface Using Leap Motion With Interval Kalman Filter and Improved Particle Filter. *IEEE Trans. Ind. Inform.* **2016**, *12*, 694–704. [CrossRef]
13. Kruse, D.; Wen, J.T.; Radke, R.J. A Sensor-Based Dual-Arm Tele-Robotic System. *IEEE Trans. Autom. Sci. Eng.* **2015**, *12*, 4–18. [CrossRef]
14. Bassily, D.; Georgoulas, C.; Guettler, J.; Linner, T.; Bock, T. Intuitive and Adaptive Robotic Arm Manipulation using the Leap Motion Controller. In Proceedings of the ISR/Robotik 2014: 41st International Symposium on Robotics, Munich, Germany, 2–3 June 2014.
15. Pititeeraphab, Y.; Choitkunnan, P.; Thongpance, N.; Kullathum, K.; Pintavirooj, C. Robot-arm control system using LEAP motion controller. In Proceedings of the 2016 International Conference on Biomedical Engineering (BME-HUST), Hanoi, Vietnam, 5–6 October 2016; pp. 109–112. [CrossRef]
16. Chen, C.; Chen, L.; Zhou, X.; Yan, W. Controlling a robot using leap motion. In Proceedings of the 2017 2nd International Conference on Robotics and Automation Engineering (ICRAE), Shanghai, China, 29–31 December 2017; pp. 48–51. [CrossRef]
17. Chen, S.; Ma, H.; Yang, C.; Fu, M. Hand Gesture Based Robot Control System Using Leap Motion. In Proceedings of the Intelligent Robotics and Applications, Portsmouth, UK, 24–27 August 2015; Lecture Notes in Computer Science; Liu, H., Kubota, N., Zhu, X., Dillmann, R., Zhou, D., Eds.; Springer International Publishing: Berlin/Heidelberg, Germany, 2015; pp. 581–591. [CrossRef]
18. Nogueira, R.; Reis, J.; Pinto, R.; Gonçalves, G. Self-adaptive Cobots in Cyber-Physical Production Systems. In Proceedings of the 2019 24th IEEE International Conference on Emerging Technologies and Factory Automation (ETFA), Zaragoza, Spain, 10–13 September 2019; pp. 521–528. [CrossRef]

19. Cueva, C.W.F.; Torres, S.H.M.; Kern, M.J. 7 DOF industrial robot controlled by hand gestures using microsoft kinect v2. In Proceedings of the 2017 IEEE 3rd Colombian Conference on Automatic Control (CCAC), Cartagena, Colombia, 18–20 October 2017; pp. 1–6. [CrossRef]
20. Kaczmarek, W.; Panasiuk, J.; Borys, S.; Banach, P. Industrial Robot Control by Means of Gestures and Voice Commands in Off-Line and On-Line Mode. *Sensors* **2020**, *20*, 6358. [CrossRef]
21. Zhang, X.; Zhang, R.; Chen, L.; Zhang, X. Natural Gesture Control of a Delta Robot Using Leap Motion. *J. Phys.* **2019**, *1187*, 032042. [CrossRef]
22. Cipolla, R.; Hollinghurst, N. A Human–Robot Interface using Pointing with Uncalibrated Stereo Vision. In *Computer Vision for Human-Machine Interaction*; Pentland, A., Cipolla, R., Eds.; Cambridge University Press: Cambridge, UK, 1998; pp. 97–110. [CrossRef]
23. Gunawardane, P.; Medagedara, N.T.; Madusanka, B.; Wijesinghe, S. The development of a Gesture Controlled Soft Robot gripping mechanism. In Proceedings of the 2016 IEEE International Conference on Information and Automation for Sustainability (ICIAfS), Galle, Sri Lanka, 16–19 December 2016; pp. 1–6. [CrossRef]
24. Devine, S.; Rafferty, K.; Ferguson, S. Real time robotic arm control using hand gestures with multiple end effectors. In Proceedings of the 2016 UKACC 11th International Conference on Control (CONTROL), Belfast, UK, 31 August–2 September 2016; pp. 1–5. [CrossRef]
25. Li, C.; Fahmy, A.; Sienz, J. Development of a Neural Network-Based Control System for the DLR-HIT II Robot Hand Using Leap Motion. *IEEE Access* **2019**, *7*, 136914–136923. [CrossRef]
26. Razjigaev, A.; Crawford, R.; Roberts, J.; Wu, L. Teleoperation of a concentric tube robot through hand gesture visual tracking. In Proceedings of the 2017 IEEE International Conference on Robotics and Biomimetics (ROBIO), Macau, Macao, 5–8 December 2017; pp. 1175–1180. [CrossRef]
27. Sarkar, A.; Patel, K.A.; Ganesh Ram, R.; Capoor, G.K. Gesture control of drone using a motion controller. In Proceedings of the 2016 International Conference on Industrial Informatics and Computer Systems (CIICS), Sharjah, United Arab Emirates, 13–15 March 2016; pp. 1–5. [CrossRef]
28. Hu, B.; Wang, J. Deep Learning Based Hand Gesture Recognition and UAV Flight Controls. In Proceedings of the 2018 24th International Conference on Automation and Computing (ICAC), Newcastle Upon Tyne, UK, 6–7 September 2018; pp. 1–6. [CrossRef]
29. Yu, N.; Xu, C.; Wang, K.; Yang, Z.; Liu, J. Gesture-based telemanipulation of a humanoid robot for home service tasks. In Proceedings of the 2015 IEEE International Conference on Cyber Technology in Automation, Control, and Intelligent Systems (CYBER), Shenyang, China, 8–12 June 2015; pp. 1923–1927. [CrossRef]
30. Cheng, L.; Sun, Q.; Su, H.; Cong, Y.; Zhao, S. Design and implementation of human–robot interactive demonstration system based on Kinect. In Proceedings of the 2012 24th Chinese Control and Decision Conference (CCDC), Taiyuan, China, 23–25 May 2012; pp. 971–975, ISSN 1948-9447. [CrossRef]
31. Liu, Y.; Zhang, Y. Toward Welding Robot With Human Knowledge: A Remotely-Controlled Approach. *IEEE Trans. Autom. Sci. Eng.* **2015**, *12*, 769–774. [CrossRef]
32. Yang, C.; Zeng, C.; Liang, P.; Li, Z.; Li, R.; Su, C.Y. Interface Design of a Physical Human–Robot Interaction System for Human Impedance Adaptive Skill Transfer. *IEEE Trans. Autom. Sci. Eng.* **2018**, *15*, 329–340. [CrossRef]
33. Krupke, D.; Steinicke, F.; Lubos, P.; Jonetzko, Y.; Görner, M.; Zhang, J. Comparison of Multimodal Heading and Pointing Gestures for Co-Located Mixed Reality Human–Robot Interaction. In Proceedings of the 2018 IEEE/RSJ International Conference on Intelligent Robots and Systems (IROS), Madrid, Spain, 1–5 October 2018; pp. 1–9. [CrossRef]
34. Zhou, H.; Hu, H. Human motion tracking for rehabilitation—A survey. *Biomed. Signal Process. Control* **2008**, *3*, 1–18. [CrossRef]
35. Vardhan, D.; Prasad, P.P. Hand Gesture Recognition Application for Physically Disabled People. *Int. J. Sci. Res.* **2014**, *3*, 765–769.
36. Parvini, F.; McLeod, D.; Shahabi, C.; Navai, B.; Zali, B.; Ghandeharizadeh, S. An Approach to Glove-Based Gesture Recognition. In Proceedings of the Human-Computer Interaction. Novel Interaction Methods and Techniques, San Diego, CA, USA, 19–24 July 2009; Lecture Notes in Computer Science; Jacko, J.A., Ed.; Springer: Berlin/Heidelberg, Germany, 2009; pp. 236–245. [CrossRef]
37. Allevard, T.; Benoit, E.; Foulloy, L. Fuzzy Glove For Gesture Recognition. In Proceedings of the 17th IMEKO World Congress, Dubrovnik, Croatia, 22–28 June 2003; pp. 2026–2031.
38. Ghunawat, M.R. Multi-point Gesture Recognition Using LED Gloves For Interactive HCI. *Int. J. Comput. Sci. Inf. Technol.* **2014**, *5*, 6768–6773.
39. Ganzeboom, M. How Hand Gestures Are Recognized Using a Dataglove. 2009. Available online: https://www.researchgate.net/publication/228702251_How_hand_gestures_are_recognized_using_a_dataglove (accessed on 19 December 2021).
40. Shotton, J.; Girshick, R.; Fitzgibbon, A.; Sharp, T.; Cook, M.; Finocchio, M.; Moore, R.; Kohli, P.; Criminisi, A.; Kipman, A.; et al. Efficient Human Pose Estimation from Single Depth Images. *IEEE Trans. Pattern Anal. Mach. Intell.* **2013**, *35*, 2821–2840. [CrossRef]
41. Guzsvinecz, T.; Szucs, V.; Sik-Lanyi, C. Suitability of the Kinect Sensor and Leap Motion Controller—A Literature Review. *Sensors* **2019**, *19*, 1072. [CrossRef]
42. Vysocký, A.; Grushko, S.; Oščádal, P.; Kot, T.; Babjak, J.; Jánoš, R.; Sukop, M.; Bobovský, Z. Analysis of Precision and Stability of Hand Tracking with Leap Motion Sensor. *Sensors* **2020**, *20*, 4088. [CrossRef]
43. Guna, J.; Jakus, G.; Pogačnik, M.; Tomažič, S.; Sodnik, J. An Analysis of the Precision and Reliability of the Leap Motion Sensor and Its Suitability for Static and Dynamic Tracking. *Sensors* **2014**, *14*, 3702–3720. [CrossRef]

applied sciences

MDPI

Article

Optimal Design of a Parallel Manipulator for Aliquoting of Biomaterials Considering Workspace and Singularity Zones

Dmitry Malyshev [1,*], Larisa Rybak [1], Giuseppe Carbone [1,2], Tatyana Semenenko [1,3] and Anna Nozdracheva [1,3]

1 International Research Laboratory Intelligent Robotic Systems and Technologies, Belgorod State Technological University named after V.G. Shukhov, 308012 Belgorod, Russia; rlbgtu@gmail.com (L.R.); giuseppe.carbone@unical.it (G.C.); semenenko@gamaleya.org (T.S.); nozdrcheva0506@gmail.com (A.N.)
2 Department of Mechanical, Energy and Management Engineering, University of Calabria, 87036 Rende, Italy
3 Department of Epidemiology, Gamaleya National Research Center for Epidemiology & Microbiology, 123098 Moscow, Russia
* Correspondence: malyshev.d.i@ya.ru

Abstract: This article presents the concept of a robotic system for aliquoting of biomaterial, consisting of a serial manipulator in combination with a parallel Delta-like robot. The paper describes a mathematical formulation for approximating the geometric constraints of the parallel robot as a set of solutions to a system of nonlinear inequalities. The analysis of the workspace is carried out, taking into account singularity zones, using a method based on the analysis of the Jacobian matrix of the mechanism and the interference of links. An optimal design procedure is proposed for the dimensional synthesis based on a criterion for maximizing the volume of the workspace, taking into account the ambiguity of the solution of the inverse kinematics. Simulation results are reported and discussed to propose a suitable design solution.

Keywords: parallel robot; workspace; non-uniform covering; optimization; singularity zones; link interference

Citation: Malyshev, D.; Rybak, L.; Carbone, G.; Semenenko, T.; Nozdracheva, A. Optimal Design of a Parallel Manipulator for Aliquoting of Biomaterials Considering Workspace and Singularity Zones. *Appl. Sci.* **2022**, *12*, 2070. https://doi.org/10.3390/app12042070

Academic Editor: Alessandro Gasparetto

Received: 10 November 2021
Accepted: 10 February 2022
Published: 16 February 2022

1. Introduction

Over the past two decades, the number of biomedical research studies, including the study of the foundations of the pathogenesis of infectious pathology, carried out all over the world, has increased [1]. The level of such research is largely determined by the availability of a certified collection of biological material. This applies equally to the development of countermeasures in the face of the threat to the global community associated with the COVID-19 [2] pandemic. The need to reduce the number of production processes with the use of manual labor is also significant for preventing contamination of equipment and work surfaces by infectious agents and, consequently, minimizing the risk of personnel infection. The more steps that can be performed with high reproducibility, the lower the level of random error there will be in the final data. As robotic hardware and computer control systems have advanced and the price has reduced, more steps in the sample preparation process have become amenable to automation and more end users are able to justify the capital investment relative to the anticipated labor savings [3].

The term aliquot comes from the word "aliquot", which means the exact measured fraction of the sample taken for analysis, which retains the properties of the main sample. Thus, the aliquot process involves dividing the whole sample of biomaterial into subfractions. The main developers of robotic equipment for aliquoting biological samples are Dornier-LTF (Germany) [4], TECAN (Switzerland) [5], and Hamilton [6]. The laboratory equipment they produce is as follows: robotic digging station PIRO manufactured by Dornier-LTF (Germany), robotic laboratory station (Freedom EVO® series) manufactured by TECAN, and automatic dispensing system (robot dispenser) Microlab STAR manufactured by Hamilton. They are designed to perform the same type of technological processes and include the main elements: a robotic module with dosing multichannel pipettes and removable tips; a robotic module with a gripper for transferring racks; and auxiliary devices (pumps, barcode readers, thermoshakers, UV emitters, detectors of the position of objects of the working surface, etc.), the number of which depends on the purpose of using the robotic station. These automated stations have a typical architecture and successfully solve the problems of aliquoting biomaterial using tubes of various sizes. For liquid dispensing, intelligent systems are used that record the moments when the pipette touches the liquid surface, the absence of liquid in the test tube, or the presence of colloidal clots that impede pipetting.

However, their significant disadvantage is the inability to aliquot serum from test tubes with different liquid phases. For the operation of the equipment, only samples that have undergone primary manual processing in the form of separating serum from a blood clot can be used.

At the same time, it is important to design a robot, taking into account the optimization of its parameters in order to achieve the best combination of design parameters. Optimal design of robotic systems is an actual and important topic [7–12].

2. Building a Model of a Robotic System

The work [13] shows the effectiveness of automated aliquoting compared to manual labor. Research in the field of application of robotic systems for liquid dosing and automation of laboratory processes for sample preparation is carried out by many scientists. Active research is being conducted at the Institute of Automation and Center for Life Science Automation, University of Rostock (Rostock, Germany). In works [14,15], the use of a cognitive two-handed robot, working in conjunction with an operator and remotely controlled, is proposed. Pipetting of liquid is carried out with an automatic hand-type pipette, widely used in laboratory practice. Each robot arm has 7 degrees of freedom and can perform any manipulations within the workspace. The disadvantage of this solution is the low speed of the end-effector and the low accuracy of its positioning. So, in [16], an integrated system based on robots is considered for automating a multistage system of laboratory research, including the movement of equipment and biomaterial samples. The possibility of integrating devices controlled through electronic interfaces such as RS232, USB, Firewire, or Ethernet into a common network is shown, which allows centralized control of the technological process.

The article discusses a new architecture and design of a robotic system, which must meet the following requirements to ensure the technological process of aliquoting:

1. The volumes of the tubes used: from 2 mL to 9 mL, the height of which is 75 and 100 mm, respectively (Figure 1).
2. The volumes of the pipettes used are from 10 μL to 5000 μL, the height of which is from 32 mm to 150 mm.
3. A rack for loading tubes with fractionated whole blood samples should hold 12 or more tubes with a volume of 9 mL.
4. The height of the manipulator workspace must be bigger than the sum of the maximum volume tube height (100 mm), maximum pipette height (150 mm), and the safe distance between the tube and the pipette. The safe distance is 100 mm. Thus, the height of the manipulator workspace should be 350 mm.
5. Initially, the liquid is divided into fractions. The movement of tubes with samples of whole blood, divided into fractions, should be carried out evenly and progressively in the horizontal direction, limited by the velocity of the mechanism and in the vertical direction—no more than 0.03 m/s, without sudden movements (to prevent the risk of breaking the integrity of the clot blood).
6. The robotic system must meet the requirements for the safety of human-robot interaction. So, all movements of the robot should be stopped when human limbs enter the workspace to avoid injury.

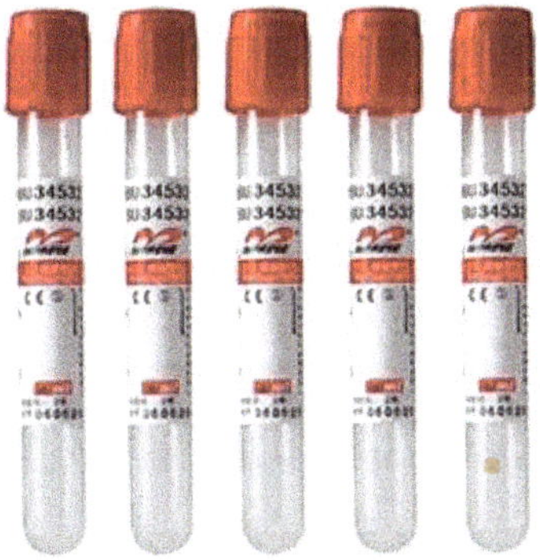

Figure 1. Round bottom blood tubes.

The flow chart of the aliquoting process is shown in Figure 2.

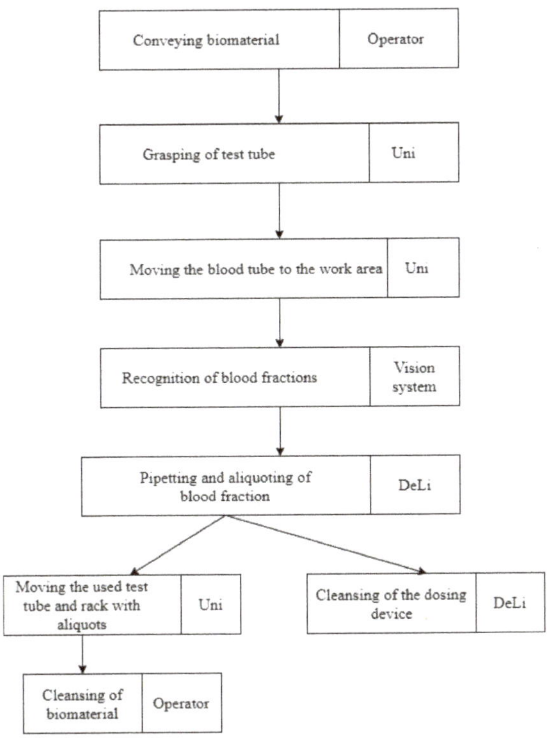

Figure 2. Aliquoting process.

Figure 3 shows a 3D model of a robotic system, which includes a parallel DeLi manipulator with 4 degrees of freedom, made based on the Delta mechanism and equipped with an excavating head for performing basic operations for aliquoting biomaterial, and a collaborative robot (Uni) with a serial structure for performing transportation operations, racks with test tubes and plates, and microtubes. Uni robot is used for the stage when liquid is divided into fractions. The DeLi robot should take only one upper fraction from a test tube and distribute it into smaller tubes, while not allowing the lower fraction (sediment) to enter. After taking the desired fraction, the DeLi robot can move at any speed available to this type of robot. The amount of the withdrawn fraction is determined using the developed technical vision system, which is not described in this work. The DeLi manipulator dispenses only one fraction of blood; therefore, the velocity for the DeLi manipulator is not limited. The Delta parallel robot topology has been chosen due to the fact that it has a better performance in terms of dynamics and accuracy as compared with serial robots. Furthermore, Delta architectures are better suited for moving along curvilinear trajectories that may arise in the proposed specific application due to different heights of test tubes and racks. This is justified by the features of the robot structure, which has low inertia and increased characteristics of speeds and accelerations, which are important when moving along circular trajectories, including those involving reverse movement of drives. This makes the Delta architecture better suited also in comparison with portal architecture.

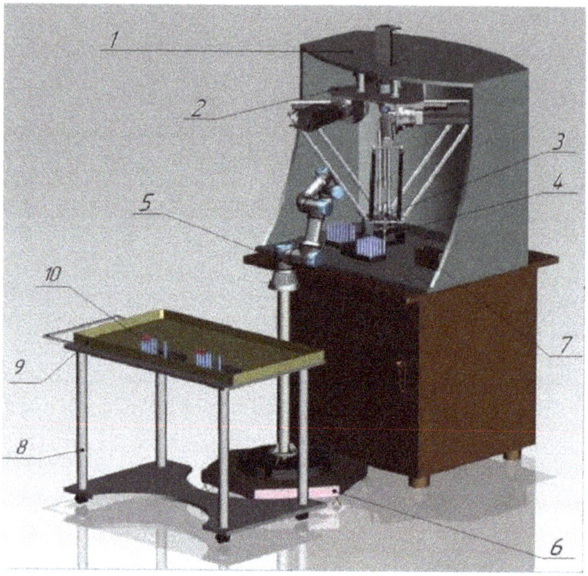

Figure 3. Three-dimensional model of the robotic system: 1—body, 2—DeLi manipulator, 3—dispensing head, 4—dispenser tip, 5—robot manipulator, 6—base of the robot manipulator, 7—workspace, 8—rack with test tubes, 9—tray for consumables, 10—trolley.

It is necessary to determine the boundaries of its workspace to select the geometric and design parameters of the robotic system. Determination of the workspace of parallel robots is much more complex than for serial robots, due to the peculiarities of structure, kinematics, and dynamics. There are various methods for determining the workspace of translational robots. A multiobjective optimum design procedure to 3 degrees of freedom (DOF) parallel robots with regards to three optimality criteria: workspace boundary, transmission quality index, and stiffness, is presented in article [17]. A kinematic optimization was performed to maximize the workspace of the parallel robot. Genetic algorithms were applied to optimize the objective function. Article [18] discusses the kinematic and geometric aspects of the 3-DOF translational orthoglide robot. New solutions are proposed to solve the inverse and forward kinematics and conduct a detailed analysis of the workspace and features, taking into account specific joint limit constraints. In [19], such methods of optimizing the workspace as the genetic algorithm and the maximum surrounding workspace are considered. In [20], the cylindrical algebraic decomposition is mentioned, which is used to calculate the robot's workspace in the projection space (x, y, z), taking into account some joint constraints. In [21], an approach to calculating the workspace of a parallel robot Delta is proposed based on the forward kinematics model. In paper [22], the proposed method is implemented for a spatial 3-DOF parallel robot, known as the Tripteron. The mechanical interference, including interference of links, interference of links with obstacles, and interference of the end-effector with obstacles, are investigated by using new geometrical reasoning. For this purpose, a new geometric method is proposed which is based on the segment to segment intersection test. This method can be well extended to a wide range of robotic mechanical systems, including, among others, parallel robots. This method is also applied in [23] for the Delta robot. However, the method considered in this article has disadvantages; in particular, the authors propose to determine the interference of segments on the auxiliary plane and not the distance between the nearest points. This makes it impossible to identify such interference of the links, in which there is no interference of the axes. The workspace for a 3-PUU Delta mechanism with translational drive joints is considered in the article [24]. The analysis of the achievable workspace can be performed using the inverse kinematics method; the workspace is expressed with three-dimensional dot clouds of points. The article [25] presents a workspace, a shared space, and features of a family of Delta-like parallel robots using algebraic tools. In articles [26,27], attention is paid to the analysis of the influence of singularity zones and interference of links. However, the articles did not consider the various configurations of the Delta robot or the influence of singularity zones on the amount of workspace.

Although manipulators of the Delta type are well studied, the development of new methods for analyzing the workspace that are more efficient in terms of increasing the accuracy of approximation and versatility of their application for various versions of their architecture is an important task. The task of determining the workspace can be solved with a given accuracy using interval analysis methods [28–30]. One of the disadvantages of deterministic interval analysis methods is a large number of elements in the covering set. Nevertheless, the article's authors proposed an approach to the transformation of the covering set, which was considered earlier in [31]. It is used to transform the covering set before rendering the stage. The interval analysis method is applied to approximate the set of solutions to a system of nonlinear inequalities that determine the constraints on the parameters of the DeLi robot links (Figure 4). The manipulator has 4 degrees of freedom and includes three RUU kinematic chains. In each of the chains, a rotary drive joint A_i is used to connect to the base, and there are two universal joints—C_i to connect to the working platform and B_i to connect two links together. The base and work platform are equilateral triangles. As an end-effector, we will consider point P—the center of the working platform.

We write the boundaries of the workspace in the form of a system of inequalities:

$$\begin{cases} \theta_i - \theta_{max} \leq 0 \\ \theta_{min} - \theta_i \leq 0 \end{cases} \tag{1}$$

where θ_i are the angles of rotation in the drive joints, and θ_{min} and θ_{max} are the minimum and maximum angles, respectively. Based on the solution of the inverse kinematics described in [32], we obtain

$$\theta_i = 2\tan^{-1}\left(\frac{-F_i \pm \sqrt{E_i^2 + F_i^2 + G_i^2}}{G_i - E_i}\right) \tag{2}$$

where E_i, F_i, and G_i are defined as:

$$E_1 = 2d\left(y_P + \frac{a-2c}{2\sqrt{3}}\right),\ F_1 = 2z_P d, \tag{3}$$

$$G_1 = x_P^2 + y_P^2 + z_P^2 + \left(\frac{a-2c}{2\sqrt{3}}\right)^2 + d^2 + 2y_P\left(\frac{a-2c}{2\sqrt{3}}\right) - e^2 \tag{4}$$

$$E_2 = -d\left(\sqrt{3}\left(x_P + \frac{2c-a}{4}\right) + y_P + \frac{2c-a}{4\sqrt{3}}\right),\ F_2 = 2z_P d, \tag{5}$$

$$G_2 = x_P^2 + y_P^2 + z_P^2 + \left(\frac{2c-a}{4}\right)^2 + \left(\frac{2c-a}{4\sqrt{3}}\right)^2 + d^2 + 2\left(x_P\left(\frac{2c-a}{4}\right) + y_P\left(\frac{2c-a}{4\sqrt{3}}\right)\right) - e^2 \quad (6)$$

$$E_3 = d\left(\sqrt{3}\left(x_P - \frac{2c-a}{4}\right) - y_P - \frac{2c-a}{4\sqrt{3}}\right),\ F_3 = 2z_P d, \quad (7)$$

$$G_3 = x_P^2 + y_P^2 + z_P^2 + \left(\frac{2c-a}{4}\right)^2 + \left(\frac{2c-a}{4\sqrt{3}}\right)^2 + d^2 + 2\left(y_P\left(\frac{2c-a}{4\sqrt{3}}\right) - x_P\left(\frac{2c-a}{4}\right)\right) - e^2 \quad (8)$$

where a is the side length of a regular triangle of the base, c is the side length of a regular triangle of the movable platform, d is the distance between A and B joints, e is the distance between B and C joints, and x_P, y_P, z_P are the coordinates of point P.

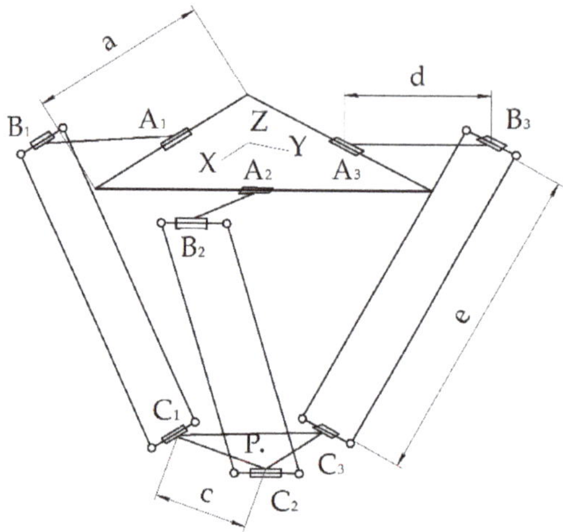

Figure 4. The structure of the DeLi manipulator.

There is an ambiguity in the solution of the inverse kinematics, when for the same coordinates of the point P there are two possible angles of rotation in the drive rotary joint for each of the three kinematic chains. An example of two possible positions of the kinematic chain $A_iB_iC_i$ is shown in Figure 5. The first position corresponds to [−] in Equation (2), the second position corresponds to [+].

Thus, there are eight possible solutions to the inverse kinematics, and, therefore, for each of them, a workspace can be determined. In this case, one should take into account the presence of singularity zones of the mechanism when hit, in which the dynamic loads on the links increase significantly, and the robot loses controllability. We consider the method proposed in [33,34], based on the analysis of the Jacobi matrix, to determine the singularity zones. The determinant of the Jacobi matrix has the form:

$$\det(J_A) = \begin{bmatrix} \frac{\partial\Theta_1}{\partial x_P} & \frac{\partial\Theta_1}{\partial y_P} & \frac{\partial\Theta_1}{\partial z_P} \\ \frac{\partial\Theta_2}{\partial x_P} & \frac{\partial\Theta_2}{\partial y_P} & \frac{\partial\Theta_2}{\partial z_P} \\ \frac{\partial\Theta_3}{\partial x_P} & \frac{\partial\Theta_3}{\partial y_P} & \frac{\partial\Theta_3}{\partial z_P} \end{bmatrix}, \quad (9)$$

where Θ_i are determined by Equation (2).

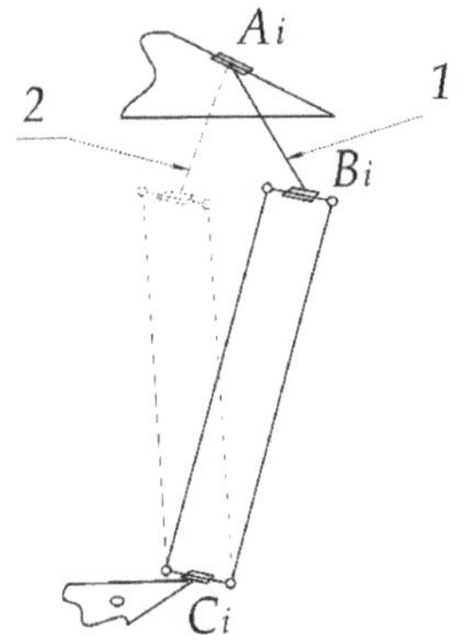

Figure 5. The ambiguity of the solution of the inverse kinematics.

In view of the cumbersomeness of the formulas obtained for each of the elements of the determinant, we present only the first of them:

$$\frac{\partial\Theta_1}{\partial x_P} = \frac{2\left(\frac{2x_p(2dz_p+s_1)}{\left(s_2-2d\left(y_p-\left(\frac{2c-a}{2\sqrt{3}}\right)\right)\right)^2} + \frac{2x_p s_2}{x10\left(s_2-2d\left(y_p-\left(\frac{2c-a}{2\sqrt{3}}\right)\right)\right)}\right)}{\left(\frac{(2dz_p+s_1)}{2x_p s_2}\right)^2+1}, \tag{10}$$

where

$$s_1 = \sqrt{4d^2{z_p}^2+4d^2\left(y_p-\left(\frac{2c-a}{2\sqrt{3}}\right)\right)^2} - {s_2}^2 s_2 = \left(\frac{2c-a}{2\sqrt{3}}\right)^2 - 2y_p\left(\frac{2c-a}{2\sqrt{3}}\right) - e^2+d^2+{x_p}^2+{y_p}^2+{z_p}^2$$

We write down the condition for the presence of singularity zones:

$$\det(J_A) = 0 \tag{11}$$

Each of the eight solutions of the inverse kinematics also corresponds to eight possible Jacobi matrices and, accordingly, their determinants. It is necessary to ensure the constant sign of the determinant of the Jacobi matrix to exclude singularity zones from the workspace of the robot. To ensure this condition, it is necessary to add one of the conditions to inequality system (1): $\det(J_A) < 0$ or $\det(J_A) > 0$, depending on the sign of the determinant. So, singularity surfaces separate the workspace into non-overlapping volumes. It is then necessary for the robot to remain within the same volume as it moves, in order to avoid crossing a singularity.

3. Synthesis of an Algorithm for Determining the Workspace

Taking into account Equations (1)–(5), an algorithm for approximating the workspace of the DeLi robot is synthesized. The workspace analysis method is a direct application of interval analysis methods, as described in Refs. [28–31].

The algorithm works with a system of inequalities written in a general form:

$$\begin{cases} g_1(x) \le 0, \\ \dots \\ g_m(x) \le 0 \\ a_i \le x_i \le b_i,\ i = 1,\dots,n. \end{cases} \tag{12}$$

The initial box Q, which includes the entire set of solutions X, is determined by the constraints of the intervals $a_i \le x_i \le b_i$, $i = 1,\dots,n$. Consider an arbitrary box B. Let m

$m(B) = \max\limits_{i=1,\ldots,m} \min\limits_{x \in B} g_i(x)$ and $M(B) = \max\limits_{i=1,\ldots,m} \max\limits_{x \in B} g_i(x)$. If $m(B) > 0$, then B contains no possible points for system (12). The proposed algorithm excludes such boxes. If $M(B) \leq 0$, then each point of the box is a possible solution. Therefore, it can be added to the coverage as an inner box. If a box cannot be excluded, it is divided into two smaller boxes if its diameter is not less than the specified precision δ. (Figure 6). The check is iteratively repeated for boxes formed after division until their size becomes less than precision δ.

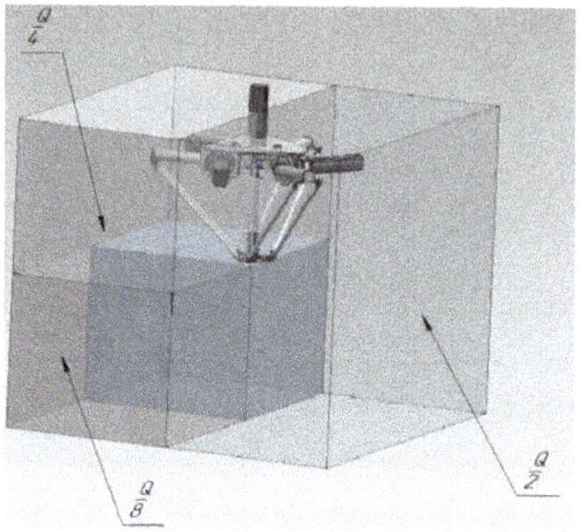

Figure 6. Division of the original box Q: $\frac{Q}{2}$—into 2 parts, $\frac{Q}{4}$—into 4 parts, $\frac{Q}{8}$—into 8 parts.

The algorithm for determining the workspace, taking into account singularity zones, is similar and differs in a large number of inequalities in the system, as well as additional conditions when checking *m (B)* and *M (B)*, taking into account the strict inequality for the determinant of the Jacobian matrix.

Equation (11) is used instead of the system of inequalities in the algorithm for determining the singularity zones. To find the set of its solutions, we calculate *M (B)*, which is calculated as $M(B) = \min\limits_{i \in 1,k} \max\limits_{x \in B} g_j(x)$, the condition for adding is that there is no inner coverage, and the condition for excluding boxes will take the form:

$$M(B) < 0 \vee m(B) > 0 \quad (13)$$

We use synthesized algorithms for numerical simulation. A program was written in the C++ programming language to implement the developed algorithms.

4. Simulation Results of Workspace Taking into Account Singularity Zones

The results of modeling the workspace were obtained for the following parameters of the DeLi manipulator: a = 450 mm, c = 200 mm, d = 150 mm, e = 230 mm. The visualization of the results of modeling the workspace is carried out by converting the covering set describing the workspace into a universal format of 3D models—an STL file. The constructed workspace without taking into account singularity zones is shown in Figure 7. Figure 7a shows the workspace in full, 7b—in section by the XOZ plane. In order to increase the performance, the processor multithreading is applied using the OpenMP package. The computation time for the approximation accuracy δ = 2 mm and the grid dimension for performing computations of the 32 × 32 × 32 functions was 67 s for each inverse kinematics case.

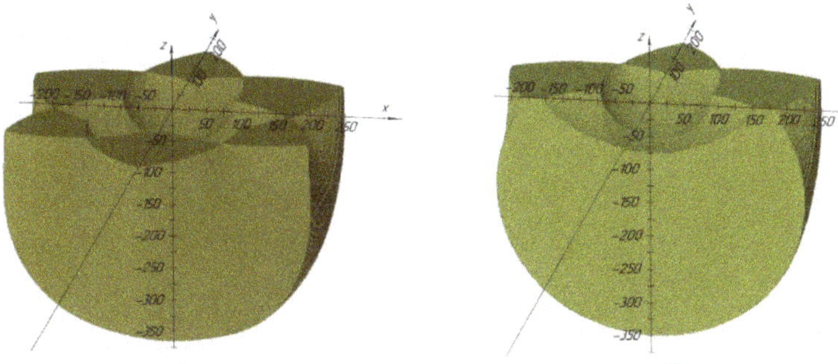

(**a**) (**b**)

Figure 7. (**a**) Volumetric workspace, (**b**) in section in the XOZ plane.

The results of modeling the workspace taking into account singularity zones are shown in Figure 8. Figure 8a shows the regions corresponding to the singularity zones when using the "+" sign in inverse kinematics (2), Figure 8b—when using the "−" sign, Figure 8c—with all eight possible combinations of the "+" and "−" signs in the inverse kinematics, Figure 8d—for all combinations within the workspace. The average computation time for an approximation accuracy $\delta = 2$ mm and a grid dimension of $64 \times 64 \times 64$ on a personal computer was 32 s.

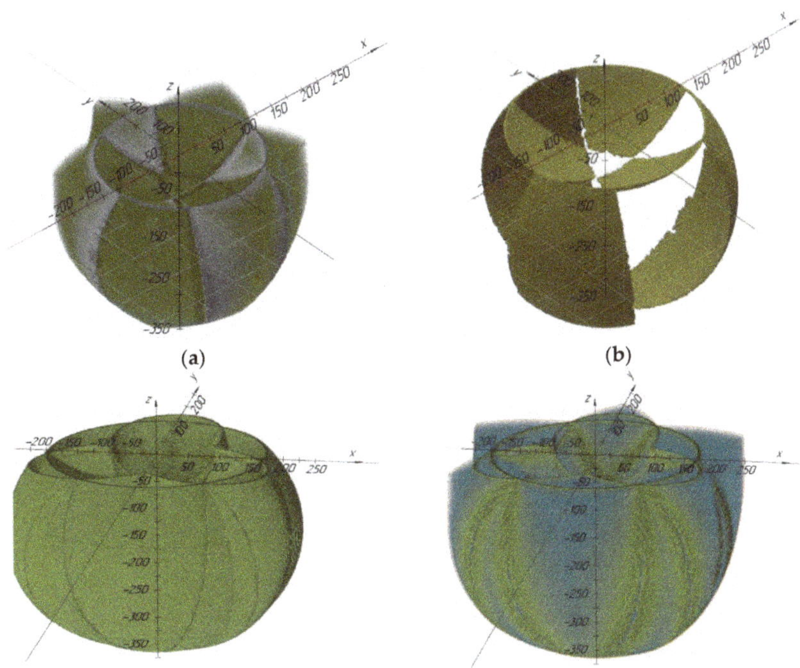

(**a**) (**b**) (**c**) (**d**)

Figure 8. Areas corresponding to singularity zones: (**a**) for "+" in inverse kinematics, (**b**) for "−", (**c**) for all possible combinations, (**d**) inside the workspace.

The simulation results show the following conclusions. The inner part of the workspace has a positive sign of the determinant of the Jacobi matrix if two or three kinematic chains have a "−" sign in the inverse kinematics. The negative sign of the determinant is obtained in the opposite cases, that is, if the "+" sign is selected for at least two chains. With this in mind, when determining the workspace, taking into account singularity zones for four possible solutions, we add the condition of the positive determinant, i.e., det $(J_A) > 0$, and for the other four, the condition of negativity, i.e., det $(J_A) < 0$.

Table 1 shows the simulation results for these cases.

Table 1. Volumes of workspaces taking into account singularity zones.

Configuration No.	Sign in Inverse Kinematics			Sign of the Jacobi Determinant	Volume of the Workspace, $\times 10^4$ mm^3		Volume Reduction in%
	θ_1	θ_2	θ_3		With Singularity Zones	Without Singularity Zones	
1	−	−	−	−	5140	4822	6.18
2	−	−	+	−	5034	4285	14.88
3	−	+	−	−	5034	4285	14.88
4	−	+	+	+	4931	4705	4.58
5	+	−	−	−	5034	4285	14.88
6	+	−	+	+	4931	4705	4.58
7	+	+	−	+	4931	4705	4.58
8	+	+	+	+	4831	4449	7.91

As can be seen from the table, out of eight options for the inverse kinematics, four groups of solutions can be distinguished: 1—all kinematic chains with "+", 2—all with "−", 3—one chain with "+" and two with "−", and also 4—two with "+" and one with "−". The first two groups correspond to one option, the second two to three options each. Considering the symmetry of the robot structure, the volume of the resulting workspace is practically the same for all variants of the group. This is confirmed by the calculated values of the volume of the workspace. The table also shows that options 2, 3, and 5, as well as 4, 6, and 7 correspond to the same volume of the workspace. Figure 9 shows the workspaces corresponding to each of the groups.

In Figure 10, it can be seen that taking into account the singularity zones led to a decrease in the size of the workspace by 4.58–14.88%, depending on the combination of possible solutions to the inverse kinematics. In addition to singularity zones, the shape and volume of the workspace are directly influenced by the interferences of the robot links with the platforms and with each other.

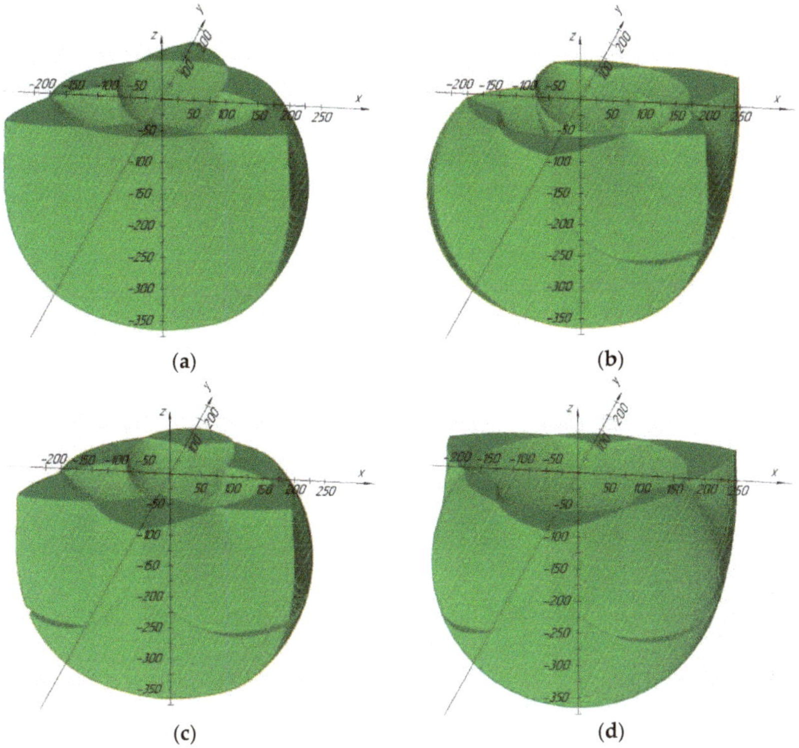

Figure 9. Workspaces, taking into account singularity zones: (**a**)—for option 1, (**b**)—for 2, 3, and 5, (**c**)—for 4, 6, and 7, (**d**)—for 8.

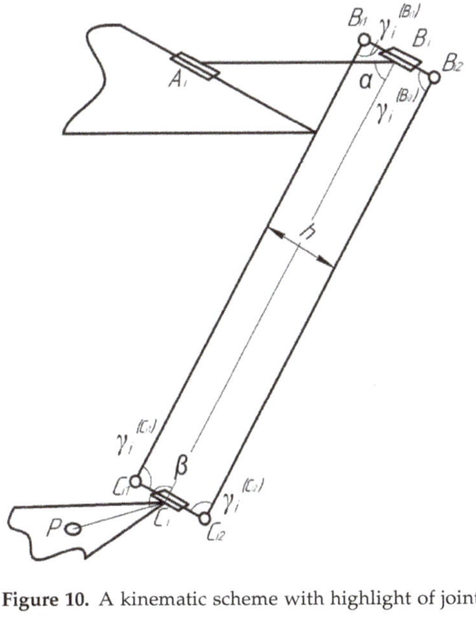

Figure 10. A kinematic scheme with highlight of joint rotation angles for a limb of the proposed mechanism.

5. Determination of the Interference of the Links of the Mechanism

The interferences of the links of the mechanism can be divided into three groups:

- Interference at small angles between links connected by joints.
- The interference of links with platforms.
- The interference of links that are not connected to each other.

The first group can be determined, taking into account the restrictions on the angles of rotation in the joints B_i and C_i (Figure 10):

$$\left\{\begin{array}{c} \alpha_i \in [\alpha_{\min}; \alpha_{\max}] \\ \beta_i \in [\beta_{\min}; \beta_{\max}] \\ \gamma_i^{(j)} \in [\gamma_{\min}; \gamma_{\max}] \end{array}\right\} \tag{14}$$

The angles can be determined using the formula for the cosines between vectors:

$$\alpha_i = \frac{(x_{Ai} - x_{Bi})(x_{Ci} - x_{Bi}) + (y_{Ai} - y_{Bi})(y_{Ci} - y_{Bi}) + (z_{Ai} - z_{Bi})(z_{Ci} - z_{Bi})}{de} \tag{15}$$

$$\beta_i = \frac{\sqrt{3}((x_{Bi} - x_{Ci})(x_P - x_{Ci}) + (y_{Bi} - y_{Ci})(y_P - y_{Ci}) + (z_{Bi} - z_{Ci})(z_P - z_{Ci}))}{ce} \tag{16}$$

$$\gamma_i^{(Bi1)} = \frac{(x_{Ci1} - x_{Bi1})(x_{Bi2} - x_{Bi1}) + (y_{Ci1} - y_{Bi1})(y_{Bi2} - y_{Bi1}) + (z_{Ci1} - z_{Bi1})(z_{Bi2} - z_{Bi1})}{he} \tag{17}$$

$$\gamma_i^{(Bi2)} = \gamma_i^{(Ci1)} = -\gamma_i^{(Ci2)} = -\gamma_i^{(Bi1)} \tag{18}$$

where $x_{A1} = z_{A1} = z_{A2} = z_{A3} = x_{B1} = 0$, $x_{A2} = \frac{a}{4}$, $x_{A3} = -\frac{a}{4}$, $y_{A1} = -\frac{a}{2\sqrt{3}}$, $y_{A2} = y_{A3} = -\frac{a}{4\sqrt{3}}$, $y_{B1} = y_{A1} - d\cos(\theta_1)$, $z_{Bi} = -d\sin(\theta_i)$, $x_{B2} = x_{A2} + \frac{\sqrt{3}d\cos(\theta_2)}{2}$, $y_{B2} = y_{A2} + \frac{d\cos(\theta_2)}{2}$, $x_{B3} = x_{A3} - \frac{\sqrt{3}d\cos(\theta_3)}{2}$, $y_{B3} = y_{A3} + \frac{d\cos(\theta_3)}{2}$, $z_{Ci} = z_P$, $x_{C1} = x_P$, $y_{C1} = y_P - \frac{c}{\sqrt{3}}$, $x_{C2} = x_P + \frac{c}{2}$, $y_{C2} = y_{C3} = y_P + \frac{c}{2\sqrt{3}}$, $x_{C3} = x_P - \frac{c}{2}$, $x_{B31} = y_{B31}$, $z_{B31} = z_{B32} = z_{B3}$.

The second group of interference includes possible interference of links with platforms A_1, A_2, A_3 and C_1, C_2, C_3. In this case, the interference with the platform C_1, C_2, C_3 is excluded by condition (14). The interference of links A_i, B_i and B_{i1}, B_{i2} with platform A_1, A_2, A_3 is also excluded by condition (1). Let us add the conditions for the appearance of the remaining interference, namely links B_{i1}, C_{i1} and B_{i2}, C_{i2} with platform A_1, A_2, A_3 (Figure 11). The first condition for the occurrence of interference is:

$$z_{Bi} \geq 0 \tag{19}$$

If condition (19) is satisfied for the joint B_i, then we calculate the coordinate points of the points D_{i1} and D_{i2}, lying on the segments B_{i1}, C_{i1} and B_{i2}, C_{i2}, respectively, which can be the interference points of the segments with the platform A_1, A_2, A_3:

$$z_{Di1} = z_{Di2} = 0 \tag{20}$$

$$x_{Di1} = x_{Bi1} + \frac{-z_{Bi1}(x_{Ci1} - x_{Bi1})}{z_{Ci1} - z_{Bi1}},\ y_{Di1} = y_{Bi1} + \frac{-z_{Bi1}(y_{Ci1} - y_{Bi1})}{z_{Ci1} - z_{Bi1}} \tag{21}$$

$$x_{Di2} = x_{Bi2} + \frac{-z_{Bi2}(x_{Ci2} - x_{Bi2})}{z_{Ci2} - z_{Bi2}},\ y_{Di2} = y_{Bi2} + \frac{-z_{Bi2}(y_{Ci2} - y_{Bi2})}{z_{Ci2} - z_{Bi2}} \tag{22}$$

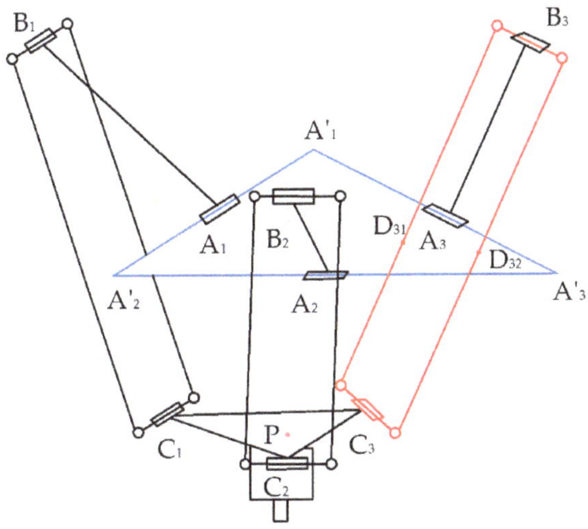

Figure 11. Interference of links with a fixed platform of the mechanism.

The platform is an equilateral triangle with points A_1, A_2, A_3 being the midpoints of the sides. The coordinates of the vertices A'_1, A'_2, A'_3 are defined as:

$$z_{A'_1} = z_{A'_2} = z_{A'_3} = x_{A'_3} = 0,\ x_{A'_1} = -\frac{a}{2},\ x_{A'_2} = \frac{a}{2},\ y_{A'_1} = y_{A'_2} = -\frac{a}{2\sqrt{3}}, \quad (23)$$

$$y_{A'_3} = \frac{a}{\sqrt{3}} \quad (24)$$

As a result, it is required to check in the two-dimensional plane XOY whether the points D_{i1} and D_{i2} belong to the triangle $A'_1A'_2A'_3$. If at least one of the points is included in the triangle, then there is interference with the platform.

We define the third group of interference using an approach based on determining the minimum distance between the segments drawn between the centers of the joints of each of the links. In [22,23], a similar condition is used, but the approach has drawbacks. In particular, the authors propose to determine the interference of the segments on the auxiliary plane, and not the distance between the nearest points. This does not allow identifying such an interference of links in which there is no interference of the axes.

The approach proposed in the current work is as follows. We construct an auxiliary plane to determine the interference of the links of the mechanism. This plane is parallel to the axis of one of the links and to which the axis of the other link belongs. In this case, the condition for the absence of interference of the links will take the form:

$$\sqrt{u_1^2 + u_2^2} > \mathrm{r}_{link1} + \mathrm{r}_{link2}, \quad (25)$$

where u_1 is the distance between the axis of a link that does not belong to the plane and the auxiliary plane, u_2 is the distance between the nearest points of the segments connecting the centers of the joints of each of the links when projecting a segment that does not belong to the auxiliary plane onto this plane, and r_{link1}, r_{link2} are the radiuses of the links.

It is worth noting the special case when the links are parallel to each other and the construction of an auxiliary plane is not required.

The algorithm for determining the interference of links is as follows:

- Determine the coordinates of the centers of the joints of each of the links.
- Draw segments 1 and 2 between the centers of the joints at each of the links, respectively.

- If the line segments are parallel to each other, then go to step 5. Otherwise, go to step 9.
- Rotate the line segments relative to the vertex of one of the line segments so that they become parallel to the OX axis.
- Determine the distance u_1 between the straight lines passing through the segments.
- Let us project the line segments onto the OX axis.
- Determine the distance u_2 between the nearest points of the line segments.
- If condition (25) is satisfied, then there is no interference. Completion of the algorithm.
- Construct an auxiliary plane in which segment 1 will lie, and to which segment 2 will be parallel.
- Determine the distance u_1 between line segment 2 and the construction plane.
- Project line segment 2 onto a construction plane.
- Determine the distance u_2 between the nearest points of segment 1 and the projection of segment 2.
- If condition (25) is satisfied, then there is no interference. Completion of the algorithm.

A more detailed approach to the definition of the third group of interference is presented in [35].

The proposed approach can be similarly applied to determine the interference between the links of the Uni and DeLi robots. In this case, it is possible to discretely separate the trajectories of the robots in the process of collaborative work of two robots. Then, it is required to determine the positions of the joints for each of the positions of the robots. Using the coordinates of the joints, one must check the intersection of all combinations of links using condition (25) and the above algorithm.

6. Analysis of the Workspace Taking into Account the Interference of the Links

The coverage obtained at the stage of defining the workspace is used to determine the positions of the end-effector inside the workspace, in which the interference of the links occurs. The coverage is divided into many boxes of equal size less than the approximation accuracy.

For the coordinate of the center of each of the boxes, conditions (14) and (25) are checked. If all the conditions are met, it is entered into a new set. The resulting set will contain boxes that correspond to the workspace without interference.

Workspace for configuration 1 according to Table 1 after excluding interference areas, as well as interference areas for a = 450 mm, c = 200 mm, d = 150 mm, e = 230 mm, $\alpha_{min} = \beta_{min} = \gamma_{min} = 10°$, $\alpha_{max} = \beta_{max} = \gamma_{max} = 350°$, D_{link} = 10 mm are shown in Figure 12. Table 2 shows the change in the volume of the workspace, taking into account the interference, without interference of links, as well as volume reduction for various configurations by taking into account the interference.

Table 2. Volumes of workspaces taking into account link interference.

Configuration No.	Volume of the Workspace, $\times10^4$ mm^3		Volume Reduction in%
	With Interference Areas	Without Interference Area	
1	4822	2457	49.05
2	4285	1741	59.37
3	4285	1741	59.37
4	4705	1183	74.86
5	4285	1741	59.37
6	4705	1182	74.86
7	4705	1182	74.86
8	4449	806	81.88

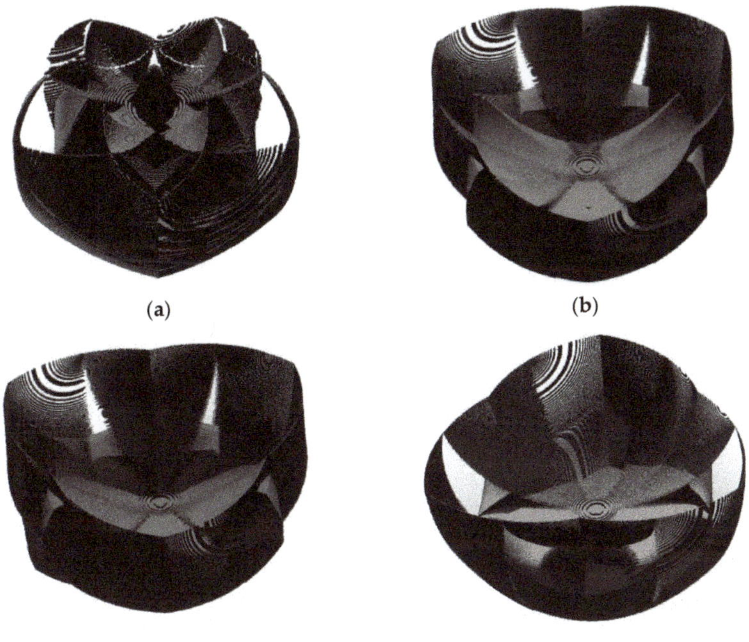

Figure 12. Workspace of the mechanism after excluding the areas of interference: (**a**) for configuration 1, (**b**) 2,3,5, (**c**) 4,6,7, (**d**) 8.

As can be seen from Table 2, with an increase in the number of kinematic chains, which corresponds to "+" in the inverse kinematics, the greater the percentage of the workspace that is excluded due to interference. This is due to the fact that in this case the kinematic chains are "bent" inward. As a result, the volume of the workspace for the first configuration is 3.05 times greater than the volume for the eighth configuration.

Visualization of the positions of the links at which they intersect occurs to verify the results of determining the interference. Figure 13 shows some of the link crossings that occur.

Figure 12 shows that the main inner part of the work area is round. Simulations were also performed for other link sizes (Table 3). The simulation results are shown in Figure 14. As can be seen from the figures, the workspace for various sizes has a round shape.

Table 3. Volumes of workspaces taking into account link interference.

No.	a	c	d	e
1	600	50	1200	1200
2	200	300	400	800
3	700	300	700	700

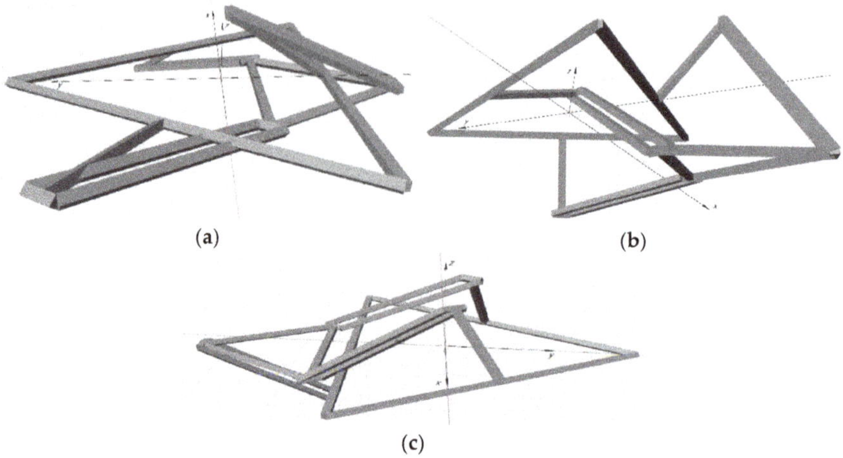

Figure 13. Examples of identified link interference: (**a**) links $B_{31}C_{31}$ and $B_{32}C_{32}$ with a fixed platform and link A_3B_3 with link $B_{32}C_{32}$, (**b**) links $B_{31}C_{31}$ and $B_{32}C_{32}$ with a fixed platform, (**c**) link A_2B_2 with link $B_{22}C_{22}$.

Figure 14. Workspace of the mechanism after excluding the areas of interference: (**a**) for configuration 1, (**b**) 2, (**c**) 3.

7. Geometric Parameters Optimization

Next, we will optimize the geometric parameters of the mechanism. The volume of the workspace, taking into account singularity zones and the interference of links, is an optimization criterion. Determination of the optimal sizes of links a, c, d, and e was carried out in several stages. At each stage, the volume of the workspace was calculated taking into account the interference of the links for various combinations of sizes. The ranges of resizing and the iteration step are reduced with each stage to reduce the computational complexity.

Numerical values are given in Table 4. The last column contains the volume of the workspace, which is a criterion for excluding or including in the next stage a certain part of the size range. Constant dimensions for modeling: h = 100 mm, D_{link} = 30 mm. The sum of the dimensions a, c, d, and e was taken to be constant and equal to 1850 mm. The amount was selected in accordance with the dimensions of the industrial robot ABB IRB 360: $a \approx 600$ mm, $c \approx 100$ mm, $d \approx 350$ mm, $e \approx 100$ mm. It should be noted that at the first stage, the computations were carried out for all configurations; however, the condition for the required volume of the workspace was fulfilled only for configurations 1 and 2/3/5. At the second stage, the volume condition was met only for the first configuration. In this regard, modeling was carried out only for the first configuration starting from the third stage.

Table 4. Stages of determining the optimal link lengths.

Stage	a_{min}	a_{max}	c_{min}	c_{max}	d_{min}	d_{max}	e_{min}	e_{max}	Step	Required Volume, m^3
1	50	1700	50	1700	50	1700	50	1700	100	1.5
2	50	320	50	320	200	800	750	1320	30	2.4
3	51	139	81	169	346	634	1020	1316	8	2708
4	60	72	82	94	469	520	1173	1227	3	27,105
5	61	65	83	87	512	519	1183	1190	1	-

The maximum workspace is reached at a = 61 mm, c = 86 mm, d = 513 mm, and e = 1190 mm. For such a ratio of dimensions, the total volume of the workspace, taking into account singularity zones, is 4.597 m^3, of which 2.714 m^3 is an area without interference of links. Thus, assuming the conditional size of the side of the base a = 1, the ratios of the sizes c, d, and e were 1.41, 8.41, and 19.51, respectively.

The dependence of the volume of the workspace for various configurations on the change in the parameters d for a = 61 mm and c = 86 mm, $e = (1850 - a - c - d)$ is shown in Figure 15. The numerical values of the workspace volume for the plot are obtained by iterative execution of the developed program for various parameters.

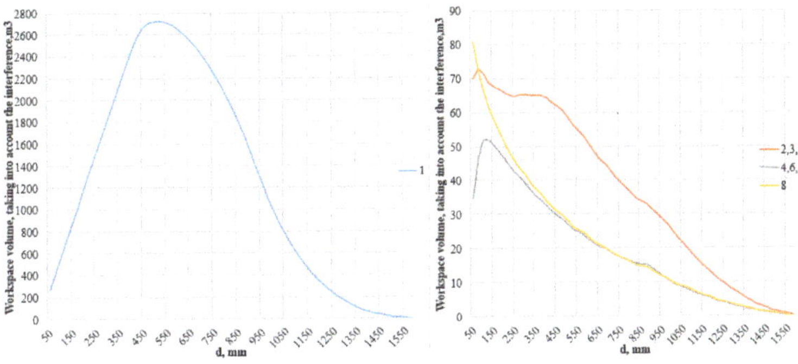

Figure 15. Dependence of the volume of the workspace on the change in the length of the links d and e.

Figure 16 shows the dependence of the proportion of areas in which interference occurs in the total volume of the workspace depending on the parameters d and e. The graphs show that for the first configuration, the fraction of interference decreases with increasing size d. For the rest of the configurations, the percentage of interference for almost all sizes d is more than 90%.

Let us perform the second stage of optimization. It takes into account the requirements of the aliquoting process. The required workspace height h_c is based on the requirements of tube and pipette sizes (350 mm). We can conclude that the manipulator workspace is round based on the previous analysis. Based on this, an optimization problem can be posed, which maximizes the radius r_c of a cylinder of a given height h_c, which can be inscribed into the manipulator's workspace. In this case, the sum of the dimensions a, c, d, and e of the manipulator should also be minimized. Therefore, the general criterion k can be written as:

$$k = \frac{a + c + d + e}{r_c} \tag{26}$$

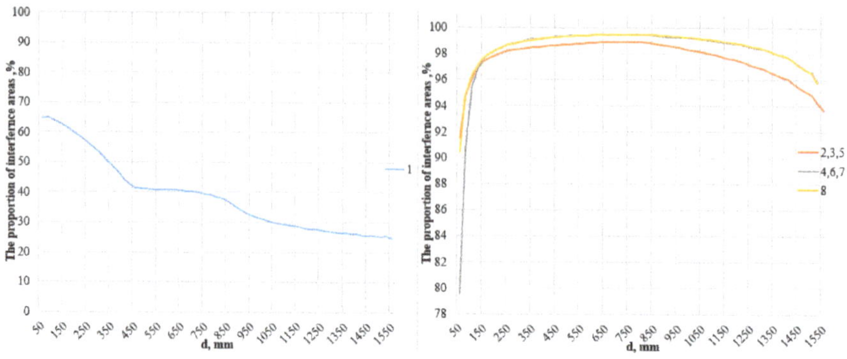

Figure 16. Dependence of the ratio of areas of interference to the workspace on the length of the links.

The optimization was performed in several stages, similar to the optimization performed above to maximize the workspace. Initial data for optimization: $a \in [30; 1500]$, $c \in [30; 1500]$, $d \in [30; 1500]$, $e \in [30; 1500]$, $h_c = 350$ mm. As a result, the optimal value of the criterion $k = 1.778$ was obtained for the following dimensions: $a = 61$ mm, $c = 87$ mm, $d = 478$ mm, $e = 1485$ mm, $r_c = 1187,29$ mm. The simulation of the workspace was performed for the obtained dimensions. Figure 17 shows that the workspace completely includes the cylinder with the calculated radius.

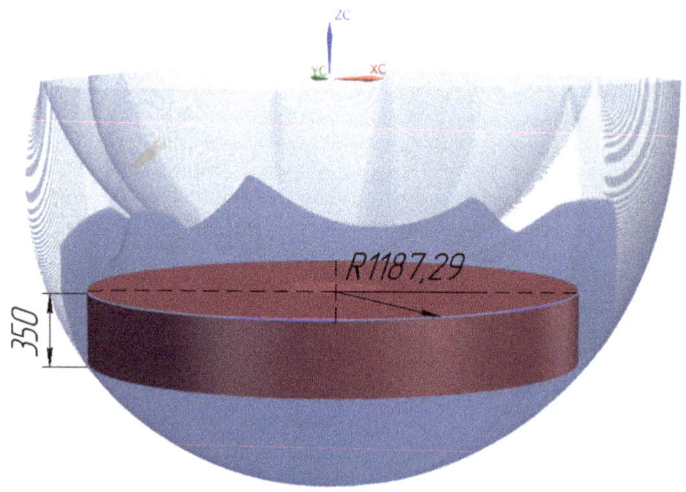

Figure 17. Workspace for optimal parameters.

An additional simulation was performed to compare the results of the two optimizations. The volume of the workspace and the maximum cylinder diameter were calculated for both optimal results. The parameters obtained during the first optimization were changed while maintaining the ratio 1:1.41:8.41:19.51. The change is necessary due to the incorrectness of comparing results that have a different sum of parameters. The sum of the parameters for the second optimization was 2111 mm. Therefore, the dimensions of the first optimization were changed so that their sum was 2111 mm. The obtained initial data and simulation results are shown in Table 5.

Table 5. Stages of determining the optimal link lengths.

Optimization No.	a	c	d	e	Sum	Volume, m^3	Cylinder Radius
1	69.606	98.133	585.375	1357.886	2111	4.024	1085.91
2	61	87	478	1485	2111	4.005	1187.29

As can be seen from the table, the workspace volume was larger as expected for the parameters obtained as a result of the first optimization and the maximum radius of the inscribed cylinder for the second optimization. However, it should be noted that the workspace volume in the first case turned out to be only 0.47% larger, while the radius for the second case was larger by 9.34%. Therefore, in this case, the radius of the inscribed cylinder is the main criterion for the problem of aliquoting. So, we can conclude that second optimization results should be currently accepted as the final results. However, a formal optimization algorithm will be planned in the future for a further optimization of the obtained results.

The aliquoting process assumes that Uni and DeLi should have a common part of workspaces. The Uni kinematic scheme is shown in Figure 18.

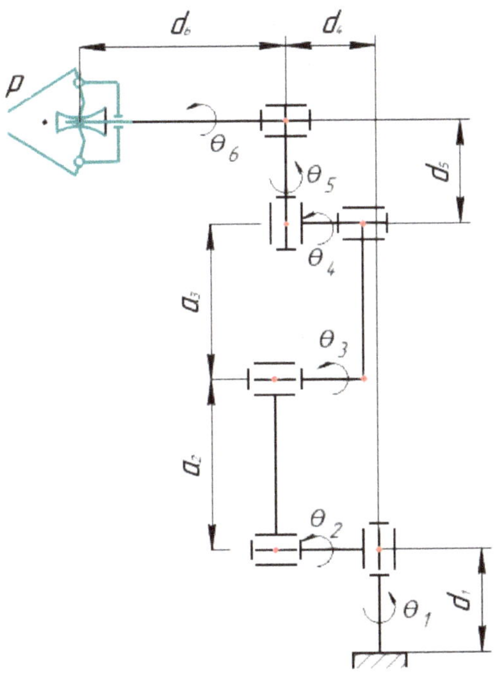

Figure 18. The kinematic scheme of Uni robot.

Workspace of Uni is calculated using D-H parameters and transformation matrices, where D-H parameters are equal to $a_2 = 240$ mm, $a_3 = 210$ mm, $d_1 = 150$ mm, $d_4 = 110$ mm, $d_5 = 85$ mm, and $d_6 = 81$ mm. The end-effector positions were calculated in discrete steps using the forward kinematics. The calculated positions were entered into the workspace array. The resulting array was exported to the STL model. Likewise, a program was written in the C++ programming language to implement the developed algorithm. The Uni workspace is shown in Figure 19.

Figure 19. The calculated Uni workspace.

Taking into account the calculated workspace of the Uni robot, we perform a simulation of a robotic system. The workspace models of robots were exported to the CAD system. The relative position of the robots was determined taking into account their workspaces and the requirements for the absence of intersections of links of mechanisms. Simulation of collaborative work of robots is shown in Figure 20. The execution of operations should be consistent: the Uni manipulator delivers a rack with test tubes to the joint workspace; then the DeLi manipulator performs the aliquoting process. The simulation confirmed the absence of intersections during collaborative work.

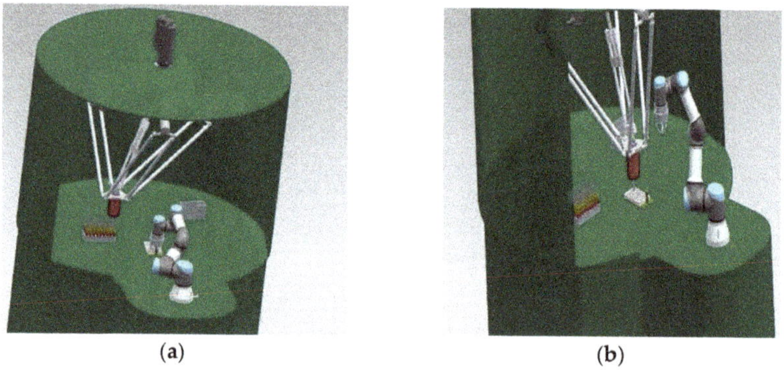

Figure 20. CAD model of RS movement with the movement of robots: (**a**) Uni moves the rack into the joint workspace, (**b**) DeLi interacts with the rack within a joint workspace.

Figure 21 shows the intersection of the DeLi and Uni workspaces. The intersection area is sufficient to accommodate a rack with 96 test tubes. Based on this, it can be concluded that the workspace of the DeLi robot and the relative position of the robots meet the requirements of the aliquoting process. The above conditions are met if the robot Uni is fixed relative to the coordinate system DeLi at a point with coordinates (0, 830, −897).

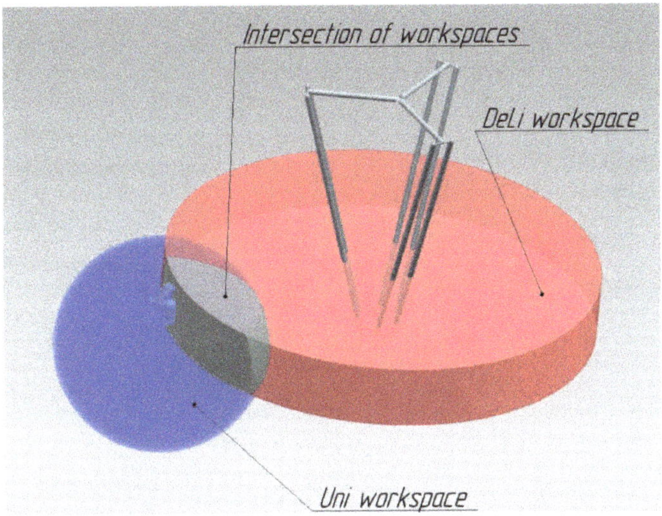

Figure 21. The intersection of the workspaces of a Uni and DeLi robot.

8. Conclusions

This article discusses a new robotic system design for aliquoting of biomaterial based on a Delta-type parallel robot with revolute kinematic joints. For the proposed robotic system, effective numerical methods and algorithms for determining the workspace, taking into account singularity zones and interference of links, have been developed and tested. An optimal design procedure has been carried out and robot design solution has been selected that achieves the maximum workspace for a given footprint size. For the selected design solution, the optimal ratio of the lengths of links a, c, d, and e relative to the link length a has been determined as being 1:1.41:8.41:19.51, respectively. The parameters were optimized taking into account the requirements of the aliquoting process. As a result, the optimal value of the criterion $k = 1.778$ was obtained for the following dimensions: $a = 61$ mm, $c = 87$ mm, $d = 478$ mm, $e = 1485$ mm, and $r_c = 1187,29$ mm.

Author Contributions: Conceptualization, D.M., L.R. and G.C.; methodology, D.M., A.N. and L.R.; software, D.M.; validation, L.R., T.S. and A.N.; formal analysis, D.M.; investigation, D.M.; resources, D.M., L.R. and A.N.; data curation, T.S. and A.N.; writing—original draft preparation, D.M.; writing—review and editing, L.R. and G.C.; visualization, D.M.; supervision, L.R. and G.C.; project administration, L.R. and G.C.; funding acquisition, L.R. and G.C. All authors have read and agreed to the published version of the manuscript.

Funding: This work was supported by the state assignment of Ministry of Science and Higher Education of the Russian Federation under Grant FZWN-2020-0017.

Institutional Review Board Statement: Not applicable.

Conflicts of Interest: The authors declare no conflict of interest.

References

1. Furuta, K.; Allocca, C.M.; Schacter, B.; Bledsoe, M.J.; Ramirez, N.C. Standardization and innovation in paving a path to a better future: An update of activities in ISO/TC276/WG2 biobanks and bioresources. *Biopreserv. Biobank.* **2018**, *16*, 23–27. [CrossRef] [PubMed]
2. ABB Robots Aid Rapid Automated Testing for COVID-19 Virus. Available online: https://new.abb.com/news/detail/75687/cstmr-abb-robots-aid-rapid-automated-testing-for-covid-19-virus (accessed on 9 November 2021).
3. Lord, H.L.; Pfannkoch, E.A. Sample Preparation Automation for GC Injection. In *Comprehensive Sampling and Sample Preparation*; Pawliszyn, J., Ed.; Academic Press: Oxford, UK, 2012; pp. 597–612. [CrossRef]

4. Robotic Digging Station PIRO. Available online: https://www.labortechnik.com/en/pipetting-robot-piro/uv-lamp-piro (accessed on 1 June 2021).
5. Freedom EVO Platform. Freedom EVO®Series. Available online: https://lifesciences.tecan.com/products/liquid_handling_and_automation/freedom_evo_series?p=tab--2 (accessed on 9 November 2021).
6. Hamilton Microlab STAR Liquid Handling System. Available online: https://www.hamiltoncompany.com/automated-liquid-handling/platforms/microlab-star (accessed on 15 March 2021).
7. Ceccarelli, M.; Carbone, G.; Ottaviano, E. Multi criteria optimum design of manipulators. *Bull. Pol. Acad. Sci. Tech. Sci.* **2005**, *53*, 9–18.
8. Bouraine, S.; Azouaoui, O. Safe Motion Planning Based on a New Encoding Technique for Tree Expansion Using Particle Swarm Optimization. *Robotica* **2021**, *39*, 885–927. [CrossRef]
9. Courteillie, E.; Deblaise, D.; Maurine, P. Design Optimization of a delta-like robot through global stiffness performance evaluation. In Proceedings of the IEEE/RSJ International Conference on Intelligent Robots and Systems, St. Louis, MI, USA, 11–15 October 2009.
10. Ben Hamida, I.; Laribi, M.; Mlika, A.; Romdhane, L.; Zeghloul, S. Dimensional Synthesis and Performance Evaluation of Four Translational Parallel Manipulators. *Robotica* **2021**, *39*, 233–249. [CrossRef]
11. Shen, H.; Yang, T.; Li, J.; Zhang, D.; Deng, J.; Liu, A. Evaluation of Topological Properties of Parallel Manipulators Based on the Topological Characteristic Indexes. *Robotica* **2020**, *38*, 1381–1399. [CrossRef]
12. Yu, G.; Wu, J.; Wang, L.; Gao, Y. Optimal Design of the Three-Degree-of-Freedom Parallel Manipulator in a Spray-Painting Equipment. *Robotica* **2020**, *38*, 1064–1081. [CrossRef]
13. Malm, J.; Végvári, A.; Rezeli, M.; Upton, P.; Danmyr, P.; Nilsson, R.; Steinfelder, E.; Marko-Varga, G. Large scale biobanking of blood—The importance of high density sample processing procedures. *J. Proteom.* **2012**, *76*, 116–124. [CrossRef]
14. Fleischer, H. Analytical Measurements and Efficient Process Generation Using a Dual-Arm Robot Equipped with Electronic Pipettes. *Energies* **2018**, *11*, 2567. [CrossRef]
15. Fleischer, H.; Drews, R.R.; Janson, J.; Reddy, B.; Patlolla, C.; Chu, X.; Klos, M.; Thurow, K. Application of a Dual-Arm Robot in Complex Sample Preparation and Measurement Processes. *J. Lab. Autom.* **2016**, *5*, 671–681. [CrossRef] [PubMed]
16. Vorberg, E.; Fleischer, H.; Junginger, S.; Liu, H.; Stoll, N.; Thurow, K. A Highly Flexible, Automated System Providing Reliable Sample Preparation in Element- and Structure-Specific Measurements. *J Lab. Autom.* **2016**, *5*, 682–692. [CrossRef] [PubMed]
17. Stan, S.; Manic, M.; Maties, V.; Balan, R. Evolutionary Approach to Optimal Design of 3 DOF Translation Exoskeleton and Medical Parallel Robots. In Proceedings of the HSI 2008, IEEE Conference on Human System Interaction, Krakow, Poland, 25–27 May 2008.
18. Pashkevich, A.; Chablat, D.; Wenger, P. Kinematics and Workspace Analysis of a Three-Axis Parallel Manipulator: The Orthoglide. *Robotica* **2006**, *24*, 39–49. [CrossRef]
19. McCormick, E.; Wang, Y.; Lang, H. Optimization of a 3-RRR Delta Robot for a Desired Workspace with Real-Time Simulation in MATLAB. In Proceedings of the 2019 14th International Conference on Computer Science & Education (ICCSE), Toronto, ON, Canada, 19–21 August 2019; pp. 935–941. [CrossRef]
20. Jha, R.; Chablat, D.; Rouillier, F.; Moroz, G. Workspace and Singularity Analysis of a Delta Like Family Robot. In *Robotics and Mechatronics*; Springer: Cham, Switzerland, 2016; pp. 121–130.
21. Bentaleb, T.; Hammache, H.; El Amine Belouchrani, M. Workspace, Accuracy Analysis and Kinematic Calibration of a "Delta" Parallel Robots. In Proceedings of the International Conference on Electronics & Oil: From theory to applications (ICEO'11), Ouargla, Algeria, 1–2 March 2011; pp. 290–295.
22. Anvari, Z.; Ataei, P.; Masouleh, M.T. The collision-free workspace of the tripteron parallel robot based on a geometrical approach. In *Computational Kinematics*; Springer: Cham, Switzerland, 2018; pp. 357–364. [CrossRef]
23. Anvari, Z.; Ataei, P.; Masouleh, M.T. Collision-free workspace and kinetostatic performances of a 4-DOF delta parallel robot. *J. Braz. Soc. Mech. Sci. Eng.* **2019**, *41*, 99. [CrossRef]
24. Liping, W.; Huayang, X.; Liwen, G.; Yu, Z. A novel 3-PUU parallel mechanism and its kinematic issues. *Robot. Comput.-Integr. Manuf.* **2016**, *42*, 86–102. [CrossRef]
25. Jha, R.; Chablat, D.; Baron, L.; Rouillier, F.; Moroz, G. Workspace, joint space and singularities of a family of delta-like robot. *Mech. Mach. Theory* **2018**, *127*, 73–95. [CrossRef]
26. López, M.; Castillo, E.; Garcia, G.; Bashir, A. Delta robot: Inverse, direct, and intermediate Jacobians. *Proc. Inst. Mech. Eng. Part C J. Mech. Eng. Sci.* **2006**, *220*, 103–109. [CrossRef]
27. Zhao, Y.J. Singularity, isotropy and velocity transmission evaluation of a three translational degrees of freedom parallel robot. *Robotica* **2013**, *31*, 193–202. [CrossRef]
28. Evtushenko, Y.; Posypkin, M.; Rybak, L.; Turkin, A. Approximating a solution set of non-linear inequalities. *J. Glob. Optim.* **2018**, *7*, 129–145. [CrossRef]
29. Merlet, J.-P. Interval Analysis and Robotics. *Tracts Adv. Robot.* **2010**, *66*, 147–156.
30. Moore, R.; Kearfott, R.; Cloud, M. *Introduction to Interval Analysis*; SIAM: Philadelphia, PA, USA, 2009.
31. Rybak, L.; Malyshev, D.; Gaponenko, E. Optimization Algorithm for Approximating the Solutions Set of Nonlinear Inequalities Systems in the Problem of Determining the Robot Workspace. *Commun. Comput. Inf. Sci.* **2020**, *1340*, 27–37.

32. The Delta Parallel Robot: Kinematics Solutions. Available online: https://www.ohio.edu/mechanical-faculty/williams/html/PDF/DeltaKin.pdf (accessed on 14 January 2022).
33. Gosselin, C.; Angeles, J. Singularity analysis of closed-loop kinematic chains. *IEEE Trans. Robot. Autom.* **1990**, *6*, 281–290. [CrossRef]
34. Gherman, B.; Birlescu, I.; Plitea, N.; Carbone, G.; Tarnita, D.; Pisla, D. On the singularity-free workspace of a parallel robot for lower-limb rehabilitation. *Rom. Acad. Ser. A—Math. Phys. Tech. Sci. Inf. Sci.* **2019**, *20*, 383–391.
35. Behera, L.; Rybak, L.; Malyshev, D.; Gaponenko, E. Determination of Workspaces and Intersections of Robot Links in a Multi-Robotic System for Trajectory Planning. *Appl. Sci.* **2021**, *11*, 4961. [CrossRef]

MDPI
St. Alban-Anlage 66
4052 Basel
Switzerland
Tel. +41 61 683 77 34
Fax +41 61 302 89 18
www.mdpi.com

Applied Sciences Editorial Office
E-mail: applsci@mdpi.com
www.mdpi.com/journal/applsci

www.ingramcontent.com/pod-product-compliance
Lightning Source LLC
Chambersburg PA
CBHW040807250926
51850CB00007BA/1102
* 9 7 8 3 0 3 6 5 6 7 1 4 3 *